Studies in Space Policy

Volume 22

Series Editor

European Space Policy Institute, Vienna, Austria

The use of outer space is of growing strategic and technological relevance. The development of robotic exploration to distant planets and bodies across the solar system, as well as pioneering human space exploration in earth orbit and of the moon, paved the way for ambitious long-term space exploration. Today, space exploration goes far beyond a merely technological endeavour, as its further development will have a tremendous social, cultural and economic impact. Space activities are entering an era in which contributions of the humanities — history, philosophy, anthropology —, the arts, and the social sciences — political science, economics, law — will become crucial for the future of space exploration. Space policy thus will gain in visibility and relevance. The series Studies in Space Policy shall become the European reference compilation edited by the leading institute in the field, the European Space Policy Institute. It will contain both monographs and collections dealing with their subjects in a transdisciplinary way.

More information about this series at http://www.springer.com/series/8167

Stefano Ferretti
Editor

Space Capacity Building in the XXI Century

Editor
Stefano Ferretti
European Space Policy Institute
Vienna, Austria

ISSN 1868-5307 ISSN 1868-5315 (electronic)
Studies in Space Policy
ISBN 978-3-030-21940-6 ISBN 978-3-030-21938-3 (eBook)
https://doi.org/10.1007/978-3-030-21938-3

This Springer imprint is published by the registered company Springer Nature Switzerland AG
The registered company address is: Gewerbestrasse 11, 6330 Cham, Switzerland

Foreword

It is a great honour and an immense pleasure for me to introduce this book on *Space Capacity Building in the XXI Century*. It is the fruit of an intense effort made by Stefano Ferretti to put together many contributions from a multitude of highly knowledgeable experts on the various aspects of space capacity building from all over the planet.

It is also a follow-up and one of the many outcomes of the outstanding workshop on "Space2030 and Space 4.0: Synergies for Capacity Building in the XXI Century", organized by the European Space Policy Institute (ESPI), co-organized by UNOOSA, with the support of ESA, on the 3rd of February 2018. This was a contribution to the—at that time upcoming—UNISPACE+50 event to celebrate the fiftieth anniversary of the first United Nations Conference on the Exploration and Peaceful Uses of Outer Space.

The ambition of this book is to provide an extensive overview of the national and international policy frameworks in relation to space capacity building and the rationales for their adoption on a global scale. The methodology applied consists of:

- Considering various examples of space capacity building efforts across different regions as implemented by a range of actors from agencies to industry, from NGOs to users;
- Presenting space capacity building programs that can empower the international community towards fully accessing all the potential benefits of space assets and data, for a sustainable socio-economic development.

Currently, new innovation models are increasingly spread across sectors and disciplines, including space, which is becoming an integral part of many societal challenges (e.g. telecoms, weather, climate change and environmental monitoring, civil protection, infrastructures, transportation and navigation, health care and education).

This book, edited by ESPI, is the first international publication that analyses the multiple challenges associated with space capacity building across the board. It assists readers in defining achievable ambitions in this matter, constructing their own technical and operational road maps involving renowned and knowledgeable stakeholders but also new private actors, NGOs and civil society, and in mapping out the available opportunities, summarizing the most appropriate programmatic options for their successful implementation.

Starting from a policy and strategy perspective, the book dives into some key areas of capacity building including innovation and exploration, global health, climate change and resilient societies. It showcases reflections from a range of senior space and non-space professionals as well as users from around the world, who kindly agreed to share their unique perspectives and solutions to make space technology and applications increasingly accessible.

The outcome is a rich mosaic where different cultural and policy approaches to space are translated into actionable programs and ideas so that space may truly benefit all of humankind.

Vienna, Austria

Jean-Jacques Tortora
Director of the European Space Policy Institute

Preface

The space sector is undergoing a paradigmatic shift, which is bringing about a rapidly evolving and increasingly complex space agenda with more participants, both governmental and non-governmental, increasingly involved in ventures to explore space and carry out space activities. As new spacefaring nations emerge, the risk of duplicating activities and missing opportunities to improve our livelihoods on Earth should be offset by promoting cooperation and the spread of best practices and cutting-edge technologies across the globe, creating common frameworks and dialogue platforms to accompany this process.

The year 2018 marked the 50th anniversary of the first United Nations Conference on the Exploration and Peaceful Uses of Outer Space—UNISPACE +50. This conference reviewed the contributions that three previous UNISPACE conferences (UNISPACE I, held in 1968, UNISPACE II, held in 1982, and UNISPACE III, held in 1999) have made to global space governance. In line with the 2030 Agenda for Sustainable Development and Sustainable Development Goals, UNISPACE+50 aimed to chart the future role of the UN Committee on the Peaceful Uses of Outer Space (COPUOS), its subsidiary bodies and the United Nations Office for Outer Space Affairs (UNOOSA). The activities of UNOOSA are an integral part of the UNISPACE+50 thematic cycle and are aimed at contributing to four pillars: space economy, space society, space accessibility and space diplomacy.

On the occasion of the 50th anniversary of the first UNISPACE Conference, the European Space Policy Institute also organized a high-level conference, together with UNOOSA and the European Space Agency. The conference was structured as a dialogue platform for a range of different actors who have a stake in future space governance and services, and this book summarizes the relevant findings and recommendations, outlining the synergies between the United Nations agenda Space2030 and the Space 4.0 strategy of the European Space Agency and providing insights into various areas and disciplines that are related to the most innovative and recent socio-economic developments impacted by evolving space policies.

The overarching framework of the book is *Space Capacity Building in the XXI Century* and it is based on the high-level contributions from the key stakeholders who will shape the space agendas and programs for the coming decades, proving useful guidance to policymakers on a global scale and ultimately benefitting citizens by fully exploiting the potential of space in addressing user needs. In fact, "Capacity-building for the twenty-first century" is the 7th Thematic Priority of UNISPACE+50 and it is the most cross-cutting aspect of the initiative, responding to the request of the various Member States of the United Nations seeking to define new innovative and effective approaches to overall capacity building and development needs, as a fundamental pillar of global space governance.

International cooperation is essential in the new space ecosystem, and therefore, the book identifies potential synergies for capacity building on a global scale, describing their guiding principles and providing concrete examples of these evolving policies in Africa, America, Asia and Europe. In doing so, the various contributions address the most prominent ongoing developments in the XXI century, focusing on four thematic priorities of UNISPACE+50:

- Global partnership in space exploration and innovation;
- Strengthened space cooperation for global health;
- International cooperation towards low-emission and resilient societies;
- Capacity building for the twenty-first century.

The readers will therefore be able to appreciate both the holistic perspectives of the high-level policies and strategies, as well as the programmatic developments in their main areas of interest, by exploring the future approaches of the diverse groups of stakeholders involved in innovation, exploration, global health and climate change. These stakeholders contribute with individual chapters to the book, describing in detail policies, strategies and programmatic actions in their areas of interest.

Readers will find this book to be a starting point to identify local and regional initiatives linking space activities and a more sustainable development of our world, drawing inspiration and identifying potential avenues to promote further collaboration and create stronger networks for space capacity building in the twenty-first century.

Vienna, Austria Stefano Ferretti

Acknowledgements

I would like to acknowledge the significant support in the conceptualization and organization of the joint conference "Space2030 and Space 4.0: synergies for capacity building in the XXI century", which took place at ESPI in preparation of UNISPACE+50, provided by the United Nations Office for Outer Space Affairs, and its Director Simonetta Di Pippo, the European Space Agency, and its Chief Strategy Officer Kai-Uwe Schrogl, and the European Space Policy Institute, its Director Jean-Jacques Tortora, and the respective teams.

A special mention goes to the co-chairs of the ESPI-UNOOSA-ESA conference for their support in the management of its four sessions and for their valuable insights in the construction of a rich narrative, which included different points of view, personal experiences and a forward looking approach to space capacity building: Natalia Archinard, Head of the Swiss Delegation to UN COPUOS (Federal Department of Foreign Affairs), Ana Ávila, Costa Rica Delegation to UN COPUOS, Yeshey Choden, Ministry of Information and Communications (Bhutan), Lorant Czaran, Programme Officer (UNOOSA), Jorge Del Rio Vera, Programme Officer (UNOOSA), Isabelle Duvaux-Béchon, Head of Member States Relations and Partnerships Office (ESA), Funmilayo Erinfolami, Scientific Staff (UN Regional Centre, Nigeria), Daniel Garcia Yarnoz, Programme Officer (UNOOSA), Beth Healey, Former ESA Medical Doctor (Spaceflight Analogue Concordia, Antarctica), Clelia Iacomino, Resident Fellow (ESPI), Shirishkumar Ravan, Senior Programme Officer (UNOOSA), Luc St-Pierre, Chief of the Space Applications Section (UNOOSA), and also to Otto Koudelka, Professor and Head of the Institute of Communications Networks and Satellite Communications (TU Graz), for providing the closing statement.

I am very grateful in particular to all the authors of the book chapters, that believed in such an enterprise, joining forces and involving many key stakeholders, looking at capacity building from various different angles, often thinking out of the box while focusing on making a meaningful synthesis for the readers, that can now be taken up by key decision-makers in developing innovative frameworks for space capacity building in the XXI century, for the benefit of all humankind. In addition,

I would like to thank those who provided insights and ideas that helped to shape this project. Among these were Zhuang Dafang, Institute of Geographic Sciences and Natural Resources Research (Chinese Academy of Sciences), Juan Garces de Marcilla, Director Copernicus Services (ECMWF), Simon Jutz, Head of Copernicus Space Office (ESA), Ramesh Krishnamurthy, Senior Advisor Department of Information, Evidence and Research, Health Systems and Information Cluster (World Health Organization), Mioara Mandea and Juliette Lambin, Innovation, Application and Science Directorate (CNES), Stéphane Ourevitch, Copernicus Support Office (EC), Valerio Tramutoli, Professor at Faculty of Engineering (Università degli Studi della Basilicata). It would be impossible to mention all the other colleagues, distinguished United Nations Committee on the Peaceful Uses of Outer Space (COPUOS) delegates, experts and NGO managers, users and citizens, industry and academia representatives, with whom I discussed ideas, challenges and innovative concepts in the three years leading up to UNISPACE+50, but it is praiseworthy that these exchanges lead to the compilation of this book, which will hopefully serve the global community at large in the years to come.

Stefano Ferretti

Contents

Acronyms

Airbus D&S	Airbus Defence and Space
AIS	Automatic Identification Satellites
ALR	Austrian Aeronautics and Space Agency
AMESD	African Monitoring of the Environment for Sustainable Development
APAC	China and other Asia-Pacific
ARISE	Agriculture Resource Inventory and Survey Experiment
ASAP	Austrian Space Application Programme
ASEAN	Association of Southeast Asian Nations
ASI	Agenzia Spaziale Italiana (Italian Space Agency)
ATV	Automated Transfer Vehicle
AVHRR	Advanced, Very High Resolution Radiometer
BDS	BeiDou Navigation Satellite System
BELSPO	Belgian Federal Science Policy Office
BHRS	Belgian High Representation for Space Policy
BIS	Business, Innovation and Skills
BMVIT	Austrian Federal Ministry for Transport, Innovation and Technology
CAD	Computer-Aided Design
CAPE	Crop Acreage and Production Estimation
CASC	China Aerospace Science and Technology Corporation
CATHALAC	Water Center for the Humid Tropics for Latin America and the Caribbean
CBERS	China–Brazil Earth Resources Satellite
CDTI	Centre for the Development of Industrial Technology
CEC	Consortium for Educational Communication
CELAC	Community of Latin American and Caribbean States
CENI	Commission Électorale Nationale Indépendante
CEOS	Committee on Earth Observation Satellites

CERSGIS	Centre for Remote Sensing and Geographic Information Services
CET	Centre for Education Technology
CFAS	Federal Commission for Space Affairs
CGWIC	China Great Wall Industry Corporation
CIET	Central Institute of Educational Technology
CILSS	Comité permanent Inter-Etats de Lutte contre la Sécheresse dans le Sahel (Ghana)
CIS	Communications, Intelligence and Security
CMA	Governing Body of the Paris Agreement
CMSA	China Manned Space Agency
CNES	Centre National d'Études Spatiales (National Centre for Space Studies) (French Space Agency)
CONAE	Argentinean Space Agency
CONCORDi	European Commission's biennial Conference on Corporate R&D and Innovation
COP	Conference of the Parties
COPUOS	United Nations Committee on the Peaceful Uses of Outer Space
COSTIND	Commission for Science, Technology and Industry for National Defense
CRESDA	Centre for Resources Satellite Data and Applications
CSA	Canadian Space Agency
CSE	Centre de Suivi Ecologique (Senegal)
CSES	China Seismo-Electromagnetic Satellite
CubeSats	Cube Satellites
DECU	Development and Educational Communication Unit
DJEI	Department of Jobs, Enterprise and Innovation
DLR	Deutsches Zentrum für Luft- und Raumfahrt (German Aerospace Center)
DoD	United States Department of Defence
DRDO	Defence Research and Development Organisation
DSC	Decision Support Center
DSCOVR	Deep Space Climate Observatory
EIB	European Investment Bank
EIF	European Investment Fund
EMEA	Europe, the Middle East and Africa
EMMRCs	Educational Multimedia Research Centres
EO	Earth Observation
EPM	European Physiology Modules
ESA	European Space Agency
ESPI	European Space Policy Institute
ESSO	Earth System Science Organization
ETC	Emergency Telecommunications Cluster
EU	European Union

EUMETSAT	European Organisation for the Exploitation of Meteorological Satellites
EUTELSAT	European Telecommunications Satellite Organization
FAA	Federal Aviation Administration
FAI	Floating Algal Index
FASAL	Forecasting Agricultural output using Space, Agrometeorology and Land based observations
FCT	Foundation for Science and Technology
FFG	Austrian Research Promotion Agency
FFL	Fondation Follereau Luxembourg
FOCAC	Forum on China–Africa Cooperation
GDP	Gross Domestic Product
GEO	Geostationary Earth Orbit
GEO	Group on Earth Observation
GEOSS	Global Earth Observation System of Systems
GERD	Gross Domestic Expenditure on R&D
GFDRR	Global Facility for Disaster Reduction and Recovery
GGIM	Global Geospatial Information Management
GmbH	Gesellschaft mit beschränkter Haftung
GMT	Greenwich Mean Time
GNI	Gross National Income
GNSS	Global Navigation Satellite Systems
GOES-R	Geostationary Operational Environmental Satellite-R
GPM	Global Precipitation Measurement (NASA and JAXA Mission)
GPS	Global Positioning System
GRACE	Gravity Recovery and Climate Experiment
GSA	European GNSS Agency
GSRT	General Secretariat for Research and Technology
HAT	Human African Trypanosomiasis/Sleeping sickness
HFA	Hyogo Framework for Action
HR	High Resolution
HSTI	Human Space Technology Initiative
I&B	Information and Broadcasting
IAA	International Academy of Astronautics
IAC	International Astronautical Congress
IAEG-SDGs	UN Statistical Commissions' Inter-agency Expert Group
IAF	International Astronautical Federation
IARI	Indian Agriculture Research Institute
IASC	Inter-Agency Standing Committee
ICE Cubes	International Commercial Experiments Service
ICF	ICE Cubes Facility
ICG	International Committee on Global Navigation Satellite Systems
ICIMOD	International Centre for Integrated Mountain Development
ICoC	Draft International Code of Conduct for Outer Space Activities
ICRC	International Committee of the Red Cross

ICS	Information and Communication Systems
ICT	Information and Communication Technology
IFIs	International Financial Institutions
IGS	International GNSS Service
IKAR	Interdepartmental Coordination Committee for Space Affairs
IMF	International Monetary Fund
INCOIS	Indian National Centre for Ocean Information Services
INTA	National Institute for Aerospace Technology
IODC	Indian Ocean Data Coverage
IOD/IOV	In-orbit Demonstration & Validation
IOs	Regional Organizations and International Organizations
IoT	Internet of Things
IP	Internet Protocol
IP	Intellectual Property
IPP	International Partnership Programme
IRNSS	Indian Regional Navigation Satellite System
ISC	International Satellite Company Limited
ISED	Innovation, Science and Economic Development
ISRO	Indian Space Research Organisation
ISS	International Space Station
ISU	International Space University
ITAR	International Traffic in Arms Regulations
ITU	International Telecommunication Union
IUCAA	Inter-University Centre for Astronomy and Astrophysics
JASON	Joint Altimetry Satellite Oceanography Network (NASA and CNES Joint Mission)
JAXA	Japan Aerospace Exploration Agency
J-PAL	Abdul Latif Jameel Poverty Action Lab
KARI	Korea Aerospace Research Institute (Korea Space Agency)
LEO	Low Earth Orbit
LULC	Land Use and Land Cover
MDA Corp.	MacDonald, Dettwiler and Associates Ltd.
MDGs	Millennium Development Goals
Melco	Mitsubishi Electric Co.
MEMS	Micro Electro Mechanical Systems
MEO	Medium Earth Orbit
MERIS	Medium Resolution Imaging Spectrometer (Sensor on board the ESA Envisat satellite)
MERLIN	Methane Remote Sensing LIDAR Mission
MESA	Monitoring for Environment and Security
MetOp	Meteorological Operational Satellite
MetOp-SG	MetOp Second Generation
MEXT	Ministry of Education, Culture, Sports, Science and Technology
MFG	Meteosat First Generation
MIT	Massachusetts Institute of Technology

MIUR	Ministry of Education, University and Research
MODIS	Moderate Resolution Imaging Spectroradiometer
MOSDAC	Meteorological and Oceanographic Satellite Data Archival Centre
MoU	Memorandum of Understanding
MSF	Médecins Sans Frontières
MSG	Meteosat Second Generation
MSI	Multi-Spectral Instrument on Board Sentinel-2. European Mission. Part of the European Union Copernicus Programme
MSS	Mobile Satellite Service
MTG	Meteosat Third Generation
NASA	National Aeronautics and Space Administration
NATO	North Atlantic Treaty Organization
NCERT	National Council of Educational Research and Training
NCSTE	China's National Centre for Science and Technology Evaluation
NDCs	Nationally Determined Contributions
NDVI	Normalised Difference Vegetation Index
NEC	Nippon Electric Company
NGA	National Geospatial-Intelligence Agency
NGO	Non-Governmental Organization
NNRMS	National Natural Resources Management System
NOAA	National Oceanic and Atmospheric Administration
NOW	Netherlands Organisation for Scientific Research
NRO	National Reconnaissance Office
NRSC	National Remote Sensing Centre
NSC	National Space Council
NSC	Norwegian Space Centre
NSO	Netherlands Space Office
OCO	Orbiting Carbon Observatory
ODA	Official Development Assistance
OECD	Organisation for Economic Co-operation and Development
OHB	Orbitale Hochtechnologie Bremen
OPEC	Organization of the Petroleum Exporting Countries
OST	Outer Space Treaty
PACE	Plankton, Aerosol, Cloud, ocean Ecosystem
PES	Payment for Ecosystem Services
POLSA	Polish Space Agency
Poseidon	Positioning, Ocean, Solid Earth, Ice Dynamics, Orbital Navigator (TOPEX/Poseidon was a joint NASA CNES Mission)
PPP	Public–Private Partnership
PRS	Public Regulated Service
PSA	Programme on Space Applications
PUMA	Preparation for the Use of MSG in Africa programme
QZSS	Quasi-Zenith Satellite System
RCM	RADARSAT Constellation Mission

RCMRD	Regional Centre for Mapping of Resources for Development
R&D	Research & Development
REDD+	Reducing Emissions from Deforestation and Degradation
RLV	Reusable Launch Vehicle
ROSA	Romanian Space Agency
Roscosmos	Roscosmos State Corporation
RSCC	Russian Satellite Communications Company
SAARC	South Asian Association for Regional Development
SAB	Space Advisory Board
SAC	Space Applications Centre
SAFs	Satellite Application Facilities
SAHEL	Sub-Saharan initiative for Telemedicine
SDGs	Sustainable Development Goals
SDP	Space for Development Profile
SERI	State Secretariat for Education, Research and Innovation
SES	Société Européenne des Satellites
SIA	Satellite Industry Association
SIET	State Institute of Educational Technology
SIG	Spatial Informatics Group
SITE	Satellite Instructional Television Experiment
SMAP	Soil Moisture Active Passive (NASA Mission)
SNC	Sierra Nevada Corporation
SpaceX	Space Exploration Technologies
SRC	Space Research Centre
SRON	Netherlands Institute for Space Research
SS/L	Space Systems/Loral
SSH	Sea Surface Height
SSO	Sun-synchronous orbit
SST	Space Surveillance and Tracking
SST	Sea Surface Temperature
STEAM	Science, Technology, Engineering, Art and Mathematics
STEM	Science, Technology, Engineering and Mathematics
STSC	Scientific and Technical Subcommittee
SWOT	Suface Water Ocean Topography (future satellite mission, jointly developed by NASA and CNES, launch planned for 2021)
TCBMs	Transparency and Confidence-Building Measures
TDP	Technology Demonstration Program
TEU	Treaty on European Union
THAAD	Terminal High Altitude Area Defense system
TM/TC	Telemetry & Telecommand
TOPEX	Topography Experiment (TOPEX/Poseidon was a joint NASA CNES Mission)
TRL	Technology Readiness Level
TRMM	Tropical Rainfall Measuring Mission (NASA Mission)

TWS	Terrestrial Water Storage
USA	United States of America
UAE	United Arab Emirates
UAV	Unmanned Aerial Vehicle
UGC	University Grants Commission
UK	United Kingdom
ULA	United Launch Alliance
UN	United Nations
UNCOPUOS	United Nations Committee on the Peaceful Uses of Outer Space
UNCTAD	United Nations Conference on Trade and Development
UNDP	United Nations Development Programme
UNDSS	United Nations Department of Safety and Security
UNFCCC	United Nations Framework Convention on Climate Change
UNGA	United Nations General Assembly
UNGIWG	United Nations Geographic Information Working Group
UNIDIR	United Nations Institute for Disarmament Research
UNISPACE+50	Conference on the occasion of the 50th anniversary of the United Nations Conference on the Exploration and Peaceful Uses of Outer Space
UNISPACE	United Nations Conference on the Exploration and Peaceful Uses of Outer Space
UNOOSA	United Nations Office for Outer Space Affairs
UNSDI	United Nations Spatial Data Infrastructure
UN-SPIDER	United Nations Platform for Space-based Information for Disaster Management and Emergency Response
USAID	U.S. Agency for International Development
VAST	Vietnamese Academy of Science and Technology
VHR	Very High Resolution
VIMS	Visible and Infrared Mapping Spectrometer
VKO	Aerospace Defence Forces
VSAT	Very Small Aperture Terminal
WFP	World Food Programme
WG	Working Group
WGP	World Gross Product
WISE	Wide-field Infrared Survey Explorer
WRS	World Radiocommunication Seminar

Chapter 1
Space Capacity Building in the XXI Century

Stefano Ferretti

1.1 Introduction

The space sector is currently entering a new era characterized by rapid technological changes, changes in social and environmental conditions, the advent of digitalization, increased access to space data and new key players.[1] Access to space technologies is becoming more widespread and enables new solutions that respond to the needs of the global community. To access these benefits fully, a holistic space capacity building program is being endorsed by a wide range of actors, in order to create a networked and decentralized model that can bring the benefits of space across the World.

Space capacity building doesn't only involve training and education, as is often assumed, but encompasses a much broader range of activities designed to equip nations with the tools needed to successfully deploy, employ and exploit space activities. It therefore also includes processes such as R&D, technology development and transfer, entrepreneurship and the creation of new business markets, developing and enforcing legal tools and policies, international cooperation and partnerships, and tailoring of space solutions to specific (or local) needs.

Within this context, the idea for this publication was born from the convergence of two streams of activities. On the one hand, all the activities related to sustainable development and the UN Agenda2030 with its sustainable development goals, and all activities, projects and initiatives organized as part of the UNISPACE+50 process and conference, fifty years after the first UNISPACE conference. On the other hand, the activities related to ESA's strategy space 4.0 for the future of the Agency in navigating the challenges that this new era of space travel and commercial space is

[1] Al-Ekabi, C., Ferretti, S. (Eds.). (2018). Yearbook on Space Policy 2016: Space for Sustainable Development. Springer.

S. Ferretti (✉)
European Space Policy Institute (ESPI), Vienna, Austria
e-mail: stefano.ferretti@esa.int

S. Ferretti (ed.), *Space Capacity Building in the XXI Century*, Studies in Space Policy 22,
https://doi.org/10.1007/978-3-030-21938-3_1

heralding, and the positions of new space private actors, space industry, NGOs, other space agencies, new space faring nations and academia.

ESPI was extensively involved in both activities and therefore was able to identify synergies and complementarities between these two activity streams that warranted further exploration. It was in this spirit that a joint conference was organized, whose outcomes and further reflections are presented in this publication, including specific contributions from senior space professionals.

1.2 ESPI's Activities on Sustainable Development and Capacity Building with UNOOSA

Regarding space for sustainable development, ESPI extensively published reports, organized conferences, consulted with stakeholders and provided key recommendations for the implementation of space programs.[2,3,4,5,6] In the year leading up to the 50th anniversary of the first United Nations Conference on the Exploration and Peaceful Uses of Outer Space, a series of ESPI activities were focused around the seven thematic priorities, as identified by UNOOSA:

TP1 Global partnership in space exploration and innovation
TP2 Legal regime of outer space and global space governance: current and future perspectives
TP3 Enhanced information exchange on space objects and events
TP4 International framework for space weather services
TP5 Strengthened space cooperation for global health
TP6 International cooperation towards low-emission and resilient societies
TP7 Capacity-building for the twenty-first century.

The 7th thematic priority (TP7), capacity building for the XXI century, is the most cross-cutting aspect of the initiative. Under this theme, UNOOSA and UN COPUOS member states are looking for new ideas, including especially:

- innovative approaches to capacity building in the space sector
- new measures for progress and development.

The ESPI activities in this framework focused on how to identify upcoming global challenges and elicit societal needs, particularly at local level, for which a dedicated

[2]Ferretti S., Feustel-Büechl J., Gibson R., Hulsroj, P., Papp A., Veit, E. (2016) ESPI Report 59 Space for Sustainable Development. ESPI, Vienna.

[3]Ferretti S (2016) Space as an enabler of sustainable development. 4th ICSD, Columbia University, New York, NY.

[4]Ferretti S., Tortora J.J., Veit E., Vernile A. (2016) "ESPI Report 60 Engaging with Stakeholders in Preparation for UNISPACE+50", ESPI, Vienna.

[5]Ferretti S., Veit E. (2016) "Space for Sustainable Development" (IAC-16-E3.1.15). Proceedings of the 67th International Astronautical Congress (IAC), Guadalajara, Mexico.

[6]See Footnote 1.

workshop with NGOs was organized at ESPI, and whose outcome was presented to the 2016 UN/International Astronautical Federation (IAF) Workshop on Space Technology for Socioeconomic Benefits: Integrated Space Technologies and Applications for a Better Society. A dedicated conference was then jointly organized by Austria and UNOOSA in Graz, looking at capacity building as a cross-cutting priority out of the 7 identified by the UN for UNISPACE+50.

The ESPI contribution on assessing and addressing societal needs of all the actors operating in the field, provides insights on the fact that the approval of the Sustainable Development Goals (SDGs) by the UN General Assembly in 2015 marks a departure point for our theoretical approach to sustainable development. The emphasis that the 17 goals place on the consideration of the long-term economic, social and environmental requirements of beneficiary communities, poses new challenges for stakeholders in the development community. To achieve these goals, multi-stakeholder, cross-cutting approaches will be necessary. Space services such as position, navigation and timing, telecommunications and earth observation have become increasingly embedded in the everyday lives of much of the global population, but their social impact reaches further still. A number of successful, demand-driven initiatives using space assets have demonstrated the often untapped potential which Space has in supporting development activities (e.g. Preparation for Use of MSG in Africa (PUMA), African Monitoring of the Environment for Sustainable Development (AMESD), Monitoring for Environment and Security in Africa programme (MESA), GMES and Africa Support Programme and other recent developments suggest that we may expect this trend to grow further in coming years).

The establishment of dialogue mechanisms between the fields, which successfully defines and prioritises user requirements, will be a prerequisite of this collaboration, and for the Space community, understanding the needs of stakeholders in the field is the first step towards achieving this objective. This contribution underlines the recommendations of a variety of stakeholders from the Space, NGO and UN communities, who participated in an ESPI workshop in May 2016, which was an opportunity for development actors to voice their views. It aimed at developing an integrated approach that collects the needs on the field and translates them into technological solutions that support sustainable development on a global scale. Discussions included how NGOs are currently using Space assets, which requirements are filled by cooperation with Space stakeholders, and how the interaction between the fields may be improved. The ESPI dialogue platform included both theoretical reflections on a range of development activities, including health, energy, humanitarian emergencies, and security, as well as the creation and management of Space-NGO programmes. Specific case studies reflect existing Space asset incorporation into development efforts in a variety of locations and environments, including islands, mountains and urban environments. Together, the stakeholders' inputs highlight the pivotal importance of Space in achieving the Sustainable Development Goals, but point out that successful cooperation between the fields depends on continuous, open dialogue to define user requirements, and the creation of Space-derived solutions tailored to specific contexts.

A major step in this direction was taken by UNOOSA, developing a space capacity index and a space solutions compendium, that would allow measuring progress and development of spacefaring nations. The United Nations/Austria Symposium on "Access to Space: Holistic Capacity-Building for the 21st Century", supported by the Government of Austria, the European Space Agency, the State of Styria, and the City of Graz, was held at the Space Research Institute of the Austrian Academy of Sciences in Graz, Austria, in September 2017. This Symposium provided a unique opportunity to discuss the contributions of space technology and its applications for sustainable development in the light of the technical, legal, political, and regulatory developments. The determination of the long term impact of an activity can be a challenging task but can be overcome by creating a set of standardized Space Capabilities indicators covering all space related domains and applications, and providing orientation to countries in the evaluation of their capabilities and prioritization of activities. This set of indicators becomes a key decision support tool for countries in space related matters, and it could also support UNOOSA in the measurement of the impact of its activities and help propose targeted initiatives by identifying gaps or areas of improvement.

The United Nations/Austria Symposium on "Access to Space: Holistic Capacity-Building for the 21st Century" was therefore a flagship event towards UNISPACE+50 Thematic Priority 7 under which UNOOSA started the pilot project on a Space for Development Profile (SDP), to create a profile of the space domain of countries, and a Space Solutions Compendium (SSC), to provide solutions to improve the indicators defined in the SDP as suggested in the UN/IAC symposium outcomes. This conference opened a dialogue on TP7 to discuss innovative approaches to capacity building in the space domain, particularly in the areas of applications and technology, policy and law including the need to measure progress and development.

The European Space Policy Institute contributions to this UN/Austria Symposium included the presentation of a research on "New Visions, Methods and Tools for Capacity-Building in the 21st Century" by the author of this chapter. This ESPI contribution, considers that the space sector is currently entering into a new era, where changes in technologies and in social and environmental conditions, such as the advent of digitalisation, increased access to space data thanks to free, full and open data policies, and new key players from start-ups, private companies and NGOs to entire spacefaring nations are available. In this context, new disciplines and methodological approaches are converging towards novel ways of conducting business and academic research, which open up new opportunities for the emergence of an holistic capacity building scenario for all humankind.

In the space sector, such a scenario could open the doors to achieving a seamless chain of innovation and exploitation of space, distributed across the world thanks, for example, to digitalisation, innovative manufacturing technologies and miniaturisation of satellites. In this way, it could become possible for a vast majority of people to access space technology, conduct research and development, design, manufacture, assemble flight hardware, finally launching it to exploit the wealth of space data and services, that respond to the needs of their communities. The well-established United Nations/Japan Cooperation Programme on CubeSat Deployment from the International Space Station (ISS) Japanese Experiment Module (Kibo) "KiboCUBE", already provides developing countries the unique opportunity to deploy, from the ISS Kibo, cube satellites (CubeSats) which they develop and manufacture.

This enhanced networked and decentralised model, which includes CubeSats, concurrent design, additive layer manufacturing, "KiboCUBE" launch and operations, could be initially implemented and validated through the Regional Centres for Space Science and Technology Education which are affiliated to the United Nations. These centres are uniquely positioned to engage in synergistic partnerships with space agencies and industries to share design best practises, convey specific needs and requirements and identify existing matching space solutions. This could further the potential for economic growth and sustainable development in the regions covered, which could benefit from technology spin-offs, a larger number of educated young people, the availability of key services in the agricultural sector, in natural resources management, in e-health and tele-epidemiology domains, ensuring a more sustainable living for all on this planet.

For the longer term, improved connectivity in rural and under-served areas could extend these opportunities to both schools and universities, which could start with co-creating and delivering dedicated Massive Open Online Courses (MOOCs) in order to enhance STEM education opportunities for the younger generations. Space Agencies, Intergovernmental Organisations, Academia and Industry, thanks to their access to advanced space know-how, could sponsor and contribute to the creation of these innovative interdisciplinary curricula, thereby preparing more citizens to fully exploit the benefits arising from innovation, space technologies and data. This new approach is fully aligned with the objectives of the UN Agenda 2030 and could contribute to the achievement of a substantial number of sustainable development goals, while also addressing TP7 of UNISPACE+50.

Notably ESPI has been at the forefront of the UNISPACE+50 preparatory activities by organizing workshops, conferences and by participating to the two UN workshops just described above. In addition, it coordinated efforts at the UNOOSA High Level Fora, which took place in Vienna and Dubai, where it presented the outcomes of the ESPI 10th Autumn Conference, organized in 2016 on the theme "Space for Sustainable Development", the contributions of which are captured in the ESPI 2016 Yearbook on Space Policy published by Springer.[7]

In 2017 ESPI contributed to additional UNOOSA activities, notably the UN/WHO/Switzerland Conference on Strengthening Space Cooperation for Global Health, concerning TP5 (Strengthened space cooperation for global health) and involving also the World Health Organization, as the specialized Agency of the UN for health, where the author of this chapter presented the research "Science, technology and applications for global health".

Another highly relevant event has been the 3rd UN/ICAO Aerospace Symposium in Vienna, where ESPI presented a brief paper on suborbital flights and contributed to the discussions on Space Traffic Management (STM). ESPI also contributed to the Open Universe initiative, organized by the United Nations Office for Outer Space

[7] See Footnote 1.

Affairs and the Italian Space Agency, with the support of the European Space Agency, highlighting the relevance of International cooperation, policy and legal aspects and the educational, social and economic benefits of open data. In fact, the Open Universe initiative aims at promoting transparency in astronomy and space science data sharing, resurfacing inaccessible data and broadening the user-base.

All these ESPI activities had the main objective of identifying innovative and effective approaches to overall capacity building by discussing infrastructure for cross-sectorial and integrated applications, while enhancing partnerships for capacity building and institutional support, including recommendations on the future evolution of the UN regional centres.

1.3 ESPI Activities with ESA

In the meantime, the European Space Agency (ESA) initiated an activity to examine future global megatrends and organized a dedicated workshop to assess them in the context of the Space 4.0 era. Analogously to industry 4.0, the fourth industrial revolution, Space 4.0 heralds a new era in the global spaceflight domain, stemming from rapid and profound changes in technology, society and the economy, coupled with the increasing entry of new actors in the space industry, both on a national and industrial scale. ESPI also significantly contributed by joining these dedicated workshops and offering insights to shape ESA's future.

The above-mentioned streams of activities, championed by UNOOSA and ESA, are inextricably linked in defining our global future and offer complementary perspectives for capacity building considerations. Therefore, the importance of linking the Space 4.0 strategy with UN policies, frameworks and objectives was identified and further elaborated as part of a joint ESA-ESPI-UNOOSA conference held at the ESPI premises in January 2018, where the involvement of private actors and NGOs in addressing societal needs through the use of space, became one of the core elements around which this joint conference took shape.

1.4 The ESPI-UNOOSA-ESA Conference

The key theme proposed for the conference was *Space2030 and Space 4.0: Synergies for Capacity Building in the XXI Century*. Through this framework, the objective was to identify potential synergies, focusing, in particular, on four thematic priorities of UNISPACE+50 that were also relevant for Space 4.0:

- Innovation and exploration;
- Space for global health;
- Climate change and resilient societies;
- Capacity-building for the twenty-first century.

A special emphasis was placed on the future approaches of the diverse groups of stakeholders involved in innovation, exploration, global health and climate change, building upon the various UN workshops cited in the previous paragraph, including the outcomes of the on-going ESA/UNOOSA cooperation on the Space Solutions Compendium.

Each session included speakers from a range of different stakeholders including governmental and international agencies, NGOs, space industry, academia and users (see Fig. 1.1). The remainder of this chapter presents a summary of the individual contributions that can be found in the following chapters of this book, highlighting key perspectives, ideas and solutions. Moreover, each session also featured a co-chair with a specific regional focus, illustrating needs and solutions implemented by different global regions involving capacity building for space. The conference was enriched by two opening keynotes given respectively by Simonetta Di Pippo, Director of UNOOSA and Kai-Uwe Schrogl, Chief Strategy Officer of ESA.

The following paragraphs present a walk-through of the main topics and ideas presented in this book by international professionals with an expertise in space capacity building endeavours.

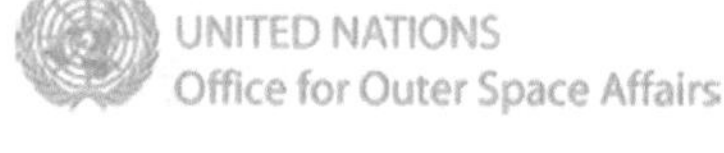

Fig. 1.1 Participant organizations to the ESPI-UNOOSA-ESA conference *Space2030 and Space 4.0: synergies for capacity building in the XXI century*

1.4.1 Opening Contributions

In her contribution, Chap. 2 of this publication, Director Simonetta Di Pippo, together with Markus Woltran and Martin Stasko (UNOOSA), highlight how the UNOOSA supports the use of space-based assets to support the three UN Frameworks Agenda 2030, COP 21 and Sendai Framework for Disaster Risk Reduction. As part of the UN, a unique organization that supports its 193 member states in fostering international dialogue and diplomacy, UNOOSA was tasked with preparing UNISPACE+50, in order to promote a Space2030 agenda, that takes into account the potential of space in supporting global sustainable development. The agenda was structured around the seven thematic priorities listed above, each contributing to four pillars: space economy, all the resources and activities that provide benefits to man-kind while exploring, understanding or using space; space society, space assets and know-how used across different sectors to benefit society; space accessibility regarding both launch capabilities and access to space data; and space diplomacy or the cooperation between nations to tackle global challenges with space. It is in this context that space capacity building is paramount to widen the pool of nations with space capabilities that can contribute to the identification and deployment of space-based solutions to improve life on our planet.

In Chap. 3 the other opening contribution is reflected, summarizing the keynote of the ESA Chief Strategy Officer Prof. Kai-Uwe Schrogl. Ulrike Bohlmann from the ESA Strategy Department, illustrates ESA's objectives and activities that will define its strategy within the space 4.0 era. Following on from the era of astronomy or space 1.0, through the space race defined as space 2.0, through space 3.0 the era of cooperation between few spacefaring nations to build the international space station, we have now arrived at space 4.0. This new space paradigm has offered ESA the opportunity to reflect on its role as the prime European Agency for space-related activities, and how it wishes to position itself in the coming years. Besides guaranteeing European access to space and presiding over the protection of its existing space infrastructure, ESA has set forth several goals that show its vocation for capacity building by maximizing the integration of space activities into the European economy and fostering a globally competitive European space sector. This will be achieved through four key actions which are to inform, interact, innovate and inspire.

1.4.2 Innovation and Exploration

The first section of this book is focused on the Innovation and Exploration thematic priority (TP1) of UNISPACE+50 and the related contributions span from Chaps. 4 to 11.

In Chap. 4, Roland Walter (University of Geneva) explains how space science knowledge is increasing together with the volume and variety of space missions, driven by a marked decrease in launch costs that is releasing the monopoly on space from a handful of spacefaring nations, and ushering in new governments, industry players, academia and private citizens. This data is an important driver of economic and scientific value of modern space endeavors. Having recognized the importance and value of this data, government agencies, scientific communities and economic development organizations are increasingly opening up their repositories to interested parties that can therefore perform research and extract value. To fully exploit this potential, considering the technological challenges and increased complexity, a new paradigm will have to be embraced. First, knowledge of the universe should be considered as part of mankind's common heritage. Developing dedicated cloud computing solutions and improving AI technology will also be paramount, as well as providing higher level interfaces to the raw data and establishing partnerships between complementary actors. This chapter discusses how UNOOSA could contribute to this paradigm, for example through its Open Universe Initiative in cooperation with the Italian Space Agency.

In the present scenario, where the space sector is coping with large changes both in technology, in the range of space actors present, and in the type of companies that are competing, there is also a need to change the way in which space missions are conceived and developed. Therefore, in Chap. 5, Stefano Ferretti discusses Disruptive R&D in the Space Sector. This chapter provides concrete examples of innovative approaches already used for R&D in the space sector and also identifies new methodologies which can be applied to space related processes at large, such as co-creation, open innovation, design thinking, innovative manufacturing, digital organizations and open service innovation. Finally, practical examples and suggestions that could be adopted by new space actors, wishing to leap-frog into the modern era, are presented.

The space industry often presents complex and time-consuming design, manufacturing and testing processes, with small production outputs that do not benefit from economies of scale. As pressures for efficiency weigh on the space sector, new technologies that improve communication and collaboration can be deployed to significantly improve the efficiency of the design phase. Chapter 6 by Antonio Martelo (DLR), focuses on the need both for space agencies and especially industry, to use new digital technologies to favor communication and cooperation, to reduce the research and development costs associated with space solutions. He focuses on Concurrent Engineering (CE), an approach used to evaluate the feasibility of new designs in a collegial manner by all those involved in the design phase. In particular, he illustrates that although many space agencies and large industrial companies have made significant investments into this technology, CE environments can be built using a modular and lean approach, making them scalable to any organization's needs, no matter how small. The lean approach is examined through different possible solutions and related costing methods. A practical implementation is presented using the example of DLR's "CE On the Go" facility, a portable CE environment currently used for educational purposes.

Not only innovation methodologies but also mega-trends in society are impacting space actors. Which are the most important mega-trends where opportunities lie for ESA to act as a broker and facilitator for new space-based opportunities, thereby providing capacity building towards its member states and beyond? Chap. 7 by Andrea Vena, Gianluigi Baldesi (ESA) and Arnaud Bossy (Frost & Sullivan) presents a study conducted together with Frost & Sullivan identifying nine key areas and related strategies that ESA should look to for its future development as it navigates the Space 4.0 era. These areas include first of all a series of innovative technologies such as space-based power, autonomous vehicles, cyber-security, artificial intelligence and blockchain. They also include health-based considerations specifically related to ageing populations, while Smart Cities were chosen as a topic which impacts societal resiliency, and finally they identified a regional focus on the African continent.

Another space agency that is leveraging its capabilities and technology for innovative capacity building in the Japanese Aerospace Exploration Agency (JAXA) In Chap. 8, Fuki Taniguchi, Hiroki Akagi and Kunihiro Matsumoto (JAXA) describe the partnership with UNOOSA to launch CubeSats from the Japanese Kibo module of the International Space Station. Through its unique features and technology, Kibo is ideally positioned to enable this type of launch on a regular basis and has already performed more than 200 launches including CubeSats from universities and research institutes. Through its partnership with UNOOSA, the program KiboCube offers this launch opportunity also to emerging countries that are members of UNOOSA, such as Kenya and Guatemala that were selected for the first two rounds. Through this program, Japan leverages its technology and space prowess to support developing nations in expanding their space development, fostering human resources and promoting new industrial space-based opportunities worldwide.

Not only space agencies but also private companies can foster capacity building through innovative uses of existing space assets. This is the case of the International Commercial Experiment Cubes (ICE Cubes) presented by Hilde Stenuit and Mauro Ricci (Space Applications Services) in Chap. 9, that is simplifying the access to the ISS for scientific experiments and technical demonstrators, supported by ESA's new policy to open up the ISS to commercial opportunities. This platform provides standardized tools and equipment, creating an end-to-end service to deploy and control experiments in microgravity with only 12 months of development and in a cost-effective manner. The service caters to fundamental and applied research, commercial R&D on life sciences and materials, in-orbit testing and validation of hardware, STEAM educational experiments and capacity building for new space actors. This chapter presents the current status of the platform, its future development and examples of successfully deployed experiments.

Within the context of space 4.0 and looking ahead at future space exploration programs, Chap. 10 by Veronica La Regina and Bernard Hufenbach (ESA) sets the context in which space innovation and exploration activities are being conducted by the Space Agencies and the private sector in the main spacefaring nations. One instrument analyzed to implement these activities in ESA is the one of partnerships between actors with complementary knowledge and capabilities. Each ESA Directorate has started implementing initiatives along these strategic guidelines, and the chapter offers concrete examples, focusing mainly on the "Innovation in Action" program set out by the ESA Space Exploration teams. This initiative sources ideas for the upcoming exploration destinations such as LEO, the Moon and Mars, following four key steps: fertilization, seeding, selection and implementation. First outcomes from this activity are presented in terms of new partnerships created.

In Chap. 11 Luciano Saccani, Director for International Business Development (Sierra Nevada Corp) presents the Dream Chaser and the unique mission planned together with the United Nations Office for Outer Space Affairs (UNOOSA) to offer United Nations Member States the opportunity to participate in an orbital space mission. Non-spacefaring countries will now have inclusive access to space, thanks to the Memorandum of Understanding signed in 2016 between UNOOSA and SNC, to coordinate a space mission for multiple United Nations Member States utilizing the Dream Chaser vehicle, developing and flying payloads for missions related to experiments in microgravity science, remote sensing or space hardware qualification. United Nations Member States are invited to provide payloads or experiments to be flown in LEO, with the requirement that they advance one or more of the Sustainable Development Goals.

1.4.3 Space for Global Health

The second section of this book is focused on the Strengthened Space Cooperation for Global Health thematic priority (TP5) of UNISPACE+50 and the related contributions span from Chaps. 12 to 19.

On the topic of how space assets and data may contribute to global health, in Chap. 12 Chiaki Mukai, Senior Advisor and Astronaut and Yoko Kagiwada (JAXA), together with Nanoko Ueda (Ministry of Foreign Affairs of Japan) explain how the International Space Station (ISS) has been extensively used by its partners to pursue research on human health that can be particularly useful in emergencies, epidemics and early warning events, strengthening capacity building in global health. The issues studied in microgravity include the weakening of bones and muscles, psychological stress and the effects of radiation. In addition, JAXA established several research partnerships, most notably with the World Health Organization (WHO) on Polio eradication, as well as other initiatives that use remote sensing to correlate environmental and epidemiologic variables.

Tele-epidemiology is a key asset also for CNES and its partners, to study, through space technology, the links between different factors that can influence the outbreak of disease epidemics both in humans, animals and plants. In Chap. 13 by Cecile Vignolles (CNES), we learn how these studies have contributed to highlighting in which ways the environment, local ecosystems and etiological agents contribute to disease epidemics. The initial concept has been applied successfully in Africa and is being further tested in different countries and for different diseases. Cooperation with local health authorities has enabled the creation of risk maps with high spatial and temporal accuracy and can significantly contribute to the prevention of epidemics, bringing health benefits to vulnerable populations.

Medical relief efforts in the field depend on the availability of critical information on the geographical location. Satellite-based data and GIS support thus have the potential to support humanitarian efforts and increase the efficiency of operations on the ground. In Chap. 14, Edith Rogenhofer, Sylvie de Laborderie, Jorieke Vyncke (Médécins Sans Frontières) with Andreas Braun (Universität Tübingen) and Gina Maricela Schwendemann (University of Salzburg), recount how, since the Ebola outbreak of 2014, this data has increasingly been used to support the activities of Médécins Sans Frontières (MSF). In addition, a large citizen science initiative called Humanitarian OpenStreetMap (HOT) was deployed whereby thousands of volunteers helped build maps bottom-up starting from satellite data and on the ground GPS readings. Space-based data has been used also to support vaccination campaigns and refugee assistance in informal camps, showing the clear benefits of the collaboration between different actors such as Copernicus, Image providers, the HOT communities, remote sensing specialists and teams on the ground.

Remote medical care solutions could be developed further thanks to space analogue research. In Chap. 15 Beth Healey, former ESA Medical Doctor at Spaceflight Analogue Concordia (Antarctica), summarizes her experience and provides an account of the scientific research conducted on "White Mars". Many technologies tested at Concordia can both enable future manned missions on planetary surfaces, but also support technology transfer activities on Earth. For example innovative Life Support Systems are today providing clean water in Morocco, in a similar fashion to the recycling of water on the International Space Station and at Concordia. Similarly, the medical and procedural know-how acquired during space analogue missions could be shared to provide tele-medicine solutions in developing countries. Finally, this contribution highlights the challenges of living and working in a remote and isolated continent, in a small international team, providing critical lessons learned for future space exploration ventures.

In Chap. 16, Stefano Ferretti and Alfredo Roma argue that in regions where local needs cannot be fulfilled by existing ground infrastructure or when these are out of service, space systems can deliver effective solutions, particularly when they are combined with airborne platforms during public health challenges, emergencies and natural hazards. Full interoperability and integration of data streams in real time among these systems allows end users to receive the information they need directly in the field, also using mobile equipment. Nevertheless, it is essential that the owners of space data allow for such an integration, so that their information could be enriched with data provided by airborne platforms, that have the added value of flexibility, modularity, accuracy and targeted response in a localized area, particularly once a critical situation evolves. Images, videos, samples of volcanic or nuclear clouds, etc., should be provided in transparent ways to the end users, irrespectively of their source, in line with an end to end scenario designed according to their needs and requirements. This data also needs to be structured through proper coordination mechanisms, to acquire, processes and store them in a fully integrated and secure manner.

Considering regional and national efforts in making the best use of space for global health, this book also presents the views of three initiatives, illustrated by representatives of the UN COPUOS expert focus group on global health, from Oceania, North America and Europe. Australia is a country where advanced research on combating global epidemics is being carried out. Their space applications program includes the use of space-based solutions for predictive modelling in disease surveillance. In Chap. 17, Chandana Unnithan, Ajit Babu and Melanie Platz explain the vision to create a global health alert system which could be useful in preventing and addressing global epidemics, as part of the Australian government's Health Security Initiative for the Indo-Pacific region launched in October 2017. The aim is to foster regional and global cooperation in order to coordinate responses to health security issues also through space assets, by leveraging on Australia's capabilities and know-how.

Canada's leadership in space technology is uniquely positioned to make a transformative impact in global health, as presented in Chap. 18 by Aranka Anema, Nicholas D. Preston, Melanie Platz and Chandana Unnithan. Innovations in robotics, satellite and radar imagery have catalyzed novel applications in healthcare within Canada and have the potential to inform the country's future directions in foreign aid. Applications of space technology are particularly well suited to Global Affairs priorities in pandemic preparedness, food and nutrition security, disease prevention and disaster response, and should form an integral part of Canada's foreign aid strategy. As these 'exponential technologies' continue to be tested, validated and scaled worldwide, Canada will need to consider how it can best position the use of digital data generated by these tools to foster accountability and transparency of investments towards the UN SDGs. Partnerships between government, private and non-profit sector will continue to be essential to achieve these goals. With political support and catalytic investments in healthtech, such as Grand Challenges Canada, the SD Tech Fund and the Innovation Superclusters Initiative, Canada aims to commercialize new technology products, processes and services that will improve global health for all.

Prof. Engelbert Niehaus from Germany has been promoting the Open Community Approach (OCA) for interfacing Space and Global Health at the UN COPUOS since several years. OCA leads to Open Educational Resources for the benefit of all humankind, making them available through Wikiversity, which serves also as a WikiJournal environment, similarly to peer-reviewed journals. In Chap. 19, Melanie Platz, Chandana Unnithan and Engelbert Niehaus explain why such a platform is ideal for medical tertiary education, offering the scientific community an open access to important resources without financial constraints, maximizing the wide dissemination of knowledge. Australia, for example, is managing the health infrastructure for many diverse communities, living in remote, distant locations from each other, thanks to an OCA based technology (Serval). In this way people can be more easily connected during emergencies, and health infrastructures needs can be assessed more rapidly.

1.4.4 Climate Change and Resilient Societies

A key thematic priority of UNISPACE+50 is international cooperation towards low-emission and resilient societies (TP6), which is addressed in the third section of this book, where relevant ideas and actions are presented in Chaps. 20 to 24.

In Chap. 20 Werner Balogh and Toshiyuki Kurino present the World Meteorological Organization and Space-based Observations for Weather, Climate, Water and Related Environmental Services. The World Meteorological Office (WMO) Space Programme links the providers (satellite operators) with the users, working in partnership with the Coordination Group for Meteorological Satellites (CGMS) and the Committee on Earth Observation Satellites (CEOS). WMO guarantees open data access to many developing countries in Africa, Asia, the Pacific, the Americas and the Caribbean, thanks to technical and financial support from satellite operators and from WMO space programme components: integrated space-based observing system; access to satellite data and products; awareness and training; space weather coordination. The future work of the WMO will be based on an Earth system approach, covering meteorology, climatology, hydrology, oceanography, seismology, volcanology, air quality, greenhouse gases, space weather and related multi-hazard and impact-based seamless services, whether over land, at sea, in the air or in space. United Nations global development agendas are closely linked to this approach, since the WMO contributes to 12 out of the 17 SDGs and, together with UN Environment (UNEP), acts as co-custodian for SDG 13 on Climate Action, including the implementation of the architecture for climate monitoring from space. The WMO Space Programme also supports the WMO Disaster Risk Reduction Programme and other related programmes that consider the impact of weather, climate and water on infrastructures and human populations.

The European Space Agency is also looking at capacity building efforts in Earth Observation, as outlined in Chap. 21 by Francesco Sarti, Amalia Castro Gomez and Chris Stewart (ESA). Satellites provide data at various scales (global, national or local), available in a continuous and consistent way for the entire Earth as a system, illustrating its evolution and the impact of climate change, which is crucial information for policy decision makers and for monitoring the proper implementation of international conventions related to environmental protection, sustainable development and mitigation of natural disasters. ESA is focusing on EO online education, e-learning (web tutorials, software, ibooks and tablet apps) and Massive Online Open Courses (MOOCs), targeting specific sectors or regions, and themes such as water management, where, for example, African users benefit of a dedicated Copernicus Data Access Cooperation Agreement. The hydrology Thematic Exploitation Platform instead is a development tool, that allows to build capacity in user communities, thanks to an open source and algorithms sharing policy. Other examples are the Sentinel-2 for Agriculture (Sen2Agri) to support the operational monitoring of agriculture at different scales, and a Space University in cooperation with the African Union, with the support of European Commission and ESA.

While assessing climate change and its effects on Earth, it is crucial to evaluate the resiliency of the environment and inhabited areas. In Europe 72% of people live in cities (80% by 2050), while globally, 50% of the world's population lives in urban areas, which are centres of economic growth, innovation and employment. In Chap. 22, Grazia Fiore (Eurisy) discusses how satellite applications are enhancing the quality of life in urban areas, particularly when faced with challenges related to infrastructures, environmental sustainability, social inclusion, and health. Eurisy carried out a survey of public administrations using satellite-based services, showing an increased resilience of urban areas to natural disasters thanks to a better predictability of floods, weather forecasts, vegetation status monitoring along water courses, slope and soil sinking risks, and air quality (i.e. temperature, pollution, presence of pollens and other allergenic substances). In case of natural hazards, rescue teams can generate post-disaster maps thanks to the International Charter "Space and Major Disasters" and the Copernicus Emergency Management Service.

Resilient societies may strongly benefit from space science, technology and its applications, which provide a wide range of solutions to global challenges related to global health, water, energy and urban development, while contributing to achieving overall economic and socio-economic sustainable development. In Chap. 23 Stefano Ferretti, Barbara Imhof and Werner Balogh illustrate how future space technologies may represent a unique opportunity for increasing sustainability on Earth. According to this strategy, which is based on the concepts of interdisciplinarity, technology spin-in and spin-off, and open innovation processes at large, sustainability could be achieved through space mission requirements and technologies with various maturity levels. The underlying idea is to integrate such space requirements into a roadmap which encompasses innovative key enabling technologies, big data, artificial intelligence systems and advanced robotics. In this way future technological developments on Earth, benefitting from an open innovation approach with space actors, will provide smart solutions to the citizens of tomorrow and open up new business sectors associated to these spin-offs. For example, a city could be seen as a spaceship system, since space habitats will include self-regenerative functions, and this perspective could also allow smart cities on Earth to become greener and more sustainable, especially in view of the population expansion and the resulting densification and increase of urban areas.

Moreover, in rural areas and in developing regions of the world, space technology offers innovative solutions to increase resiliency to climate change and related public health challenges. In Chap. 24 Shubha Sathyendranath (Plymouth Marine Laboratory), Anas Abdulaziz, Nandini Menon, Grinson George, Haley Evers-King, Gemma Kulk, Rita Colwell, Antarpreet Jutla, and Trevor Platt, discuss how Earth Observation could help to address the challenges faced, for example, by the communities of Vembanad Lake in Kerala, India, and the surrounding low-lying regions, due to frequent flooding, especially during the monsoon season. The floods of 2018 were exceptionally catastrophic, with up to a million people displaced, 400 deaths and widespread damage to public infrastructure, private homes and natural ecosystems. Copernicus data offers the possibility to reach a cell-phone application with a dynamic sanitation map for the region, to protect the population from some of the potential consequences of these natural disasters, such as cholera and other vector-borne diseases. The work also extends to nearby coastal waters and communities, and to beach safety from a human health perspective, involving the use of satellite data to map water quality and water clarity, ultimately supporting the development of more resilient societies. Public engagement and related capacity building is seen as the way forward to a sustainable future.

1.4.5 Capacity Building for the XXI Century

The final chapters from 25 through 33 address specifically capacity building efforts at large in terms of frameworks and international approaches, capturing recommendations around UNISPACE+50 thematic priority 7.

Around the time when the UN Agenda 2030 was launched, ESA had started an exercise to identify how existing space assets were being used and could further be used to support sustainable development. The Agency therefore started developing a catalogue, which is now available online, of ESA space solutions for each SDG. This activity is in line with UNOOSA's activities to share the benefits of space at large, while identifying opportunities towards achieving the UN Agenda 2030, and thus became a key contribution to the UNOOSA Space Solutions Compendium. Chapter 25, authored by Isabelle Duvaux-Bechon, Head of Member States Relations and Partnerships Office (ESA), describes the creation of this Catalogue of activities, its content and proposed use for capacity building as well as its future evolution in the context of addressing global challenges.

The United Nations created regional centers, following the second UNISPACE conference, in order to provide local capacity building activities in different regions that were not traditionally familiar with space activities. They operate to promote and outreach the benefits of space science and technology at all levels of society. Funmilayo Erinfolami (UN Regional Center, ARCSSTEE, Nigeria) reflects in Chap. 26 on how the UN regional centres can enhance their existing roles to meet the capacity building needs of the XXI century. From the perspective of the Nigerian regional center, ARCSSTEE, she illustrates the activities currently carried out to benefit a wide range of stakeholders, from preschoolers to government officials. Moreover, key partnerships are highlighted between the regional center and societal platforms such as NGOs, technology hubs, mentoring programs and social media platforms.

In Chap. 27, Servatius van Thiel (EU European External Action Service), examines the European Union's Space Policy and particularly its efforts to contribute to the UN's Sustainable Development Goals through capacity building initiatives, one of the three main policy objectives of the Council of Ministers, alongside autonomous European space capacity and a secure and safe space environment. Regarding sustainable development, the EU is already ensuring availability of space-based data and tools to facilitate its access and use. The chapter provides examples of important initiatives that the EU has initiated within Europe and in Africa and Latin America to foster the use of its data for transport, environmental monitoring and sustainable development. Moreover, it presents several interesting suggestions for additional activities, to be created in cooperation with UNOOSA and ESA, that can foster space capacity building towards a sustainable economic growth, leveraging key EU assets and capabilities.

The European Space Agency is similarly working towards the development of integrated applications, using satellite data, connectivity, positioning and other space technologies that can concur to the solution of societal problems. In Chap. 28, Roberta Mugellesi (ESA) presents the concept of Space Applications. These are solutions that make simultaneous use of different space services and technologies. Space applications combine different space systems (Earth observation, navigation, telecommunications, etc.) with airborne and ground-based systems to deliver solutions to local, national and global needs. The chapter discusses the different actors within ESA that are in charge of the building blocks for space applications, and the different programs that external firms in ESA member states can leverage to support the feasibility study, development and market application of integrated applications.

An international and non-profit organization that has a concrete and vast experience with space capacity building at global level, is the Space Generation Advisory Council (SGAC). Clementine Decoopman and Antonio Stark present its mission, activities and the challenges it has overcome in Chap. 29. SGAC, also known as SpaceGen, was created during the first UNISPACE conference to perform space capacity building among 18–35-year-olds, eliciting their ideas and solutions on space policies and strategies in order to supply space actors with the new generation's point of view. Through their decentralized organization with representatives in different countries and 13000 members, SGAC's capacity building activities can also be tailored to local needs and cultures. They include outreach events and conferences, as well as year-round project groups that produce research and position papers on specific themes. This work is enabled through the engagement of a large team of volunteers and the use of modern digital tools. Challenges of internal and external communications are also addressed with their relevant solutions. The SGAC therefore presents an interesting example to be followed by other international organizations and non-profit organizations.

In order to collaboratively develop space projects at global level, Concurrent Engineering can represent a key methodology and tool. In Chap. 30 Massimo Bandecchi (ESA) describes insights and achievements gained during 20 years of managing the ESA Concurrent Design Facility (CDF). It highlights the benefits of three specific aspects and applications of the concurrent design approach that can become assets for innovative capacity building: Education, Interoperability, System of Systems (SoS) architecting. Following several years in which the CDF was used predominantly for internal purposes, ESA started sharing its know-how about Concurrent Engineering as well as the software and the models that have been developed in the facility with the rest of the international community including National Agencies and institutions, space industry and academia that in turn developed their own CDFs. This has fostered educational opportunities, with theoretical and practical sessions including the design of CubeSats. The chapter further stresses the importance of interoperability considerations between systems and across firms and organizations to create a "global virtual design facility". Recent evolutions in technology have also enabled CDFs to be used for system-of-systems architecting through simulations of complex mission or disaster scenarios of which practical examples are illustrated.

Space capacity building includes also addressing the new technological trends that are making data and information increasingly available and open. This trend should be embraced by the space sector, where almost all datasets produced are a result of public funding and can therefore be considered public goods. In Chap. 31, Paolo Giommi, Gabriella Arrigo, M. De Angelis, S. Di Ciaccio, S. Iacovoni (ASI), U. Barres De Almeida (Centro Brasileiro de Pesquisas Físicas, CBPF), J. Del Rio Vera, S. Di Pippo (UNOOSA), A.M.T. Pollock (Sheffield University), present the Open Universe initiative, under the auspices of UNOOSA, to make all space science datasets available across society through the consolidation, standardization and expansion of data-based services. ASI therefore presents the prototype web portal it has developed, that centralizes access to a number of scientific data repositories. This prototype will be used to provide learning, to elicit requirements for Open Universe and to engage with non-space actors who will be able to discover the benefits of space science datasets. Cost/benefit considerations are then expanded upon, with recommendations of how data producers may adjust future programs to include open data concepts from the outset with minimal additional costs.

Connectivity can be considered an essential element for delivering innovative services and supporting space capacity building efforts. The Space Strategy for Europe underlines the role satellites can play to provide cost-effective solutions to improve connectivity for Europe's digital society and economy as part of future 5G networks, where numerous applications and services using space data will require uninterrupted connectivity. A recent wave of technological innovation and disruptive business models has led to a growing demand for new mobility services, integrating space and the transport sector at large: autonomous and connected cars, railways, aviation and unmanned aerial vehicles. Chapter 32 by Stefano Ferretti, Ludwig Moeller (ESA), Jean-Jacques Tortora (ESPI) and Magali Vaissière (ESA) identifies cooperative frameworks to encourage the interworking of satellite and terrestrial technologies in support to the transport sector, providing perspectives of the business and institutional communities and the roles of the various stakeholders involved in the development of satellite-based 5G networks and of future transportation vehicles. It focuses on markets where satellite communications provide strong value-add and complementarity to terrestrial 5G solutions and outlines the importance of innovative public-private partnership models.

1.5 Conclusion

In Chap. 33 Stefano Ferretti provides some conclusive recommendations and key messages on the theme *Space2030 and Space 4.0: synergies for capacity building in the XXI century.* Some contributors highlighted their specific regional focus, which is well illustrated and captured by the expressed needs, while the ESPI-UNOOSA-ESA conference participants discussed potential solutions that could be implemented by different global actors and regions through space capacity building initiatives. In the Innovation and Exploration area, Yeshey Choden (Bhutan) presented the initiative taken by a country with limited space experience, to launch their first satellite into orbit. This initiative was supported by their young monarch, that is driving the country towards new technological developments. The satellite, Bhutan-1, was successfully built and launched through a partnership with Japan, an advanced spacefaring nation, that was able to provide knowledge, tools and technology for Bhutan to reach this goal.

The Space for Global Health discussion included the story of Beth Healey (UK) who spent a year in the Concordia base in Antarctica as a Medical Doctor. Antarctica is an uninhabited continent that represents an opportunity to perform advanced R&D that can benefit future space programs involving the presence of humans on other celestial bodies and planetary settlement. In this case capacity building regards the development of techniques, procedures and human understanding necessary to redefine the boundaries of space travel as we know it today, through an international cooperation involving several developed countries.

Within the session on climate change and resilient societies, Ana Avila (Costa Rica) explained how thanks to specific and targeted policies, the country has developed a green economy based on renewable energies and natural resources preservation. This illustrates how the common values shared by a nation, of environmental respect, can be channeled through the appropriate policies to create new opportunities for industry.

Finally in the Capacity Building for the XXI century part, Funmilayo Erinfolami (Nigeria) illustrated a scenario with a fast-growing economy with the strong presence of the UN in the form of its regional center. The country is considering to develop an end-to-end service for sustainable development which is serviced by private (spinoff) companies, using space as a tool to connect the needs of the population with startups that can fulfill these needs, thereby offering also economic opportunities to their citizens.

This book includes all the contributions described in this chapter, provided by the ESPI-UNOOSA-ESA conference participants and by other invited contributors, as a further elaboration and explanation of the ideas, programs, issues and challenges related to space capacity building, grouped by theme. Putting all these contributions together, this book presents outlines of potential programs to be implemented on a global scale, as well as examples of concrete activities that have been initiated for space capacity building in the XXI century. By providing an overview of the national and international policy frameworks associated with space capacity building and the rationales for their adoption on a global scale, it offers also a tool to key decision makers on the way forward, including all different regions of the world and implementing activities with a range of actors, from agencies to industry from NGOs to new space companies. It ultimately highlights space capacity building programs that can empower the international community towards accessing all the benefits that space assets and data can offer to the economy and to society.

Chapter 2
The Case for a United Nations Vision for Outer Space Activities

Simonetta Di Pippo, Markus Woltran and Martin Stasko

Abstract Since the dawn of the space age in the late 1950s, humanity has come a long way in exploring and understanding space. Indeed, the very perception of outer space has changed dramatically in over six decades. Each and every year, the importance of space-based data, technologies, services and applications grows hand in hand with our increasing dependency on the benefits of utilizing outer space. The applicability of today's space technology is incredibly broad and it enables us to monitor the dynamics of the most striking challenges of our time, including climate change, disasters or threats to sustainable development.

2.1 Background

Since the dawn of the space age in the late 1950s, humanity has come a long way in exploring and understanding space. Indeed, the very perception of outer space has changed dramatically in over six decades. Each and every year, the importance of space-based data, technologies, services and applications grows hand in hand with our increasing dependency on the benefits of utilizing outer space. The applicability of today's space technology is incredibly broad and it enables us to monitor the dynamics of the most striking challenges of our time, including climate change, disasters or threats to sustainable development.

In 2015, the United Nations and the international community gathered to address these challenges collectively and designed the 2030 Agenda for Sustainable Development (2030 Agenda) and its seventeen Sustainable Development Goals (SDGs), with the aim of improving the lives of people worldwide. Space technologies and applications play a significant role in support of the goals and are fruitful tools, not only for tackling the targets enshrined in the document, but also in monitoring progress towards achieving them. This central importance of space for our society

---

S. Di Pippo (✉) · M. Woltran · M. Stasko
United Nations Office for Outer Space Affairs, Vienna, Austria
e-mail: simonetta.di.pippo@unoosa.org

S. Ferretti (ed.), *Space Capacity Building in the XXI Century*, Studies in Space Policy 22, https://doi.org/10.1007/978-3-030-21938-3_2

reinforces the need to preserve it as "global commons", which has to be protected through one joint vision.[1]

Space has become a cornerstone of our modern society as we collectively extend our presence in space with new satellites, telescopes, spacecraft and space stations. As a result, there are more actors in space than ever before, with not only more states and space agencies becoming involved in such activities, but at the same time the private sector has become a strong presence and competitor on the market. There are over 70 space agencies in the world, which, together with the private sector, launched more than 450 satellites in 2017 alone,[2,3] bringing the total number of operational satellites to more than 1,700.[4,5] The space sector has also become a major driver of the economy with its annual value of USD 330 billion expected to rise in the future due to fast-paced progress within the sector.[6]

To address the growing complexity of outer space activities, the United Nations Committee on the Peaceful Uses of Outer Space (COPUOS) mandated the United Nations Office for Outer Space Affairs (UNOOSA) to organize a milestone event (UNISPACE+50) to take place in June 2018 as a special segment of its 61st session. Similarly to the three main global United Nations frameworks—the 2030 Agenda for Sustainable Development, the Paris climate change agreement and the Sendai Framework for Disaster Risk Reduction 2015–2030—space, as an important part of our daily lives, requires a "Space2030" agenda in order to address challenges in the use of space technology and applications by everyone, everywhere, in support of these agendas. UNISPACE+50 served as an appropriate event to promote this and to gather inputs from the widest possible cross-sectoral audience from all levels of society.

The United Nations General Assembly, in its resolution A/RES/72/79,[7] emphasized the significance of UNISPACE+50, at which concrete deliverables and outcomes are to be concluded for presentation to the General Assembly in the form of a resolution to be considered at its seventy-third session. This includes the Space2030 agenda, together with its implementation plan, for strengthening the contribution of space activities and space tools to the achievement of the global agendas and to the

[1] Arévalo-Yepes C, Froelich A, Martinez P, Peter N, Suzuki K (2010) The need for a United Nations space policy, J. Space Policy 26: 3–8.

[2] Todd D (2018) Final score for 2017: 466—a new record for the number of satellites attempted to be launched in a single year. https://goo.gl/AQZkNy. Accessed 3 April 2018.

[3] Outer Space Objects Index (2018) Online Index of Objects Launched into Outer Space, Vienna, Austria. http://www.unoosa.org/oosa/osoindex/search-ng.jspx?lf_id. Accessed 28 March 2018.

[4] UCS Satellite Database (2017) Union of Concerned Scientists, Cambridge MA. https://goo.gl/x8q8LD. Accessed 3 April 2018.

[5] Outer Space Objects Index (2018) Online Index of Objects Launched into Outer Space, Vienna, Austria. http://www.unoosa.org/oosa/osoindex/search-ng.jspx?lf_id. Accessed 28 March 2018.

[6] Di Pippo S, Kofler R, Woltran M (2016) Global Space Governance and the Future of Space. Paper presented at the 67th International Astronautical Congress, Guadalajara, Mexico, Madison, 26–30 September 2016.

[7] Resolution of the United Nations General Assembly adopted at seventy-second session, UN General Assembly Document, A/RES/72/79.

long-term development concerns of humankind, based on the peaceful exploration and use of outer space, in line with the report of COPUOS A/72/20.[8]

2.2 The Office for Outer Space Affairs

The United Nations (UN), as a global organization that brings together 193 member states, serves as a facilitator to address common challenges. From its very beginning in 1945, one of the main purposes of the United Nations has been to "achieve international co-operation in solving international problems of an economic, social, cultural, or humanitarian character and in promoting and encouraging respect for human rights and for fundamental freedoms for all without distinction as to race, sex, language, or religion".[9] The 2030 Agenda for Sustainable Development, the Sendai Framework for Disaster Risk Reduction and the Paris Agreement on climate change represent three main global frameworks for addressing the common challenges humanity is facing, and space technologies, applications, data and services play a crucial role in both implementing and monitoring progress in tackling the issues elaborated in those documents. In this regard, UNOOSA, as the main UN agency on space matters, assists United Nations Member States and facilitates the coordination of UN activities using space technology to improve lives around the world.

The Office has been working towards bringing the benefits of space to everyone around the globe by promoting international cooperation in the peaceful uses of outer space and by facilitating the use of space science and technology for sustainable development. In its capacity building and outreach efforts, UNOOSA aims to help Member States in building their space capacity and capability in order to bring benefits from utilizing space to their citizens. With the adoption of the 2030 Agenda for Sustainable Development, the SDGs have provided an additional framework for the work of the Office. UNOOSA has aimed, through its activities and initiatives, to promote and facilitate the use of space for the fulfilment of the SDGs.

In its role as a global facilitator, UNOOSA has many unique responsibilities, serving as the secretariat to COPUOS and its subcommittees and relevant working groups, as the executive secretariat to the International Committee on Global Navigation Satellite Systems (ICG) and to the Space Mission Planning Advisory Group (SMPAG). The Office also has responsibility for implementing the United Nations Secretary-General's responsibilities under international space law and maintaining the United Nations Register of Objects Launched into Outer Space. In addition, the United Nations Platform for Space-based Information for Disaster Management and

[8]Report of the Committee on the Peaceful Uses of Outer Space on its sixtieth session, held in Vienna from 7 June to 16 June 2016, UN General Assembly Document, A/72/20.

[9]United Nations, Charter of the United Nations, 24 October 1945, 1 UNTS XVI, art. 1, para. 3. http://www.un.org/en/sections/un-charter/chapter-i/index.html. Accessed 31 March 2018.

Emergency Response (UN-SPIDER), falling under UNOOSA, has been developing solutions to address the limited access of developing countries to specialized technologies in regards to managing disasters.

The Office's Programme on Space Applications (PSA) has made substantial progress in deepening knowledge and experience of space applications around the world. Activities and initiatives under PSA, including capacity building, education, research and development support, and technical advisory services, have helped to reduce the space divide, that is, the gap in space capabilities between spacefaring and non-spacefaring countries.

2.3 UNISPACE Conferences

The three UNISPACE conferences held in 1968, 1982, and 1999 were organized with the aim of examining the practical benefits of space science and technology, and their applications, data and services, for humanity, with a special focus on the needs of developing countries in line with development agendas in support of gaining benefits at all levels of society—national, regional, and global. Serving as a platform for exchanging views and new ideas related to space exploration and practical applications of such technologies and as forums to foster international cooperation, the implementation of the outcomes and decisions taken at UNISPACE conferences have succeeded in delivering a wide range of economic, social and technological benefits to humankind.[10]

UNISPACE I (1968) recommendations led to the establishment of the Programme on Space Applications (PSA), with a focus on the provision of country capacity building, education, research and development support and technical advisory services to reduce the gap between the industrialized and developing countries. The establishment of routine space operations together with the resounding success of UNISPACE I paved the way for the second and third UNISPACE conferences.

UNISPACE II (1982) addressed concerns of maintaining outer space for peaceful purposes and preventing an arms race. At the centre of discussions also was the strengthening of the United Nations' commitments to promote international cooperation by bringing the benefits of space to developing countries to build up their indigenous space capabilities via the PSA. In line with these discussions, UNISPACE II recommended the expansion of the PSA and a broadening of its mandate, and, as a result, the United Nations-affiliated Regional Centres for Space Science and Technology Education were established.

UNISPACE III concluded with the adoption of the Vienna Declaration on Space and Human Development, providing the foundation for a strategy to address global challenges of the XXI century. UNISPACE III shaped UNOOSA for its future

[10]Di Pippo S, Kofler R, Woltran M (2016) Global Space Governance and the Future of Space. Paper presented at the 67th International Astronautical Congress, Guadalajara, Mexico, Madison, 26–30 September 2016.

and laid the basis for thematic discussions in COPUOS. The conference also lead to the establishment of the United Nations Platform for Space-based Information for Disaster Management and Emergency Response (UN-SPIDER), the International Committee on Global Navigation Satellite Systems (ICG), and the creation of the International Charter on Space and Major Disasters and the Space Generation Advisory Council.

2.4 Unispace+50

The UNISPACE conferences succeeded in moving the political space agenda and mandates of COPUOS and its Subcommittees forward, to keep pace with global developments and space activities. In 2015, COPUOS mandated the Office to organize a milestone event to take place in June 2018, as a special segment of its 61st session, to celebrate the 50th anniversary of the first United Nations Conference on the Exploration and Peaceful Uses of Outer Space (UNISPACE I), held in Vienna in 1968. The international community gathered to consider the current status as well as the future role of the Committee and of the Office as important players in the field of global space governance and in shaping the evolution of the space sector (Fig. 2.1).

UNISPACE+50 was a timely opportunity to discuss the future of the space domain in light of the dynamic developments of the past years, as well as the challenges to international cooperation, sustainability, safety and security of outer space, stemming from the evolution of the sector in the six decades of the space age.

The UNISPACE+50 roadmap aimed to define concrete outputs of space activities for the development of nations, taking into account the evolving and complex space

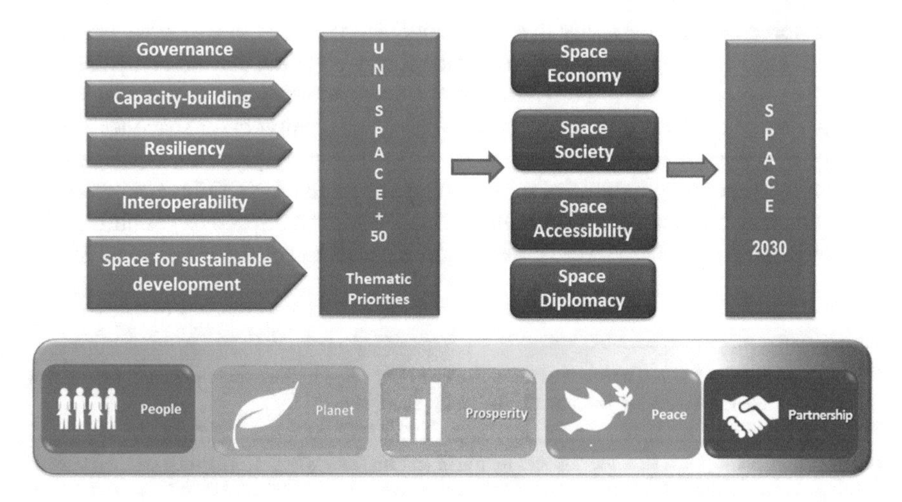

Fig. 2.1 UNISPACE+50 Process (©United Nations Office for Outer Space Affairs 2020. All Rights Reserved)

agenda. The Committee therefore adopted the following seven key thematic priorities of UNISPACE+50 to be addressed and implemented on a global scale, working with all relevant stakeholders and also as a link to the 2030 Agenda for Sustainable Development[11]:

1. Global partnership in space exploration and innovation;
2. Legal regime of outer space and global space governance: current and future perspectives;
3. Enhanced information exchange on space objects and events;
4. International framework for space weather services;
5. Strengthened space cooperation for global health;
6. International cooperation towards low-emission and resilient societies; and
7. Capacity building for the twenty-first century.

These UNISPACE+50 thematic priorities and their long-term deliverables will align with the three main global agendas adopted under the auspices of the United Nations: the 2030 Agenda for Sustainable Development, the Sendai Framework for Disaster Risk Reduction and the Paris Agreement on climate change. The discussions from UNISPACE+50 should enable nations, and all other actors, with the opportunity to build their unique capacities in space exploration and innovation.

UNOOSA, in accordance with the plan of work of UNISPACE+50[12] and the preparatory work towards the event, held initiatives corresponding to each of the thematic priorities, hosted by different States members of the Committee, to present to Member States and other space-related regional and international stakeholders each the thematic priority, their objectives and deliverables, and to engage stakeholders further in the implementation process.[13] The recommendations and official results of these activities have been injected into the UNISPACE+50 process through reports on each thematic priority.

The space arena is currently in a three-dimensional phase in which we are experiencing an increasing number of government space actors, private companies are becoming a major force in the sector, and the socioeconomic impact of space technology and spin-offs is ever growing. In order to effectively raise awareness of the current concerns of the space community as well as to improve understanding of the challenges and opportunities of today's space sector, UNOOSA elaborated the following four pillars: 'Space Economy', 'Space Society', 'Space Accessibility' and 'Space Diplomacy'. These were presented in 2016 as part of the High Level Forum:

[11]Report of the Committee on the Peaceful Uses of Outer Space, Fifty-ninth session (8–17 June 2016), UN General Assembly Document, A/71/20, paras. 296–297.

[12]Note by the Secretariat (2015) Fiftieth anniversary of the United Nations Conference on the Exploration and Peaceful Uses of Outer Space: theme of the sessions of the Committee on the Peaceful Uses of Outer Space, its Scientific and Technical Subcommittee and its Legal Subcommittee in 2018, UN General Assembly Document, A/AC.105/L.297.

[13]Note by the Secretariat (2016) UNISPACE+50: Thematic priorities and the way ahead towards 2018, UN General Assembly Document, A/AC.105/2016/CRP.3.

Space as a Driver for Socio-Economic Sustainable Development in its outcome document "The Dubai Declaration".[14] These pillars aim to enable discussions, as part of a holistic approach towards UNISPACE+50 and beyond, that address the cross-sectoral impact of integrating economic, environmental, social, policy and regulatory dimensions of space in pursuit of global sustainable development.[15]

"Space economy" can be defined as "the full range of activities and use of resources that create and provide value and benefits to human beings in the course of exploring, understanding and utilizing space". Space exploration and activities are on the frontier of scientific endeavours and these drive innovation, which brings economic benefits to the countries pursuing such undertakings. Space activities allow the creation of new markets but also strengthen current markets at all levels of society. Capacity building and awareness-raising of space activities are of utter importance, especially for developing countries, to bring economic growth and social well-being, and to improve the quality of life. Space technologies and resulting spin-offs deliver direct, tangible, and long-lasting benefits in various aspects of human life, even with modest investments. By addressing space economy, we aim to expose the vast applicability and usefulness of outer space activities for the growth and sustainable development of all nations.

"Space society" refers to a society in which nations and governments carry out their core duties and functions, while making the best use of space technologies and space-based services and applications that benefit society. Space is a source of inspiration but also sparks curiosity in the minds of people, which then leads to innovation and technological progress for the society as a whole. This pillar addresses the exchange of information and technologies to advance and contribute to the peaceful uses of outer space. International cooperation is required for successful and ubiquitous development and progress, and for bringing more countries and their people into the space sector.

"Space accessibility" aims to address the access of communities and decision-makers to space data, technologies and applications, and to utilize these for the benefit of society. Through this pillar, the focus is on providing enhanced access to space and its assets for both commercial and scientific endeavours, as well as open and free access to space-based data and information provided by such infrastructure. In order to bring more countries into the space domain, we must ensure that capacity building activities and consequent improvements in the utilization of space technology, applications, services and data are driven forward. As an international community, we must ensure that people around the globe have access to such benefits and that space does not constitute another area of social inequality.

---

[14]Dubai Declaration (2016), High Level Forum: Space as a Driver for Socio-Economic Sustainable Development, 24 November, Dubai, United Arab Emirates.

[15]Di Pippo S, Gadimova S, Woltran M (2016) Four Pillars to address the Future of Space. Paper presented at the 67th International Astronautical Congress, Guadalajara, Mexico, Madison, 26–30 September 2016.

"Space diplomacy" refers to cooperation among nations based on equal engagement and mutual respect in addressing the mutual goal of tackling the global challenges, while nurturing existing and forging new constructive partnerships to achieve this goal. Space diplomacy embraces both "space for diplomacy", that is, cooperation in space to improve international relations, and "space in diplomacy", that is, the use of space for peaceful purposes for improving international relations. In order to protect our future in space, there is a need to focus on and enhance the long-term sustainability of outer space activities by addressing the growing number of actors, space objects, space debris and threats from outer space. Moreover, in the six decades of the space age, the international community has achieved agreement on five treaties and five principles as a constructive approach towards sustainable and effective space governance. The pillar of "space diplomacy" can therefore be seen as a driver for a peaceful atmosphere of mutual respect and trust among actors in the space domain, and for collaborative advancement in using outer space for peaceful purposes.

In order to raise awareness about the usefulness and benefits of space, there is a need to introduce new communications methods and tools to explain the importance of space for society and its broad applicability in virtually all sectors. Despite growing efforts, there is still a continuing problem in the space sector as it often remains obscure to the general public when it comes to "selling itself". Space is one of the fastest growing industries and its future prospects are bright, with expectations that its economic value will more than triple in the coming decades.[16] Space's potential and recent developments need to be made publicly known for all to cherish, as with public support more investments are justifiable, which in turn brings economic growth and innovation to countries around the world. Building future communication strategies around the four pillars can be seen as a necessary first step in this direction, while also keeping in mind the possibility of new initiatives and activities to bring an increasing number of people into the STEM fields, and to the space sector in particular.

2.5 The Path Towards the "Space2030" Agenda

In the 2017 resolution A/72/79, the General Assembly emphasized the significance of UNISPACE+50, at which "concrete deliverables and outcomes are to be concluded for presentation to the General Assembly in the form of a resolution to be considered at its seventy-third session. This includes the "Space2030" agenda, together with its implementation plan, for strengthening the contribution of space activities and space tools to the achievement of the global agendas for addressing the long-term

[16]Sheetz M (2017) The space industry will be worth nearly $3 trillion in 30 years, Bank of America predicts. https://goo.gl/N9ooSf. Accessed 3 April 2018.

development concerns of humankind based on the peaceful exploration and use of outer space".[17,18]

The Space2030 agenda has to outline a comprehensive and inclusive long-term vision for space as one of the key drivers for development, while also addressing the capability of states in utilizing and taking advantage of the benefits stemming from space technologies, applications, data and services. These must be made available to the broadest possible user base worldwide, in order to bridge the space divide.

Similar to the digital divide, which is the gap in accessing information and communication technologies, including the internet, the space divide is the gap that divides the world in terms of the benefits and opportunities of the space domain. Space technologies and applications should not represent another field in which developing and non-spacefaring nations struggle to catch up with those that have already mastered space capabilities. In order to reduce this divide, it is necessary to provide new opportunities and enhance existing ones for accessing space.

The Space2030 agenda, as a forward-looking agenda, should focus on creating initiatives and programmes that make space accessible for everyone, everywhere, with a special focus on non-space actors. To broaden the number of stakeholders in the space community, there is a need to support emerging space nations by facilitating international cooperation in space exploration and innovation. The growing number of actors will inevitably influence the way we operate in space and therefore challenges to the safety, security and sustainability of outer space activities will be addressed. The agenda will also examine paths to strengthen international efforts in the effective use of space-based data, science, technology and applications for the monitoring of climate variables and disasters, and for supporting the achievement of the goals and targets related to sustainable development.

The Space2030 agenda will aim to develop policies and practices shared by all spacefaring nations, allowing every nation that desires and has the capability to use and to explore space, to carry out such activities for peaceful purposes, with respect to the long-term sustainability, safety and security of outer space. The main aim will be to provide a long-term vision for the contribution of space activities and tools to the achievement of internationally agreed global frameworks. It will be crucial to underline the mutual benefits of space, to identify and tackle all obstacles to international cooperation, and to facilitate the benefits stemming from utilization of outer space.

Space itself serves as a strategic domain for the functioning of our society but its importance and its uses have to be shared globally in order to provide everyone with an equal opportunity to take advantage of the broad range of space technologies and applications. Space is not subject to national appropriation but is a "global commons", which we need to protect through one joint vision. Space is a limited resource for the benefit of humankind and thus the ongoing task of the international community is to

---

[17]Resolution of the United Nations General Assembly adopted at seventy-second session, UN General Assembly Document, A/RES/72/79.

[18]Note by the Secretariat (2017) The "Space2030" agenda and the global governance of outer space activities, UN General Assembly Document, A/AC.105/1166.

prove its capacity to react to the changes and swift developments within the sector, with the aim of strengthening the governance of outer space.

2.6 Conclusion

The United Nations, stemming from its unique role as a global facilitator in world affairs, is irreplaceable by any other international or regional organization. It brings together 193 Member States in its quest to improve living conditions around the world by identifying and tackling challenges to sustainable development, security and international peace, human rights, and international law. Space technology and applications have for years been used as important tools in the United Nations system in addressing these issues. As space infrastructure carries significant relevance to, and constitutes an integral part of, efforts towards achieving the targets of the SDGs, UNISPACE+50 underlines the need to address developments and new realities in the space domain to set the optimal pathway on a global scale for the future of space activities.

Unimpeded access to and freedom of utilization of space for peaceful purposes has been the foundation of global space governance for decades, similar to the laws governing the open seas. Space activities in the past involved very few actors with limited opportunities for others to join the path; however, evolution in recent years has brought many countries and non-state actors into the sector, resulting in a growing complexity within the sector and with more possibilities for cooperation.

There is only one planet on which humanity can currently live and we are dependent on its limited resources, making it our duty to jointly protect it and to address the challenges we are facing as a species. Space science and technology are inextricably connected to our efforts in addressing these issues and success can only be achieved through cooperation and collaboration. Building partnerships has proven to be an effective way to pool resources, knowledge, staff or skills, and not only allows actors to invest less with higher returns, but also allows new actors—and therefore new experiences and innovations—to enter the field. The growing importance of international cooperation to deal with shared problems clearly indicates the great potential of sharing responsibility and in undertaking that which would be beyond the capacities of a single state or other entity.

The livelihood of people worldwide can be improved, but the space-faring nations, together with all relevant entities, need to stimulate those nations with limited or no capabilities and support them on their path to becoming relevant space actors in the future. Space technology, applications, data, and services not only help monitor the Earth but also create economic opportunities and new markets, serve as a source of inspiration and innovation, and support decision-making processes that directly influence sustainable development on our planet. That is why outer space must be kept open for everyone, with the support of the United Nations, as it has been for decades. Like-minded actors must come together to facilitate the development of space activities in countries worldwide, with the aim of bringing the benefits of space to all humankind. Only if international cooperation and partnerships are nurtured and cherished can all the challenges to humanity be truly addressed in an effective and successful manner.

Chapter 3
Space 4.0

Ulrike M. Bohlmann

Abstract At the ESA Council meeting at Ministerial Level 2016 in Lucerne, the Ministers of the ESA Member States adopted the landmark resolution "Towards Space 4.0 for a United Space in Europe". This chapter provides insights into the concept of space 4.0 and the way ESA understands and implements its role in this current era. Building on its 50 years of experience as Europe's Space Agency, conceiving and implementing with great success exciting and societally beneficial space projects, ESA has embarked on its journey through this new era. To capture ESA's conception of its own role in Space 4.0, the Agency has coined its own motto of Space 4.0i with four verbs to portray its main lines of action: inform, innovate, interact and inspire. This chapter not only depicts and deciphers the Leitmotiv of Space 4.0i and its components, but also provides concrete examples of how ESA implements its role for the benefit of the European citizen, society and economy.

3.1 What Does Space 4.0 Stand for?

Space 4.0 has been coined as a term to describe the context in which we currently undertake space activities. It builds upon the foregoing eras.

From the early days, the sky was contemplated by astronomers, who discovered the first patterns of celestial movements and mechanics. Space was an object of study that exerted a certain fascination, which is still a driving force for scientists today. The skies pointed the way to navigators and allowed for the first practical applications making use of the guidance from the lights from above. In the era of Space 2.0, the superpowers engaged in a first space race, with the political rivals investing heavily with the aim of establishing technological superiority for reasons of national security

U. M. Bohlmann (✉)
European Space Agency, Paris, France
e-mail: Ulrike.Bohlmann@esa.int

S. Ferretti (ed.), *Space Capacity Building in the XXI Century*, Studies in Space Policy 22,
https://doi.org/10.1007/978-3-030-21938-3_3

and national prestige, which were major factors in Cold War dynamics overall.[1] The acceleration of developments led to huge leaps in technology that left a legacy to build on in nearly all fields related to space activities and also sowed seeds in education, innovation and the economy at large. The international arena has undergone some radical changes since then, the world order is considered to be multi-or unipolar with the rise of the concepts of globalisation and global governance. The third era, Space 3.0, is therefore best characterised by the term "cooperation" as exemplified by the conception and realisation of the International Space Station as the greatest international project of all time, in which the United States, Russia, Japan, Canada and Europe are joining forces.

We are now witnessing the advent of a new all-embracing era: Space 4.0. Space is evolving from being the preserve of the governments of a few spacefaring nations to a situation in which there is increased engagement from governments around the world and in which we are witnessing the emergence of private actors, with commercialisation a growing feature. Space 4.0 represents the evolution of the space sector towards a new space era, characterised by a new playing field. This era is unfolding through inclusive interaction between governments, the private sector, society and politics. The term Space 4.0 echoes, obviously the concept of "Industry 4.0" to which it is constructed in analogy. Industry 4.0 refers to the on-going fourth industrial revolution encompassing both manufacturing and services.[2] Industry 4.0 includes the utilisation of contemporary automation, big data, data exchange and manufacturing technologies. New digital technologies, new business models, and smart integrated services are disrupting the traditional value chains. As industry is entering this new era revolutionising design, production and management mechanisms, space as a sector follows this trend, since innovation in space technologies and applications does not exist in isolation but is in fact very strongly linked to the innovation dynamics in other fields and parts of society. This aspect obviously becomes ever more marked as the space sector matures and grows more and more intertwined with other technology fields, adding an element of consumer goods manufacturing processes to spacecraft manufacturing while maintaining its other more strategic features. Many of the new general-use technologies—such as artificial intelligence, advanced robotics and 3D printing—have great potential for applications in space. Reciprocally, "Industry 4.0" is empowered by numerous features of how space activities are conducted, such as system approaches, extreme reliability or remote operations in harsh environments. The benefits of this evolution are palpable: the space sector can contribute to reshaping industry and the economy through knowledge policies and promote a society of innovation, with the trend towards growing integration of space and non-space technologies and processes. Integration in applications has the potential to concern and support many different aspects of daily life, as for example allowing the citizen to

[1]See: Ulrike M. Bohlmann and Moritz Bürger, "NewSpace—Putting an end to national prestige and accountability?" Proceedings of the 61st Colloquium on the Law of Outer Space, September 2018, publication upcoming.

[2]See: Klaus Schwab, The Fourth Industrial Revolution, Penguin 2017.

play a more important role in resources management via smart networks or providing for an increasingly effective governance via e-governance models.

3.2 How Does ESA Embody Its Role in Space 4.0?

ESA builds on its 50 years of experience as Europe's Space Agency, conceiving and implementing with great success exciting and societally beneficial space projects. Its founding instrument, the Convention for the Establishment of a European Space Agency[3] has proven its adaptability over the years, providing ESA Member States with the appropriate framework to carry out its mission.

Taking into account the space 4.0 era as described above, and the growing diversification as well as complexification of the space sector, ESA and the European Commission, together addressing 30 European Member States, on 26 October 2016 in a joint statement on shared vision and goals for the future of Europe in space,[4] defined the following overarching goals:

"*We thus have the goals to*:

- *maximise the integration of space into European society and economy, by extending the use of space technologies and applications to support public policies, providing effective solutions to the big societal challenges faced by Europe and the world;*
- *foster a globally competitive European space sector, by supporting research, innovation, entrepreneurship for growth and jobs, seizing larger shares of global markets; and*
- *ensure European autonomy in accessing and using space in a safe and secure environment, by consolidating and protecting its infrastructures.*

These goals shall rest on the solid foundation of excellence in science, technology and applications, expressed through an environment of outstanding education and skills and a thorough knowledge base."

To capture ESA's conception of its own role in Space 4.0, the Agency has coined its own motto of Space 4.0i with four verbs to portray its main lines of action: inform, innovate, interact and inspire.

3.2.1 ESA Informs

The facilitation of the exchange of scientific and technical information pertaining to the fields of space research and technology and their space applications found

[3]Opened for signature on 30 May 1975 and entered into force on 30 October 1980, text available at: https://download.esa.int/docs/LEX-L/ESA-Convention/20101200-SP-1317-EN_Extract_ESA-Convention.pdf.

[4]Joint statement on shared vision and goals for the future of Europe in space by the European Union and the European Space Agency, available at http://ec.europa.eu/DocsRoom/documents/19562/.

already a prominent place in the ESA Convention, namely in its Article III.[5] For decades, ESA has been sharing vast amounts of information, imagery and data with scientists, industry, media and the public at large via digital platforms such as the web and social media.[6] In a time where data has been labelled "the gold from Space",[7] ESA has taken the decision to facilitate, via an Open Access policy for ESA's information and data, the broadest use and reuse of the material for the general public, media, the educational sector, partners and anybody else seeking to utilise and build upon it. Through its digital agenda, ESA strives to increase the value of space information for all of its partners so as to arrive at a more transparent, immediate and collaborative cooperation with its Member States and delegations, to enhance science collaboration, to supply citizen science projects, and to enhance its free and open data policies.

3.2.2 ESA Innovates

ESA intends to drive Space 4.0 through its innovation strategy and is carried by the desire to assure a seamless grid of innovation including spin-off, spin-in and co-development activities, encompassing innovation in both technology and processes. ESA's programmes—ranging from space science, human spaceflight and exploration over Earth observation, space transportation, telecommunication and navigation to operations and technology—live from and for innovation. In addition, ESA develops concepts for extending the implementation of public private partnerships stimulates the European industry to develop innovative missions, technologies and services for commercialisation. Technical expertise and business-development support are provided to an ever-expanding network of ESA Business Incubation Centres—more than 500 start-ups have been supported already—so as to inspire entrepreneurs to turn space-connected business ideas into commercial ones and promoting a sustainable ecosystem of innovation-led start-ups and new ventures.

3.2.3 ESA Interacts

Since its inception, ESA has been all about cooperation and interaction. Providing for and promoting cooperation among European States in space research and

[5]See Footnote 3.

[6]For an overview of the historical evolution of ESA's Earth Observation policies, see: Ulrike Bohlmann and Alexander Soucek, From "Shutter Control" to "Big Data" Trends in the legal treatment of Earth observation data, in: Christian Brünner et al. (eds.), Satellite-Based Earth Observation, Springer, upcoming 2018.

[7]Margarita Chrysaki, Space: Still an important Matter of National Prestige? Available at http://www.europeanbusinessreview.eu/page.asp?pid=1820.

technology and their space applications, is ESA's stated purpose.[8] Cooperation with international partners, third States or other international organisations benefits of ESA's special nature and global position: ESA has programmatic ties with partners that have a varying political and economic background and can take on the role of a catalyst and enabler of international cooperation across the globe. Interaction with industry is not limited to the classical space industry, but is establishing partnerships also with non-space sector entities on differing levels: Together with its Member States, ESA supports the cooperation between European academia and research centres with industry so as to further encourage innovation in European space activities, their sustainability, and to fully exploit the benefits from space-based systems and the resulting data for the benefit of European society. Intensified interaction with academia contributes to nurturing an aptly trained and skilled workforce, ultimately contributing to Europe's overall knowledge-base and competitiveness. An emblematic example of ESA's interaction with the general public, is the Citizens' Debate on Space for Europe, an event organised in the run-up to the last Ministerial Council to gather opinions and ideas to help develop and nurture the future strategy for space in Europe: In 22 States, a total of about 2000 persons representing a broad diversity of citizens debated space issues during a day-long event in September 2016. This consultation exercise, on an unprecedented scale, was organised in all ESA Member States simultaneously, following the same structure and addressing the same set of questions.[9] The results of this direct interaction were fed directly into the strategic debate and found their way into the considerations on how to best meet societal challenges and put space at the service of the citizen.

3.2.4 ESA Inspires

The contemplation of the nightly starry sky has from the early days inspired generations of artists, poets and explorers. Building on this fascination and the innate human curiosity to find out more, space programmes have, since their inception, motivated children, teenagers and young adults to pursue a career in the STEM sector in general, and in the space sector in particular. Modern day heroes, such as astronauts and flight directors, provide living and tangible examples. ESA emphasises this inspirational aspect with the organisation of outreach activities and a unique education programme,[10] that addresses school children, teachers and Universities, each at their level. Offers concerning formal education, with teacher training and educational resources to support the curriculum in an innovative way via the ESERO (European Space Education Resources Offices) network are combined with Hands-on activities as well as informal education offers. The ESA Academy provides specific training courses and hands-on projects, such as Fly your Satellite or the REXUS/BEXUS

---

[8]See Article II of the ESA Convention.

[9]For results, see: https://esamultimedia.esa.int/docs/corporate/ESA-2_EN_BAG_30-11-3.pdf.

[10]http://www.esa.int/Education.

projects involving the launch of sounding rockets and stratospheric balloons, as well as the use of educational facilities, such as the Concurrent Design Facility. However, beyond the direct use of space as a subject or as the context of education, ESA aspires to inspire more: challenging endeavours, that are conceived for the long term, such as for example the Rosetta mission, the development of which started in 1992, and the start of which had to be moved from the originally foreseen date in early 2003 to March 2004, the trajectory of which had to be modified to accommodate a different target comet than originally foreseen, which performed an amazing number of manoeuvres during its ten year journey to finally arrive in the orbit of comet 67P/Churyumov–Gerasimenko and put a lander on its surface, are the most beautiful models for audacity, tenacity, resilience and the satisfaction to finally reach a goal after intense effort. The inspiration these missions spark is priceless.

Another ESA vision to spark inspiration and ideas is the Moon Village, a global exploration activity on the surface of the Moon, which has already attracted worldwide attention as well as the interest of space agencies, industry and science. It could be a successor to the International Space Station and bring together different actors, capabilities and interests and be a pit-stop or test-bed for further deep space exploration activities, reachable in a reasonable time frame. The vision of a Moon Village with its immense potential for innovation will provide opportunities for the emergence of new paradigms in science, private/commercial activities, human/robotic interaction, involvement of the general public with immersive technologies and ways of crowd-shaping. It offers a huge source of inspiration that can be seized upon for education, culture and public outreach.

3.3 Charting the Way Ahead

The foregoing chapter depicted and deciphered the Leitmotiv of Space 4.0i, under which ESA plays its role: Space 4.0i, and its components allow ESA to emphasise the main features we are pursuing in our programmes, based on the mandate given to ESA by its Member States over the years. It embodies the vision of ESA as THE space agency for Europe, an institution that constantly informs, innovates, interacts and inspires for the benefit of the European citizen, society and economy.

While new programmes and projects are continually being developed and the Agency continues to respond flexibly to challenges and opportunities ahead, this strategic approach will herald ESA's way into the future, the upcoming Ministerial Conference in 2019 and beyond.

Part I
Innovation and Exploration

Chapter 4
Space Science Knowledge in the Context of Industry 4.0 and Space 4.0

Roland Walter

Abstract Astrophysics deals with important questions characterizing human nature. Most of our knowledge was obtained in the last century. The data resulting from these explorations, collected in space or on ground, are of great value, encoding our knowledge of the Universe, of our past and of our future. With the emergence of clouds and deep learning, high level interfaces to these data could evolve towards providing direct access to knowledge. The United Nations Office for Outer Space Affairs can play a role towards a consensus in establishing a code of conduct on data handling and promoting the data and knowledge on the Universe as an intangible world cultural heritage.

4.1 Introduction

Astrophysics deals with the most important questions characterizing human nature. Where do we come from? What is the future of humanity? How did life begin? Is there life elsewhere? Our understanding of the Universe develops fast, although with detours and surprises. Most of our knowledge, from the source of energy in stars, their assembly in galaxies or the structure and history of the Universe was obtained in one century, largely thanks to the advances of technology. Astronomy has always been and is still a main driver of space science and exploration.[1]

Space is among the greatest adventures of mankind, largely the adventure of our generation. Space is dangerous, spectacular, with fascinating phenomena and full of opportunities. The data resulting from explorations, collected in space or on ground, is of great value, as it encodes our knowledge of the Universe, of our past and of our future. Astronomy and space therefore hold a special place in science communication,

[1] The 22 states parties to this convention (2010) Article V Activities and Programmes. in: Convention for the establishment of a European Space Agency. ESA Communications, ESA-SP-1317, 7th edition December 2017. Available at http://www.esa.int/About_Us/Law_at_ESA/ESA_Convention. Accessed 19 March 2018.

R. Walter (✉)
University of Geneva, Geneva, Switzerland
e-mail: roland.walter@unige.ch

S. Ferretti (ed.), *Space Capacity Building in the XXI Century*, Studies in Space Policy 22,
https://doi.org/10.1007/978-3-030-21938-3_4

because of the fundamental questions tackled, providing an exceptional link between science and society. Astronomy and space are also inclusive by nature: all humans are sharing the same spaceship.

The United Nation Committee on the Peaceful Uses of Outer Space is now considering space as a driver for sustainable development,[2] promoting the use of space data to foster not only collaboration, science and innovation but also inclusion. Providing straightforward interfaces to complex data and their analysis could make the process of generating science available to the society and education at large, a fundamental step to transmit the values of science and to evolve towards a knowledge society, worldwide.

High-energy astrophysics space missions pioneered and demonstrated how powerful legacy data sets can be for generating new discoveries,[3] especially when combined with data from different research infrastructures, and analysed in ways that the original researchers could not have anticipated. Nowadays the success of a research infrastructure is not only measured by the answers it provides to important scientific questions but also by the information it makes available to a wide community.

Space agencies,[4] science communities,[5] economic and development organisations,[6] recognizing that preservation and access to science data are central to their missions, have enforced the policy of having data becoming publicly usable. Since about 40 years the astrophysics community[7] develops standards for sharing data and making them available, mostly within specialized communities.

The business model of astronomers is however changing as the increase of data complexity and the decrease of the cost of launch services are calling for adaptations.

The data volume and computing power necessary to operate the upcoming astrophysics research infrastructures will increase by a factor of hundred in the next decade (Fig. 4.1). Even when Moore's law is considered, the complexity will increase by

---

[2]The Committee on the Peaceful Uses of Outer Space (2018) Draft resolution on space as a driver for sustainable development. United Nations A/AC.105/C.1/L.364, 9 February 2018. Available at https://cms.unov.org/dcpms2/api/finaldocuments?Language=en&Symbol=A/AC.105/C.1/L.364. Accessed 19 March 2018.

[3]White, Nicholas E. (2012) From EXOSAT to the High-Energy Astrophysics Science Archive: X-ray Astronomy Comes of Age. NASA Technical Report GSFC.ABS.7468.2012. Available at https://ntrs.nasa.gov/search.jsp?R=20120016404. Accessed on 19 March 2018.

[4]The 22 states parties to this convention (2010) Article III Information and Data. in: Convention for the establishment of a European Space Agency. ESA Communications, ESA-SP-1317, 7th edition December 2017. Available at http://www.esa.int/About_Us/Law_at_ESA/ESA_Convention. Accessed 19 March 2018.

[5]Committee on Issues in the Transborder Flow of Scientific Data (1997) Bits Of Power, Issues in Global Access to Scientific Data. The National Research Council, National Academy Press ISBN 0-309-05635-7. Available at http://www.nap.edu/read/5504/ Accessed 19 March 2018.

[6]The Secretary-General of the OECD (2007) OECD Principles and Guidelines for Access to Research Data from Public Funding. Available at http://www.oecd.org/sti/sci-tech/38500813.pdf). Accessed 19 March 2018.

[7]E.g. the data and documentation commission (https://www.iau.org/science/scientific_bodies/commissions/B2/) and the FITS working group (https://fits.gsfc.nasa.gov/iaufwg/iaufwg.html) of the International Astronomical Union.

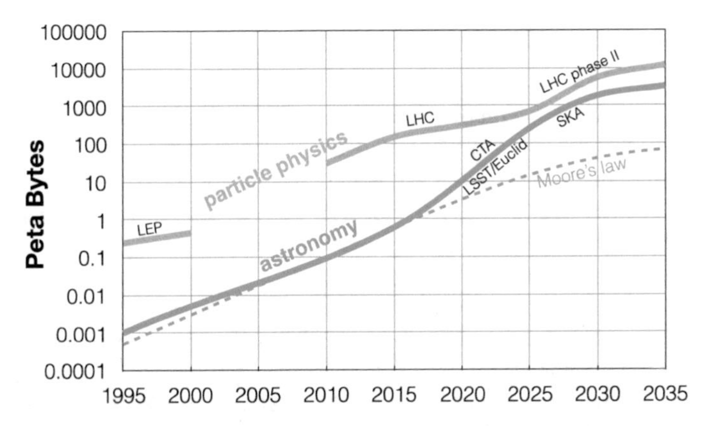

Fig. 4.1 Evolution of the data volume produced by astronomy experiments compared to that produced by the main particle colliders of the European Organisation for Nuclear Research (CERN). LEP and LHC stands for the Large Electron-Positron collider and the Large Hadron Collider. The acronyms of the astronomical experiments are mentioned in the text

a factor of ten. Astrophysicists are entering in the big data era and need an evolution of their business model. A centralisation towards clouds, providing storage, computing, and some standardisation is taking place in the industry and in science.[8] Handling data flows exceeding human insight, and accessing knowledge at a level of abstraction higher than the data themselves requires new tools, in particular artificial intelligence, which is at the core of the *fourth industrial revolution.*[9]

The cost of launch services is reducing drastically, from >10 USD down to about 1 USD per gram brought to low Earth orbit,[10] and has the potential to shift the space leadership from the few spacefaring nations to numerous space actors, industries, academia, and citizens. The multiplication of these actors makes their coordination, and definition or adoption of standards for data and knowledge management and sharing, more challenging. This transition was christened *Space 4.0* at the 16th ministerial council of the European Space Agency.[11]

[8]The European Commission (2018) Implementation Roadmap for the European Open Science Cloud. SWD(2018) 83 final. Available at http://ec.europa.eu/research/openscience/pdf/swd_2018_83_f1_staff_working_paper_en.pdf#view=fit&pagemode=none. Accessed 19 March 2018.

[9]The World Economic Forum (2018) Fourth industrial revolution. Available at http://www.weforum.org/agenda/archive/fourth-industrial-revolution. Accessed 19 March 2018.

[10]Falcon heavy can lift 63800 kg to low Earth orbit for 90 M$ (http://www.spacex.com/about/capabilities), and 30% less using reusable components (https://twitter.com/elonmusk/status/726559990480150528).

[11]The European Space Agency (2016) What is space 4.0? Available at http://www.esa.int/About_Us/Ministerial_Council_2016/What_is_space_4.0. Accessed 19 March 2018.

4.2 Evolution of Knowledge Management

Interfacing the knowledge on the Universe should be considered at different levels: data, analysis, interpretation and knowledge. The volume of details and information decreases along that sequence while the amount of study, reduction and computing increases. The level of abstraction increases from analytic to synthetic and common sense or even wisdom (Fig. 4.2).

The first digital archives of astrophysical data were created in the 1970s (see footnote 3) and developed to provide access to detector and high-level data resulting from a standard data analysis and characterising the observed targets. Data were made accessible first through interfaces dedicated to specific missions. Later, generic portals were adopted along with the world-wide web in the 1990s.[12] Eventually virtual observatory[13,14] protocols were introduced ten years later to help automated data discovery and sharing worldwide. Virtual observatory protocols were also defined to integrate high-level analysis services.

Data mining and knowledge discovery concepts were introduced in the 1990s and made clear that data analysis was an iterative and interactive process as science questions and methods evolve over time. The successive data preparation, modelling and evaluation steps indeed depend on the scientific question tackled and on the user needs.

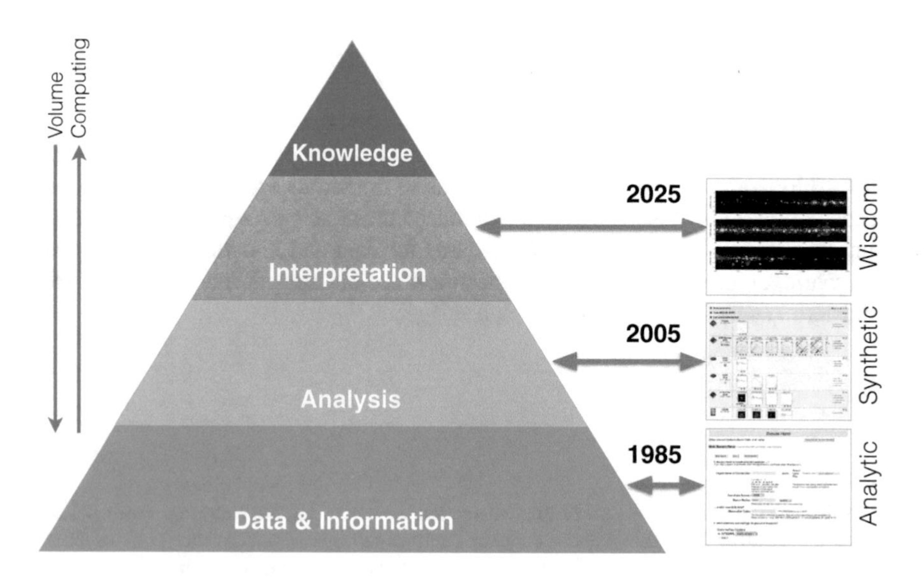

Fig. 4.2 Evolution of knowledge management in astronomy

[12]The World Wide Web Consortium (2000) A Little History of the World Wide Web Web. Available at https://www.w3.org/History.html. Accessed 19 March 2018.

[13]Szalay A.S. and Brunner R.J. (1998) Astronomical Archives of the Future: a Virtual Observatory. Available at https://arxiv.org/pdf/astro-ph/9812335.pdf. Accessed 19 March 2018.

[14]Benvenuti, P. (2001) The Astrophysical Virtual Observatory. STECF Newsletter 29,4, Available at http://adsabs.harvard.edu/abs/2001STECF..29....4B. Accessed 19 March 1018.

Several interfaces appeared from 2005 onwards using such ideas for specific astronomical missions, allowing to analyse data on-the-fly based on user specific parameters. These tools effectively interface users with the analysis rather than with the data and provide improved flexibilities for non-specialists. Some of these systems allow to remotely analyse[15, 16] observation data sets while others,[17] closer to data mining, start from transformed pre-processed data and allow to quickly generate mission-wide products suitable for various types of analysis and science goals.

As analysis software, specific to an experiment, was rarely maintained over decades, it was practically impossible to repeat analysis, or to extend it to new targets and science goals, especially when an experiment or space mission was concluded and detailed know-how had vanished. This is a fundamental limit of services, like the virtual observatory, based on pre-calculated archive holdings.

The latter services could provide an analytic access to mission data to everyone if suitable interfaces were provided. Technically, interfaces to such services could be built by extending virtual observatory standards to the needs of remote processing. These services could easily run on virtual machines on clouds, simplifying software maintenance and portability. If designed properly, they can also easily be moved between institutions, if the need arises.

Machine learning algorithms started to have an impact in the early 2000s. Deep learning can find interesting results on data sets without the need to write specific software. Supervised machine learning is particularly good to classify data and can be used to search, understand voices, recognize spam e-mails, act as a virtual helpdesk, or drive cars. Deep learning can learn, from the multiple translations available in international organizations, how to automatically translate texts. It can also be used to recognize galaxies, classify detector events, detect gravitational waves, etc. Countless new applications are developed. Generic deep learning algorithms usually outperform specific algorithms or even human classification. Deep learning outperforms humans on go, chess, poker and soon for all sort of activities.[18] Machine learning will progress further, generating games, movies on-demand or rewriting this article, in the future.

As training of neural networks is computationally expensive, the industry is developing hardware for accelerated computing, such as Graphical Processing Units (GPUs) or other specialized processors concentrating thousands of computing cores

[15]Pence, W.D. and Chai P. (2007) The HEASARC HERA system. ADASS XVI. ASPC 376, 554, Available at http://adsabs.harvard.edu/abs/2007ASPC..376..554P. Accessed 19 March 2018.

[16]The UK Swift Science Centre (2014) Swift XRT products generation system. Available at Leicester University http://www.swift.ac.uk/user_objects/. Accessed 19 March 2018.

[17]Walter R., et al. (2010) INTEGRAL in HEAVENS. Available at https://pos.sissa.it/115/162/pdf. Accessed 19 March 2018.

[18]Grace, K. et al. (2017) When Will Artificial Intelligence Exceed Human Performance? Available at http://arxiv.org/abs/1705.08807v2. Accessed 19 March 2018.

in a single chip. The combination of these software and hardware technologies is transforming our societies in many areas and even "who we are as human beings".[19]

Machine learning algorithms will be used to analyse the data collected by future research experiments studying the Universe, especially these characterised by large and complex data sets. Such techniques can be used for data analysis (e.g. performing data calibration, selection, etc.) or data interpretation, providing tools to allow the study of abstractions such as astronomical objects characteristics or their distribution in the Universe. Some examples of the use of neural networks in the analysis or interpretation of astronomical data, which are relevant for the next generation of instruments, are given below.

The Cherenkov Telescope Array[20] is the next generation ground-based observatory for gamma-ray astronomy. It will be built incrementally from 2019 and 2025, will start conducting scientific observations in 2022 and will be made of more than 100 telescopes measuring the signature of very high energy photons and particles producing flashes of blue light when interacting with the Earth's atmosphere. These flashes are recorded by cameras imaging the sky up to a billion times per second and generating about 10 Peta Bytes of data per year. Separating flashes generated by photons or particles is a key aspect of the data analysis and is performed efficiently by deep neural networks (Fig. 4.3).

The space mission Euclid[21] from the European Space Agency and the Large Synoptic Survey Telescope[22] will observe billions of galaxies in the infrared and visible light with the primary goal (among others) to map the geometry of the Universe through the study of weak gravitational lensing, correlations, supernovae, etc. Machine learning techniques are very efficient to improve galaxy images, to measure shear of these images created by weak gravitational lensing of dark matter on the line of sight, and to detect strong gravitational lenses.[23] They will be a technique of choice to interpret the Euclid data.

The Square Kilometre Array (SKA)[24] is planned to become the world's largest radio telescope array, with over a square kilometre of collecting area spread across continental scale in Australia and South Africa. High level, three dimensional images,

[19]Schwab, K. (2018) Shaping the Fourth Industrial Revolution. World Economic Forum. Available at https://www.weforum.org/agenda/2018/01/the-urgency-of-shaping-the-fourth-industrial-revolution/. Accessed 19 March 2018.

[20]The Cherenkov Telescope Array Observatory (2016) Exploring the Universe at the Highest Energies. Available at https://www.cta-observatory.org. Accessed on 19 March 2018.

[21]See "Euclid: an ESA mission to map the geometry of the dark Universe" (http://sci.esa.int/euclid), Euclid was approved in 2011 and should be launched in 2021.

[22]See "Large Synoptic Survey Telescope" (https://www.lsst.org). LSST is being built and its first light is planned in 2020.

[23]E.g. "Generative Adversarial Networks recover features in astrophysical images of galaxies beyond the deconvolution limit" (https://arxiv.org/abs/1702.00403) & "Deep Convolutional Neural Networks as strong gravitational lens detectors" (https://arxiv.org/abs/1705.07132) & "Hopfield Neural Network deconvolution for weak lensing measurement" (https://arxiv.org/abs/1411.3193).

[24]The SKA Organisation (2018) The Square Kilometre Array, exploring the Universe with the world's largest radio telescope. Available at https://www.skatelescope.org. Accessed 19 March 2018.

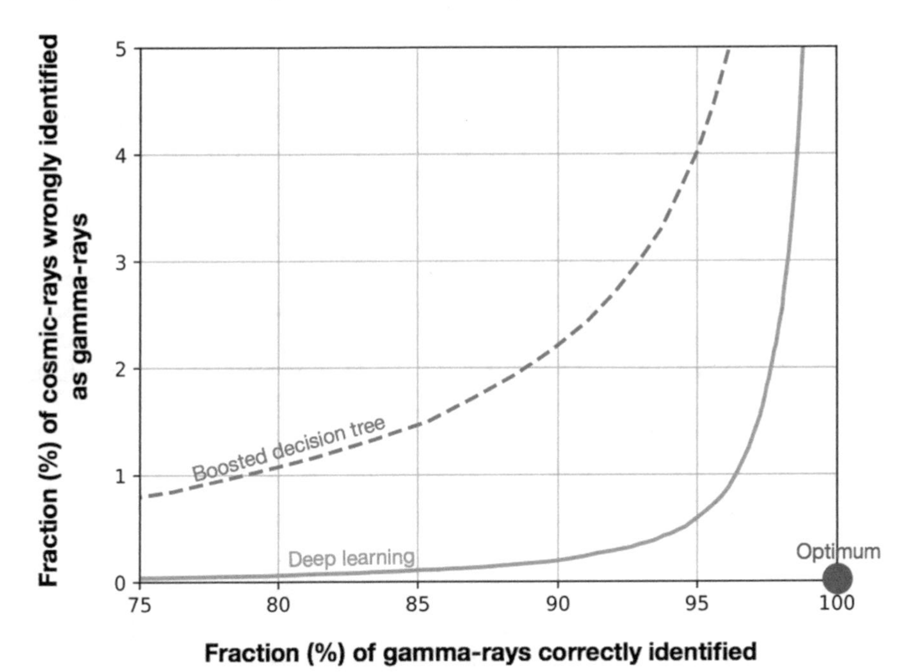

Fig. 4.3 Comparison of the efficiency of deep learning with classical techniques to separate gamma-ray from cosmic-ray images (for the CTA northern array in the energy band 0.3–1.8 TeV). The red point shows the optimal goal i.e. a perfect separation. The curves show the separation efficiency of particular techniques for continuous values of a selection parameter. The blue (dashed) curve is the state of the art obtained with classical methods (boosted decision trees on parameters extracted from the images). The orange curve was obtained applying deep learning techniques on the same selection of events, providing much improved results and not requiring any prior interpretation of the data, contrasting with classical methods

to be made available daily to the users, will be one Peta Byte in size and each include information for up to a million radio-sources. These volumes and source numbers are too large to be analysed by humans. Neural networks will be mainstream to classify astronomical objects or find specific features[25] e.g. to give access to statistical information on the ensemble of detected sources.

These experiments, observing a large fraction of the sky, will produce large data flows and provide information for many additional scientific studies beyond their primary goals. Deep learning will be very efficient in handling data sets exceeding human grasp. As the value of data lies in its use, offering interfaces to interpret this data is fundamental. The idea is to shift from the current paradigm (obtaining data) to allow users to directly build more abstract and deeper understanding. This can be done in various contexts for instance by selecting (allowing users to specify learning

[25] Aniyan, A. and Thorat, K. (2017) Classifying radio galaxies with convolution neural network. ApjS 230, 20. Available at https://arxiv.org/abs/1705.03413. Accessed 19 March 2018.

samples), cleaning (e.g. using generative adversarial networks), or defining measures and extracting statistics on them from many data sets.

Such interfaces are not yet available and will be very helpful to scientists and could be extended to other users. Neural networks are also at the base of natural language interfaces which could be adapted to the handling of statistical and scientific data.[26] Even the public could get access to that knowledge if suitable interfaces to interpretation (catalogues, source characterisations, models) are made available. Data2Dome[27] (supported by the European Southern Observatory), which displays astronomical data and models in planetariums in real time, or the ESASky web interface[28] (supported by the European Space Agency), providing access to data from space astronomy missions on the entire sky, are very interesting initiatives in this context and could be extended to the needs of scientific interpretation.

4.3 Challenges

Even in astronomy, a subject appealing to the public, making data and knowledge available and used widely is a challenge and will be even more so in the future because of the complexity and size of the data and of the multiplication of providers. This is also true in technologically advanced economies, which may face a rift between people benefitting or challenged by the implications of artificial intelligence. People need the opportunity to be exposed to artificial intelligence techniques, to understand one of the driving force of the economy and the responsibilities and ethical questions[29] associated.

The current paradigm (providing access to data) suffers from several difficulties which make data hard to use beyond the circle of specialists. The expansion of that circle will benefit from resurfacing the data (a significant fraction of the data is not publicly accessible) and analysis tools (which are often not usable, not accessible, not documented, too complex, or, even worse, lost or not maintained). Ways to improve the current paradigm are discussed elsewhere in this volume.[30] If these actions will be beneficial for the scientific community, they may not be good enough for the society, effectively maintaining a digital divide between specialists and outsiders.

---

[26]Neelakantan A. et al. (2017) Learning a Natural Language Interface with Neural Programmer. ICLR 2017. Available at https://arxiv.org/abs/1611.08945. Accessed 19 March 2018.

[27]IPS and Data2Dome (2017) Data2Dome bringing together astronomy data providers, science centre professionals and software vendors to advance the state of the art in big data visualisation. Available at http://www.data2dome.org. Accessed 19 March 2018.

[28]The European Space Agency (2018) ESAsky. Available at http://sky.esa.int/. Accessed 19 March 2018.

[29]Brundage M., et al. (2018) The Malicious Use of Artificial Intelligence: Forecasting, Prevention, and Mitigation. Available at https://arxiv.org/abs/1802.07228. Accessed 19 March 2018.

[30]Giommi, P., et al. (2018) The Open Universe Initiative. Space Capacity Building in the XXI Century. pp. 375–384.

If one wishes the data to be used beyond the circle of specialists, to transmit the values of science, for inclusion and sustainable development, improvements are needed at different levels, from the use of the underlying infrastructure to more abstract user interfaces. Accompanying the shift or paradigm requires actions: a few are suggested below.

The knowledge on the Universe is a universal heritage: The data and the knowledge collected on the Universe should be considered as a heritage of the humanity, available and preserved for all the intelligent species of the Universe.

Science clouds: Data exploitation will require increasing computing power and possibly less specific software if deep learning continues to develop. Scientists need access to centralized and cost effective computing infrastructures, especially GPUs, and to standards for the provision of services. The European Commission is proposing to interconnect public and private data infrastructures in Europe to develop a European Open Science Cloud.[31] For this purpose, a set of standards and infrastructures to support science are being created. If successful, these standards should also be the base of the data and software resurfacing efforts mentioned above.

Affordable computing: Opening data exploitation to the science community at large (beyond the circle of the specialists), to education and the public requires, at some level, freely available computing. The public should be allowed to interface with the data and their interpretation, using free services as a return for public funding. Scientists not belonging to an experiment should manage to access sufficient resources to allow using the data at all levels, if required.

Interfacing analysis, interpretation and knowledge: Interfacing data does not provide the level of abstraction required by non-specialists. Services can be built to provide higher level interfaces and to make the data and the products of their analysis available to modern analysis techniques. Interfaces could be built up to a level where they can provide information, user driven analysis and knowledge usable by scientists, education and the public.

Partnership between actors: In a world where space data providers will be more independent and numerous, the question of standardisation of data and software becomes more challenging. Open science clouds can help this process at least for publicly funded experiments. The integration of privately funded research experiments might be helped by public-private partnerships for data exploitation, and at least by the availability of open standards and codes of conduct.

Mitigating the artificial intelligence divide: Artificial intelligence will have a large impact on science and societies, acting positively and negatively. People in general, and education, need to be confronted to these techniques. Providing proper interfaces could allow at least students at various levels to experiment with modern analysis techniques on astrophysics knowledge, especially the knowledge derived from space experiments.

---

[31]The undersigning stakeholders (2017) European Open Science Cloud Declaration. Available at https://ec.europa.eu/research/openscience/pdf/eosc_declaration.pdf#view=fit&pagemode=none. Accessed 19 March 2019.

The United Nations Office for Outer Space Affairs (UNOOSA), promoting the cooperation between states and space actors, can play an important role towards a consensus, for instance in establishing a code of conduct on data handling in partnership with the scientific community. It could also raise the importance of establishing and coordinating infrastructures capable of inclusion at the level of the states, international organisations and the economy. Finally, it could take the opportunity of UNISPACE + 50 to promote the data and knowledge on the Universe as an intangible world cultural heritage.[32]

It was a pleasure to participate to the conference on Space 2030 and Space 4.0 synergies for capacity building in the XXI Century, co-organized by the European Space Policy Institute and by the UNOOSA on the 3rd of February 2018, to hear points of view on topics from a broad perspective and to share the above considerations.

Acknowledgements Figure 3 was created with the help of a grant from the Centre for Advanced Modelling Science at the University of Geneva, computing at the Swiss National Supercomputing Centre (project cad03) and advice from the Swiss Data Science Centre. Figure 1 includes materials from http://www.isdc.unige.ch and https://www.cta-observatory.org

[32]The 176 states parties to the convention (2003) Convention for the Safeguarding of the Intangible Cultural Heritage. Available at https://ich.unesco.org/en/convention. Accessed 19 March 2018.

Chapter 5
Disruptive R&D in the Space Sector

Stefano Ferretti

Abstract The Space sector is entering into a new phase which is characterized by the digital revolution as well as by industry 4.0, while it is becoming increasingly relevant to address the global challenges associated with the United Nations Agenda 2030 for Sustainable Development (Stefano, F. (2016). Space as an enabler of sustainable development. In *Proceedings of the 4th International Conference on Sustainable Development*. September 21–22, 2016, New York, USA). In this new context, it is essential that Space Agencies refocus their activities and programmes in such a way that they can properly innovate, inform, inspire and interact, while sustaining and improving development and economic growth in their member states (Stefano, F., Imhof, B., Balogh, W. (2016). Future space technologies for sustainability on earth. In *Proceedings of the 67th International Astronautical Congress*, Guadalajara, Mexico), and offering new opportunities for cooperation with new private actors (e.g. Space X, Planet, etc.), which are entering into their traditional ecosystem, disrupting both the traditional technological and business models. A new framework is therefore arising with the goal to ensure a seamless chain of innovation, enabling more cooperation between academic institutions and research establishments together with industry and end-users, in order to allow for uninterrupted, rapid development from idea to product or service and sustain competitiveness in an efficient manner, also through new funding schemes and commercial partnerships for dedicated activities, that could inspire citizens while addressing challenging space projects for the decades to come. The chapter provides concrete examples of innovative R&D approaches in the space sector and intends to identify key methodologies, which can be applied to space related processes at large, such as co-creation, open innovation, design thinking, innovative manufacturing, digital organizations, open service innovation (Stefano, F., Moeller, H. L., Tortora, J. J., Vaissiere, M. (2017). Space and SATCOM for 5G-European transport and connected mobility. In *Proceedings of the 68th International Astronautical Congress (IAC)*, Adelaide, Australia; Stefano, F. (2015). Future applications and benefits of human space flight innovations. In *Proceedings of the 66th International Astronautical Congress*, Jerusalem, Israel; Ferretti, S., Hulsroj, P. et al. (2016). *Europe in the future and the contributions of Space* (ESPI Report 55),

S. Ferretti (✉)
European Space Policy Institute (ESPI), Vienna, Austria
e-mail: stefano.ferretti@esa.int

S. Ferretti (ed.), *Space Capacity Building in the XXI Century*, Studies in Space Policy 22,
https://doi.org/10.1007/978-3-030-21938-3_5

Published in January 2016, ESPI, Vienna). In its conclusion it summarizes a recent initiative which was launched by the European Space Policy Institute to engage with the most relevant stakeholders that could contribute to the definition of the synergies between the European Space 4.0 strategy and the United Nations 2030 agenda (Ferretti, S., Feustel-Büechel, J., Gibson, R., Hulsroj, P. (2016). *Andreas Papp and Elisabeth Veit "Space for Sustainable Development"* (ESPI Report 59). Published in June 2016, ESPI, Vienna), to address capacity building in four thematic areas: innovation and exploration, space for global health, climate change and resilient societies and capacity building in the XXI century.

5.1 Introduction

5.1.1 Society and the Space Sector

In what has been merely a blink of an eye from an evolutionary perspective, humanity has become a spacefaring civilisation and has developed space technologies, applications and services that have become ever more important for the functioning of society. Despite all our steps forward, or perhaps because of them, we live now in an age of disruption. This is exciting in the sense that our world is changing faster than ever before in history. While the digital revolution is still in the midst of transforming society, recent breakthroughs—such as 3D printing—are already starting to shake up and radically alter things in a fashion we could not have imagined a few years ago. At the same time, other game-changing technologies—such as nano and biotechnology—are maturing as a result of sustained innovation efforts that might create entirely new revolutions in the near future. But the situation is not all positive. Since the challenges we are facing are both immense and ubiquitous, we cannot be complacent. In agriculture for instance, it is predicted that the global supply of food for the next fifty years will have to be as big as all harvests combined in the thousands of years since civilisation began with the invention of agriculture. Similarly, the challenges in the areas of energy, urban growth, security and sustainability are vast and will require adequate solutions, soon.

In light of opportunities and challenges like these, and considering the nature of scientific and technological progress as it has unfolded in modern history, it is more than likely that space assets will have a vital role to play in the future. But the very same dynamic of opportunities and challenges will also affect space itself. With new technologies and changing markets having the potential to alter the way we have been using space, the position of Europe will be less secure than in past decades.

As more players enter the space arena and establish ambitious space programmes, innovation and competition will also become more likely to change the status quo that has characterised the post-Cold War era in space. So while it is clear that society in the future will be radically different in some respects, the same will be true for space.

5.1.2 What Is Changing?

The Space sector is entering a new phase which is characterized by this digital revolution as well as by the industry 4.0 transformation. This occurs while it is becoming increasingly relevant to address the global challenges associated with the United Nations Agenda 2030 for Sustainable Development.[1]

The way satellites are manufactured, launched and operated is evolving rapidly, and Space industries are also transitioning towards connected and reconfigurable factories, in line with the industry 4.0 paradigm. They will soon make extensive use of cloud based systems, edge-computing, internet of things, block-chain and advanced robotics. In fact, all these new technologies allow for the implementation of much more efficient manufacturing systems, which are also very flexible and reconfigurable.

In the new space ecosystem, which is particularly competitive vis-a-vis new entrants, it is becoming highly relevant to study the factors which can favour the migration of the traditional space industries towards innovative automated and interconnected factories. But this digital revolution is also posing several new challenges, which require a strategic approach, in the cybersecurity domain.

In the meantime Space applications are becoming increasingly relevant to address the global challenges associated with the United Nations Agenda 2030 for Sustainable Development.[2] These challenges could be addressed at best following an accurate assessment of societal needs and available space technologies. In doing so, it is important to consider that these needs can be endogenous, when related to societal developments, and exogenous, when they are associated to the changing environment (e.g. climate change and natural disasters).

In general, technological progress is allowing for increasingly better living conditions for the most developed countries. This aspect has to be considered in the global context, leading to the question of how quickly developing countries will increase natural resources demands to equalize their living standards.

Space technology and applications can be particularly helpful in ensuring a sustainable socio-economic growth globally, by supporting solutions that avoid excessive and unbalanced consumption of resources and protect our planet.

To achieve this goal, Earth Observation satellites and the ever growing space based services, already provide a significant number of key indicators for monitoring the Earth's environment (e.g. Essential Climate Variables—ECVs), which can only be obtained from Space.

More in general, Space technology can provide lessons, from human spaceflight missions, to create more sustainable living conditions as population increases on Earth. In this context, all the major Space agencies have initiated technology transfer programs in various domains and it would be important to strengthen policies that

[1]Stefano, F. (2016). Space as an enabler of sustainable development. In *Proceedings of the 4th International Conference on Sustainable Development*. September 21–22, 2016, New York, USA.

[2]See Footnote 1.

sustain innovation best practises and ensure the proper dissemination of the know-how acquired in the new space ecosystem.

5.2 New Paradigms

5.2.1 The European Landscape

The European Union and its Member States still occupy a key position in the global arena, in terms of economic growth, innovation and competitiveness, despite the many global challenges they are facing. The elaboration of a long term strategy which focuses on the European citizens and their needs (e.g. democracy, equality, employment, security…) is the pre-requisite for Europe to maintain its unique position.

Combining its ability to deliver innovative solutions in a number of technological fields, Europe, thanks to the European Space Agency and European industry, has created important space infrastructures, among which the Galileo and Copernicus programmes, in addition to the contribution to the International Space Station.

Moreover, large-scale R&D investments such as the space programmes of the European Space Agency and the European Union Horizon 2020 are boosting innovation across European countries.

The potential for innovation and economic growth within Europe can be significant if orientations are made in the European Commission Multi-annual Financial Framework and at the next ESA Council at Ministerial Level. They can greatly influence the quality of life for European citizens and Europe's global role in the decades to come.

Europe can leverage its position in the global context by adopting a new model for innovation, growth and sustainable development by placing citizens and their needs at the centre of the concerns and by enabling a sustainable foreign development based on a leap-frog approach that fully exploits new R&D of technologies.

Space programmes will represent a fundamental building block, providing data that enable the provision of new services which can be tailored to specific needs of citizens or geographical areas. New space programmes should be designed around these needs and become an integral part of end-to-end integrated service provisions using new models. This also implies paradigm shift from a scientific and operational focus towards the inclusion of business requirements responding to citizens' current and especially future needs.

5.2.2 Space Agencies

In this new context it is essential that Space Agencies refocus their activities and programmes in such a way that they can properly innovate, inform, inspire and interact, while sustaining and improving development and economic growth in their member states.[3]

Therefore Space Agencies nowadays consider innovation as an fundamental and overarching theme, which enables them to continue being at the forefront in the exploration of other planets with astronauts and robots, in the scientific discoveries of the Universe, and in the provision of services and applications in the context of a rapidly evolving space sector worldwide.

The European Space Agency defines this new landscape as Space 4.0, and is addresses its mission, in this global context, at European level. The ultimate goal is to open new frontiers, by developing substantial technical know-how that can enable challenging future space missions.

However, in the new space ecosystem, innovation is also about creating economic value that could attract investments and spur new businesses. In this changing environment, Space Agencies are preparing themselves to create new opportunities for cooperation with new private actors (e.g. Space X, Planet, etc.), which are entering into their traditional ecosystem, disrupting both the traditional technological and business models.

In fact, these new space companies are already adopting an agile approach, which includes the implementation of Commercial Off-The-Shelf (COTS) technology, rapid prototyping, concurrent engineering, and innovative manufacturing processes (ALM). All these new approaches are challenging the status quo, favouring innovative Research and Development practises which disrupt the existing models.

Creating partnerships in this ecosystem becomes key for all the actors to remain competitive, while being innovative. The proposition is to go beyond the traditional boundaries of the space sector, by joining forces with other sectors in an open innovation environment, that allows the incorporation of new technologies for the manufacturing and integration of complex space systems and could spin-off space technologies and services to other industries (e.g. verticals), creating additional value by the means of integration.

5.2.3 The Relation Between Disruptive R&D Processes and the Space Sector

Traditionally, space research and innovation has been highly confidential and performed by space agencies, few corporations and specialized research centres. More recently an increasing number of new space actors are adopting open approaches to

[3]Stefano, F., Imhof, B., Balogh, W. (2016). Future space technologies for sustainability on earth. In *Proceedings of the 67th International Astronautical Congress*, Guadalajara, Mexico.

innovation, and it is becoming recognized that valuable discoveries and ideas can be generated in various external contexts.

Considering that sharing of know-how could enables reaching complex goals more efficiently and effectively, different space actors begin to cooperate, aligning their strategies to the most recent innovation models.

5.3 Innovation Models

5.3.1 Open Innovation in Space

Open innovation implies publicising requirements and roadmaps for technology, and providing incentives for a wide range of actors to contribute to the solving of problems and to making new discoveries. In human space exploration, this approach could be particularly successful as space is a field which draws great enthusiasm and which interests people from a variety of backgrounds and cultures.

In this framework, it would be highly beneficial to draw from the forefront of technological developments on Earth, in order to identify existing solutions that can be spun-in by space industry to support space exploration. To achieve this, the first necessary step is to create technology maps, identifying the technology sectors and within these the specific solutions which can benefit space expeditions. From this initial analysis, a shortlist of companies, that have expertise in or that are currently developing these technologies, can be compiled. In order for the identified technologies to be one day suitable also for space travel, these companies should then be involved in a long term spin-in process with the space agencies coordinating the mission requirements.

Starting from as early as possible in the technology development lifecycle, analyses should be made in order to identify which specific requirements should be incorporated into the technologies in order to one day be able to qualify them with a minimum effort. One example of this is in the field of electronics, where apparatuses necessitate particular efforts to be space-qualified. If solutions are foreseen in the early product design phases, no re-engineering efforts will be needed if they are finally selected for use on board a spacecraft.

For the commercial companies participating in this process, the motivation could come from the chance to advertise their contribution to the crewed exploration of our Solar System, which could give them a great return in terms of image and branding. For the space industry, the opportunity to spin in the latest technologies in the space development programs at their conception will allow for significant cost reduction and faster implementation.

The fast changing world around us is showing promising new ways of doing, from advanced manufacturing with additive layer manufacturing, to advanced robotics and artificial intelligence technologies. The creation of partnerships with private companies entering these new markets should allow for qualification of the technologies already for use on Earth, with the support of the traditional industrial system integrators that will then take over the products for flight certification and space utilisation.

Finally, for the space agencies who are planning and coordinating such missions, the platform of potential suppliers for the mission will be greatly enhanced, providing more flexibility in solutions to adopt for the final mission scenario.

However, this process will not only bring benefits to industry. In fact, funding from space programs together with the support of the expertise of the many space engineers and planetary scientists employed by high tech industries and agencies, could accelerate the development of the selected technologies with respect to the normal commercialisation cycles they would follow in their industry.

5.3.2 *Co-creation*

Co-creation is defined as a process where the "joint creation of value include both the company and the customer, allowing the customer to co-construct the service experience to suit their context".[4]

In the Space sector this concept is closely linked with the principles of open innovation and risk management. In fact, instead of the entire risk being placed on space agencies and therefore the taxpayers, a more open approach to space innovation encourages the creation of public-private partnerships where risks are shared with industrial actors, who have incentives to be more efficient and further shorten development cycles in order to commercially benefit from R&D efforts. In this spirit, ESA has recently released the source code for its Concurrent Design Facility, enabling any private company or research centre to re-create in-house the concurrent design processes used in advanced prototyping and design, and also creating a community of CDF users to share knowledge.

5.3.3 *Innovative Manufacturing*

The possibilities opened up by innovative manufacturing technologies, such as Additive Layer Manufacturing are extremely wide and range from building small components up to entire structures, starting from a variety of base materials, such as feedstock, metals, composites, glass and concrete.

Both NASA and ESA are currently looking into the possibilities offered by this technology and are carrying out tests inside the Microgravity Science Glovebox of

[4]Prahalad, C.K., Ramaswamy, V. (2004). Co-creation experiences: The next practice in value creation. *Journal of interactive marketing, 18*(3), 5–14.

the International Space Station, while conducting on Earth feasibility studies for the construction of planetary bases making use of in situ resources, such as Moon regolith.

Some advantages offered by this technology are that it allows the construction of large structures in Space, freeing them up by the constraints provided by the uplift mass and volumes of the launch vehicles, and by the structural requirements of gravity. The other key point is that robotic systems can entirely print such structures without the need of human intervention, opening up entire new business and technical opportunities associated with in-Space just-in-time production, such as new components, spare parts, cubesat, antennas, truss structures, and habitats.

Potential applications on Earth will provide sustainable and effective solutions in various environments, some of which are already being investigated like building houses in deserted areas and reconstructing underwater coral barriers. The benefits of these developments could significantly contribute to a manufacturing revolution on Earth in various fields, such as architecture and engineering, materials processing and nanotechnology, science and robotics, offering new shapes, new materials, new design solutions where they are needed at the time they are needed.

5.3.4 Design Thinking

In fact, this argument should be made not only for products, but is extremely relevant also for high-tech services. Some space applications and services developed for Earth use may find an excellent use also on other planets. The feasibility of such services should be already investigated, taking into consideration design evolutions and hardware upgrades of the payloads to be eventually flown in support of future manned planetary missions. Examples of these can be the mapping of the surface of the planets in order to assess the energy harvesting capacity, both of solar thermal, photovoltaic and geothermal energy, telemedicine services and verifying the communication capabilities, bandwidth, and procedures to be adopted in order to deliver effective medical support to the crew.

5.4 Moving Forward: Institutional Plans and Commercial Partnerships

5.4.1 The Institutional Plans

A new plan has been put in place by the European Space Agency, to ensure a seamless chain of innovation, enabling more cooperation between academic institutions and research establishments together with industry and end-users.

Innovative Research and Development activities in the Space domain mainly relate to two major fields: technology development and space based services. Space programs are usually quite rich in terms of technology developments, which are combined with a sense of discovery and exploration, that usually push the boundaries of knowledge even further compared to other high-tech sectors.

ESA recognized in 2016 that it is therefore important that this know-how is made available also to other industrial sectors, in order to "multiply their societal, economic and technological impact". At the same time, it is crucial that Space Agencies recognize the importance of spinning-in technologies from other fields and industrial sectors, potentially becoming seeds of innovation in the development processes of space systems.

In the field of space services, ESA develops space infrastructures (Navigation, Telecommunications, Earth Observation) that deliver information, communication networks and data, that allow industrial actors to generate innovations, such as new applications, products, and services responding to market needs. In order to ensure a seamless chain of innovation, the Agency is putting in place new mechanisms, to ease the access by space and non-space companies, enhancing openness, cooperation, interaction, information and knowledge sharing, flat organizational structures.

Space nowadays requires to move a step forward, in order to allow for uninterrupted, rapid development from idea to product or service and sustain competitiveness in an efficient manner, also through new funding schemes and commercial partnerships for dedicated activities, that could inspire citizens while addressing challenging space projects for the decades to come.

New space companies are developing products and services in a very fast and efficient manner, in various fields (launchers, telecommunications, earth observation, etc.). Could it be the innovation model and the disruptive R&D that are the winning factors and not a specific technology or business niche? An interesting research question would be to identify what companies like Blue Origin, SpaceX, OneWeb, Planet and Rocketlabs have in common.

Certainly, disruptive R&D needs to become part also of institutional programs, and the establishment of partnerships with commercial actors, which are already implementing new models of innovation in their processes, can become the key factor to enable a leap frog in global space activities.

5.4.2 *Building Space Capacity in the XXI Century*

New innovation models are increasingly spread across sectors and disciplines, including Space, which is becoming an integral part of many societal activities (e.g. telecoms, weather, climate change and environmental monitoring, civil protection, infrastructures, transportation and navigation, healthcare and education). Space 4.0 ambitions are to place space at the heart of the successful evolution of Europe and the full implementation of the United Nations Agenda 2030 worldwide. The outcomes of the conference "Space2030 and Space 4.0: synergies for capacity building in the

XXI century", recently co-organized by the European Space Policy Institute (ESPI) and the United Nations Office for Outer Space Affairs (UNOOSA), with the support of the European Space Agency (ESA), point to the timeliness of a fresh look at the wider perspectives and strategies to be implemented in future space programmes.

The key findings and recommendations of this platform, involving stakeholders and representatives of new space actors and civil society, map out available options and identify the ideal conditions for their successful implementation. Innovative frameworks, partnerships and collaborations in this ecosystem have been explored, posing special attention to improving the dialogue with civil society and other sectors, to make them aware of the potential of space, and to the creation of new mechanisms to identify, collect and process user needs, in order to design, implement and fully exploit future space programmes.

In this context, potential synergies between the UN agenda Space2030 and the ESA Space 4.0 strategy have been identified, focusing on four thematic priorities of UNISPACE + 50: Global partnership in space exploration and innovation; Strengthened space cooperation for global health; International cooperation towards low emission and resilient societies; Capacity building for the twenty first century. Through these themes, the interplay and dependencies amongst key actors are identified, and a special emphasis is placed on future approaches of the diverse groups of stakeholders involved, leveraging the existing space infrastructure, institutions and networks while reinforcing and expanding their scope and effectiveness in ensuring that space becomes an important driver for sustainable development.

For example, future integrated services will capture new citizen needs and target sustainable development goals, creating unprecedented opportunities for Europe worldwide. This will be enabled by Copernicus, Galileo and by the emergence of new SATCOM infrastructures for 5G, serving new markets by providing broadband connectivity to rural areas.

Space is therefore increasingly becoming the link among systems of systems, and its enabling function may represent a key element actively contributing to a sustainable future on Earth.

5.5 Conclusion

In the previous paragraphs, a paradigm shift within the industrial space sector has been presented, applying open innovation theories to spin-in technologies and encouraging spin-off from space developments, in the framework of future space programs.

This could potentially lead to an industrial renaissance not only in the traditional spacefaring nations, but see the concrete involvement, at various levels, of many nations in space endeavours, finally making its success seen as a truly global achievement.

The creation of a system-of-systems where internal and external interfaces are well defined and shared among the various mission stakeholders since the preliminary design phases, will strengthen the cooperation potential among international partners and will increase mission flexibility.

Building challenging future missions could be eased by effectively drawing all the benefits of current experiences, making use of space R&D capabilities and outside expertise and knowledge from other sectors.

Increasing this knowledge exchange, applying innovative design methodologies, will both significantly reduce development efforts and allow for new frontiers to be reached, both on Earth and in Space.

This paper was presented at the 69th International Astronautical Congress, 1–5 October 2018, Bremen, Germany. www.iafastro.org.

Chapter 6
Innovative Space Systems Design—Methods and Tools

Antonio Martelo

Abstract Within the context of the current trend in automation and data exchange in manufacturing technologies, which is leading to a digital transformation, Concurrent Engineering presents itself among the most productive methods that are in use today to study the feasibility of new design and for the early design phases of space missions and systems. This chapter introduces Concurrent Engineering (CE) and Concurrent Engineering Centres (CEC's) to the reader, and discusses the options different organizations have to create their own CEC's and apply CE, with a specific focus on the applicability of a lean approach. It also addresses the ways in which CE is evolving, and which are the upcoming trends in the field, covering Collaborative Engineering as an increasingly pursued goal, and the challenges that need to be considered.

6.1 Introduction

The design and manufacturing of space systems is a complicated, costly and time-consuming task. The extreme environment of space which a spacecraft must withstand, the difficulty of servicing on-orbit platforms, and the ad-hoc nature of the majority of missions are only some of the elements that drive these factors.

When compared to other industries, the space industry cannot build economies of scale due to the relatively low product output and the large R&D required by many missions; in spite of this, the pressure to reduce costs and increase efficiency is the same as in any other industry.

Current trends in automation and data exchange in manufacturing technologies have opened a door to a digital transformation process that can add value and improve

---

A. Martelo (✉)
German Aerospace Center (DLR), Bremen, Germany
e-mail: antonio.martelo@dlr.de

S. Ferretti (ed.), *Space Capacity Building in the XXI Century*, Studies in Space Policy 22,
https://doi.org/10.1007/978-3-030-21938-3_6

the industrial value chain, alongside other benefits. All this falls under the umbrella of what has become known as "Industry 4.0".[1]

Within the framework of Industry 4.0 and its analogous Space 4.0,[2] new methods, processes and technologies are being introduced with the objective of making the whole process of developing and manufacturing products and services more efficient.

Hypothesis such as open innovation[3]—that focuses on creating more value through cooperation and the free exchange of ideas—the Design to Produce[4] process –which intends to create a seamless production chain starting from the digital design of components and systems—or techniques like 3D printing—which support rapid prototyping and can in the future provide manufacturing capabilities in space—are only a few of the concepts and technologies that are being implemented by organizations today.

The main common threads of many of these innovative techniques and processes are an increased level of communication and collaboration, obtaining a higher innovation footprint, reducing wasted effort and/or resources, and focusing on the needs of the end user.

Among the most productive methods that are in use today to study the feasibility of new designs and for the early design phases of space missions and systems, a particularly interesting process is Concurrent Engineering (CE). CE is a recognized and increasingly used methodology that exemplifies all these aforementioned aspects; CE can benefit any organization independently from its size, support rapid prototyping and potentially be a technology leapfrog in small, dynamic organizations.

6.2 Concurrent Engineering (CE)

Concurrent Engineering is a process focussed on optimising engineering design cycles, which complements and partially replaces the traditional sequential design-flow by integrating multidisciplinary teams that work collectively and in parallel, at the same site, with the objective of performing a feasibility analysis or a high-level design in the most efficient and consistent way possible, right from the beginning.

Working within a guided process, the concurrent access of all experts to a shared database, and the direct verbal and media communication between all subsystem experts, are the defining characteristics of CE studies. The presence of the customer

[1]European Parliament (2015). Industry 4.0 Digitalisation for productivity and growth. http://www.europarl.europa.eu/RegData/etudes/BRIE/2015/568337/EPRS_BRI(2015)568337_EN.pdf [accessed 22 March 2018].

[2]European Space Agency https://m.esa.int/About_Us/Ministerial_Council_2016/What_is_space_4.0 [accessed 22 March 2018].

[3]Chesbrough, Henry W. (2003). The Era of Open Innovation. Sloan Management Review, 44(3): 35-41.

[4]European Space Agency http://m.esa.int/Our_Activities/Space_Engineering_Technology/Design_2_Produce [accessed 22 March 2018].

Table 6.1 Reported benefits of concurrent engineering (M. Lawson, H.M. Karandikar, A Survey of Concurrent Engineering, J. Concurrent Engineering: Research and Applications. 2 (1994) 1–6.)

Development time	30–50% less
Engineering changes	60–95% less
Scrap and rework	75% reduction
Defects	30–85% fewer
Time to market	20–90% less
Field failure rate	60% less
Service life	100% increase
Overall quality	100–600% higher
White-collar productivity	20–110% higher
Return on assets	20–120% higher

throughout the study, and the increased communication flow within the team, produces fast and consistent results, with a customer that knows exactly what they are getting and why, since they have been part of the decision making process throughout.

Effective implementation of CE can benefit organizations in a number of ways, including greater customer satisfaction, reduced costs, increased quality and reduced design rework and development time (see Table 6.1).

6.3 Concurrent Engineering Centres (CEC's)

The successful implementation of CE requires the integration of three main elements: a work process that encourages effective teamwork, well-coordinated multidisciplinary teams, and an infrastructure that supports the necessary activities and promotes effective communication.

The infrastructure component can be found in the aerospace industry as work spaces under the common denomination of "Concurrent Engineering Centres", although other conventional denominations include "Concurrent Design Centre", or "Concurrent Engineering Facility". A non-exhaustive map of CEC's around the world is shown in Fig. 6.1.

There are many CEC's worldwide, some in space agencies and research institutions like NASA,[5] ESA[6] or DLR[7] (see Fig. 6.2 and Fig. 6.3), and others inside private organisations and universities. A number of these centres, such as the Concurrent

[5]Smith, Jeffrey L. (1997) Concurrent Engineering in the Jet Propulsion Laboratory Project Design Center. 98AMTC-83.

[6]Bandecchi, M.; Melton, B.; Gardini, B. The ESA/ESTEC Concurrent Design Facility. Proceedings of EuSEC 2000, pages 329-336.

[7]Martelo, A.; Jahnke, S.S.; Braukhane, A.; Quantius, D.; Maiwald, V.; Romberg, O. (2017) Statistics and Evaluation of 60+ Concurrent Engineering Studies at DLR. Paper presented at the 68th International Astronautical Congress (IAC), Adelaide, Australia, 25-29 September 2017.

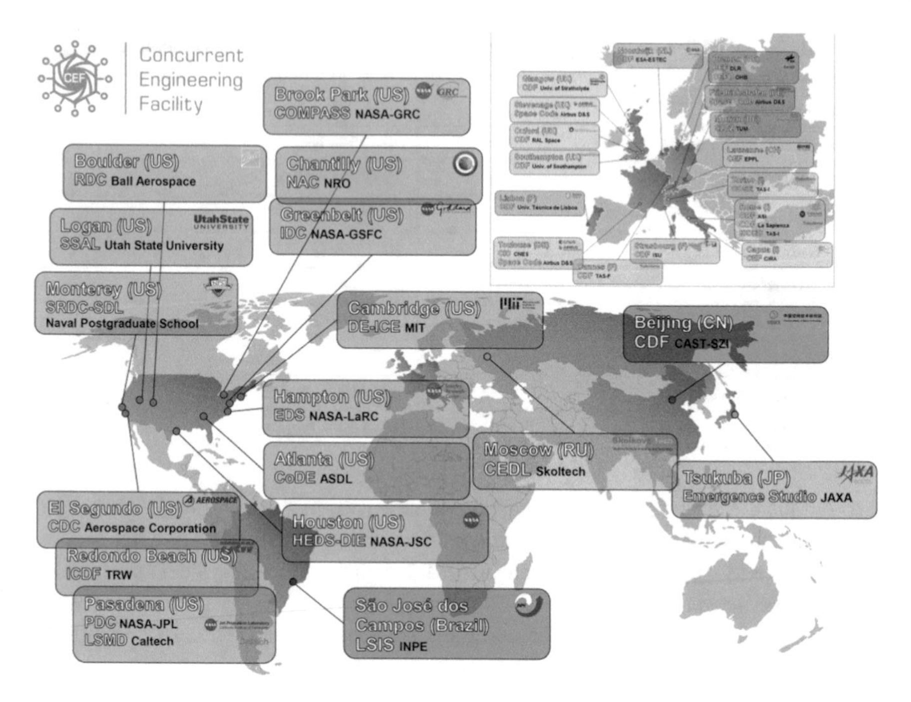

Fig. 6.1 Concurrent engineering centres

Fig. 6.2 Concurrent design facility at ESA-ESTEC

Fig. 6.3 Concurrent engineering facility at DLR Bremen

Design Facility (CDF) at ESA or the Concurrent Engineering Facility (CEF) at DLR, can be considered large facilities, but not all CEC's follow the same approach. Usually, different organizations arrange their infrastructure according to their needs; they typically provide an integrated environment for the team to work together, as well as tools that facilitate the design process and provide a framework for the exchange of information between team members. CEC's are scaled according to each organization's needs in regards to physical space as well as in processing capabilities and technical sophistication.

As with almost everything else in a CEC, the software tools at the disposal of the CE study participants depend on the work carried out in the facility and the specific organisations work structure. Common tools used in the space industry and in many CEC's include computer-aided design software such as CATIA, or the Systems Tool Kit (STK) for mission analysis, but open-source software and self-developed tools with similar functionality are also frequent due to cost concerns or as the means to adapt the tools to the needs of the organisation.

Independently of what other tools may be available, the one essential software application for a CEC comes in the form of an integrated design environment.

As the CE methodology requires access to a shared pool of information and a distributed software methodology (i.e. simultaneously accessible and editable), the use of a centralised model which can be accessed at the same time by all the technical team members, and monitored by the systems engineer, makes a Model Based System Engineering (MBSE) approach ideally suited to the task. There are a number of MBSE implementations throughout the CE community which are Open Source, such as ESA's OCDT[8] (Open Concurrent Design Tool) or DLR's Virtual

[8]Open Concurrent Design Tool: ESA Community Open Source Ready to Go!

Satellite[9,10,11] as well as commercially developed tools available for new centres that do not want to create their own. These tools, albeit working under similar principles, are implemented in all manner of different platforms (e.g. EXCEL spreadsheets or Eclipse RCP), and with different approaches (e.g. repository based or browser based), which provide parties interested in commencing CE activities with many interesting options to accommodate their needs.

Many universities, and some organizations, delve into CE following a lean approach, limiting expenses while still benefitting of many of the advantages of applying CE in their projects. This is typically done through the use of laptops connected to a Local Area Network, and by using free or open source software to support their design, and can prove to be a helpful way for organizations that would like to implement Concurrent Engineering into their overall process.

6.4 CE—A Lean Approach

Whereas an intermediate sized facility such as the CEF might provide around fifteen work stations and host anywhere between twenty and thirty participants in a CE study, larger facilities such as the CDF can double this. There is, of course, a point of diminishing returns, since the focus of CE is in increasing communication between experts and there is a limit as to how many people can efficiently communicate within a concurrent activity. The sizing of a CEC, therefore, many times is not only motivated by the needs of CE activities, but also by the multi-purpose use that these facilities are given (i.e. workshops, presentations, design reviews). Organisations aiming to develop their own CEC must consider their own needs and available resources, and tailor the facility in consequence.

Organisations that opt for a lean approach will, as a rule, face certain limitations in terms of connectivity and processing capability; they will be commonly built with portable technologies which cannot replicate the robustness of a server-based backbone network, but they will profit in terms of flexibility and reduced costs—both in setup and maintenance–. In terms of scalability, a smart use of current technologies can also benefit organisations aiming to implement a lean facility; for example, at present, a commonly used solution to streaming between a computer and mobile device to screens is the use of wireless devices such as Clickshare or Airtame. These,

[9]Schaus, V.; Fischer, P.M.; Lüdtke, D.; Braukhane, A.; Romberg, O.; Gerndt, A. Concurrent Engineering Software Development at German Aerospace Center -Status and Outlook-., SECESA 2010, Lausanne, Switzerland, 2010, 13 – 15 October.

[10]Gianni, D.; Schaus, V.; D'Ambrogio, A.; Gerndt, A.; Lisi, M.; De Simone, P. Interface Management in Concurrent Engineering Facilities for Systems and Service Systems Engineering: A Model-based Approach, CIISE2014, INCOSE Italian Chapter Conference on Systems Engineering, Rome, Italy, 2014, 24–25 November.

[11]Virtual Satellite,
http://www.dlr.de/sc/en/desktopdefault.aspx/tabid-5135/8645_read-8374/, [accessed 22 March 2018].

combined with a large screen or a projector, can provide a simple, low-cost solution for sharing information visually with the team.

A lean facility together with the use of Open Source or self-developed tools is a remarkably economical way of initiating the implementation of concurrent engineering and, since there are a number of them in the CE community at large, it should be relatively easy to find tools that integrate well into the overarching work process of any organisation.

Furthermore, the CE community is very open regarding the processes that each organisation uses, which are typically tailored to the exact objectives and way of working of the corporation or institution. This provides a good starting point for fledging CE practitioners to develop their own tailored process, benefiting from tools, ideas and lessons learnt by other more experienced organisations.

6.5 Costing and Implementation of a Lean CEC

As an example of what could be an interesting first implementation of CE for a small organisation or university, DLR has developed a portable CEC, dubbed "CE on the go", which has been used for educational purposes.

Using Virtual Satellite—DLR's centralised model-based software—very limited hardware, and the same in-study process followed in the CEF, university students have been able to experiment a CE experience in their own campus. The setup necessary for such a "CE on the go" implementation would be along the lines of what is presented in Table 6.2, based on European prices (they can vary geographically). This estimation would provide an extremely affordable setup which, if somewhat limited in terms of processing capability, could be a first step for any organisation that wishes to start applying CE to their benefit. Of course, additional costs derived from the use of other resources (i.e. electricity, use of office space, etc.) or personnel costs need to be assessed within each organisation.

Table 6.2 Estimation of lean CE implementation

Software licenses	
Open Source (e.g. Virtual Satellite, OCDT, OpenOffice, FreeCAD, NASA's GMAT)	No cost
Hardware requirements	
Router	<100€
Central processor	100–200€
Switch	<100€
Computers (8–12)	~4.000€
Cables and miscellaneous	<200 €

6.6 Evolution and Upcoming Trends in CE

As efficient as Concurrent Engineering has proven to be it continues to evolve in a number of different ways, aiming to support larger and more complex designs as well as increase the productivity beyond its current role in feasibility analysis and early design.

Although many times used indistinctly from Concurrent Engineering, the concept of Collaborative Engineering involves a wider context and some clear differences. Collaborative Engineering is similar to CE in that it uses many similar underlying ideas, such as increasing communication and availability of data to all the experts through the use of modern technologies as well as centralised models and databases. The biggest differences are found in the lack of concurrency and co-location, which means that while in CE everyone works at the same time and in the same physical space, in Collaborative Engineering experts can work from anywhere and at any time through the use of communication technologies (i.e. internet) and similar software application as the ones used in CE (e.g. centralised models and information repositories).

While the use of CE in later stages of the product design lifecycle is complicated due to the exponential growth of the size of the team involved and the amount of parameters that have to be handled—which require increasing amounts of time to perform calculations, simulations, and making informed decisions—Collaborative Engineering increases the productivity in these design phases compared to a traditional centralised sequential engineering process.

The application and development of Collaborative Engineering tools and processes is therefore an increasingly widespread trend, sometimes based on previous CE activities. This opens the door for organizations to search for ways to combine both concepts and benefit from an increased efficiency in all stages of the design process.

Another topic of discussion, especially in the space industry, is the need on international collaboration and coordination in the pursuit of larger (i.e. more costly and complex) missions. The best example of such a mission would be the International Space Station (ISS), with a production and operation cost of over 100 billion dollars[12] and involving 15 countries in its development, which almost no single country would have been capable or willing to undertake on its own. In the future, missions such as Moon and Mars settlements, or the sending of large spacecraft to the outer planets of the solar system, will surely require multiple space agencies and a large number of companies to collaborate in their design and integration. This could be facilitated through the subdivision of tasks amongst working groups which could be coordinated through multi-CEC projects; different workgroups, each operating from a CEC, could work in a specific system or subsystem following a CE process, with

[12] Office of the Inspector General. NASA's efforts to maximize research on the international space station. National Aeronautics and Space Administration, Report no.: IG-13-019 (8 July 2013). http://oig.nasa.gov/audits/reports/FY13/IG-13-019.pdf [accessed 22 March 2018].

periodic discussion sessions with the other CEC's via web-based videoconferencing systems and an overlaying Collaborative Engineering work structure.

At a smaller scale, a similar exercise could be performed bilaterally between a large facility and a lean facility to facilitate collaborations between entities. This opens the door for small organizations to benefit from the expertise and infrastructure of the larger organisation, and for the larger organisation to benefit of the specific knowledge and ideas of other institutions. This could, of course, be a perfect way of supporting capacity building through the use of internet technologies, and facilitate knowledge exchange.

6.7 Challenges

The challenges of implementing CE in an organization, which are not technical or institution-specific, have been implicitly assessed through an analysis of the potential cost of a lean facility, and the description of larger facilities. As has been discussed the sizing of a CEC, the tools and process used, and most other aspects depend of the needs of an organisation, but in almost every organisation there are benefits to be had in terms of productivity through the use of CE. The possibility of starting out with a lean approach, building experience and adapting processes and work structures over time, and then evolving the infrastructure as the need arises could be a good way to go for most small organizations.

Organizations that might not see the need to evolve their lean setup, but which might be interested in collaborating with other CEC's that can support them in aiming for more complex projects, will do well to consider the necessary telecommunications infrastructure (e.g. internet connectivity and transmission capacity, technological compatibility with other CEC's –including, for example, the use by all parties of compatible data models under a common standard such as ECSS-E-TM-10-25). Logistic aspects can also be an issue especially for intercontinental connectivity between centres, but this is an inevitable challenge that can only be addressed through proper time management.

Another element that should always be considered are the potential political barriers which might impede or limit international collaboration. Knowledge transfer between organisations can be a point of contention, but all the more so between countries where export control has to be considered.

6.8 Conclusion

Today, a number of innovative techniques, technologies and processes are being implemented by companies and institutions with the goal of increasing productivity and developing original products and services.

Concurrent Engineering is one such methodology, producing better results faster, with less effort and cost, and therefore becoming increasingly ubiquitous in the space industry. The whole emphasis of Concurrent Engineering is to connect ideas, know-how, and facilitate communication, which facilitates the generation of innovative ideas and knowledge exchange between experts, while increasing team cooperation and providing more cohesive designs.

While the collocated nature of Concurrent Engineering activities make the use of a specific physical space necessary, and the effective application of the process requires software tools that enable the team to have a centralised database and data model, it is important to note that there are economic solutions which can facilitate any organisation when establishing their own Concurrent Engineering Centre, no matter its size.

Organisations such as ESA, DLR, and others within the Concurrent Engineering user community, continue pushing the boundaries and developing processes and Open Source tools which eliminate the need for interested parties to develop their own before they can benefit from Concurrent Engineering. Additionally, following a lean approach, a low-cost set-up can be implemented which may be sized up later if necessary, or collaborations with larger CEC's can be established so as to benefit from the expertise and infrastructure of these organizations.

Both for organizations aiming to build their capacity in space, and for those that already have them but continue to work under a centralised engineering paradigm, Concurrent Engineering and Collaborative Engineering can be an accessible, yet effective way of increasing productivity and innovation, and therefore worthy of consideration.

Chapter 7
Exploring Threats and Opportunities Through Mega Trends in the Space 4.0 Era

Andrea Vena, Gianluigi Baldesi and Arnaud Bossy

Abstract In recent years, every citizen of planet Earth has experienced fundamental technological, societal and economic transformations. Furthermore, Space has evolved from being the domain of the governments of few spacefaring nations to an entrepreneurial area where we are witnessing the strong emergence of private actors and the increased engagement from emerging countries around the world. How should Space Agencies sense and seize opportunities and threats today? The European Space Agency with the support of Frost & Sullivan has launched an initiative based on the analysis of Mega Trends. Mega trends are defined as the large-scale, sustained forces of development—such as urbanisation, connectivity and convergence—set to change the future world. Each mega trend is made up of numerous sub-trends that, in turn, are likely to affect particular economic and social sectors, including space. This initiative has several objectives: sharing global mega trends across the Agency and with stakeholders, identifying and analysing opportunities for and threats to the space ecosystem and focussing on the role of ESA to mitigate/avoid threats and catch opportunities. A dedicated ESA team for future perspectives has been set up, with representatives from all parts of the Agency, including young trainees selected through an internal competition. Furthermore, some representatives of the European aerospace ecosystem have also been invited to participate to the discussions. A trend analysis has been performed with the support of Frost & Sullivan, market leader in monitoring the tendencies likely to impact our societies, economies, organisations and private lives. This chapter will present the initiative in detail and share the findings: 9 Opportunity Areas and 4 Transformation Pillars out of the total 84 ideas collected during the two specific workshops and a dedicated "call for ideas".

A. Vena (✉) · G. Baldesi
European Space Agency, Paris, France
e-mail: Andrea.vena@esa.int

A. Bossy
Frost & Sullivan, Paris, France

S. Ferretti (ed.), *Space Capacity Building in the XXI Century*, Studies in Space Policy 22,
https://doi.org/10.1007/978-3-030-21938-3_7

7.1 Introduction[1]

In recent years, fundamental technological, societal and economic transformations have impacted the life of people around the planet at unprecedented speed and with profound consequences. In addition, space-based infrastructures and services have ceased being the exclusive domain of the governments of a few spacefaring nations, and have become an entrepreneurial area where the role of emerging countries as well as the appearance of private capital and entrepreneurs have given a new impetus to the sector and opened new opportunities.

As a first-class actor in the space domain for more than fifty years, the European Space Agency (ESA) is addressing this new context, which it has named the Space 4.0 era, through the evolution of its activities, programmes and internal organisation, to be in line with the new dynamic of the space sector and anticipate user needs and seize the related opportunities.

This paper describes how ESA has addressed threats and opportunities linked to the evolution of the global context, in particular paying attention to the Mega Trends which are likely to influence the social, economic and technological landscape of the world in 15-20 years.

7.2 ESA in the Space 4.0 Era

Space 4.0 identifies the evolution of the space sector into a new era, characterised by an increasingly interconnected and dynamic playing field, where the interaction among multiple and diverse space actors around the world, e.g. private companies, academia, industry and citizens, generates new needs and opportunities.

This era is unfolding through interaction between governments, private sector, society and politics. Space 4.0 is analogous to, and is intertwined with, industry 4.0, which is considered as the unfolding fourth industrial revolution of manufacturing and services.

In line with its purpose,[2] The European Space Agency aims at leading the European space sector through the transformations required to adapt to the Space 4.0 era in order to foster its global competitiveness and fully integrate its benefits into the European society and economy.

---

[1]This article summarises the work performed by the European Space Agency with the support of Frost & Sullivan, company leader in monitoring the tendencies likely to impact our societies, economies, organisations and private lives. **Andrea Vena** heads the Corporate Development Office in the Strategy Department of the European Space Agency, within which **Gianluigi Baldesi** is Senior Corporate Development manager.

[2]According to its Convention of 1975, "The purpose of the Agency shall be to provide for and to promote, for exclusively peaceful purposes, **cooperation among European States** in **space research and technology** and their **space applications**, with a view to their being used **for scientific purposes and for operational space applications systems** [...]".

This requires, on one side, the consolidation of a sustainable space sector closely connected with the fabric of society and economy. For this to happen, space must be safe, secure and easily and readily accessible, and built on a foundation of excellence in science and technology—broadly and continuously over time.

On the other side, it also requires the Agency to be adapted to Space 4.0 and to be ready to accept the challenges and seize the opportunities ahead, in particular increasing its flexibility and agility. Besides its traditional functions as a space agency covering all the space domains, ESA aims at evolving and becoming the broker, the promoter and the facilitator of the creation and development of space-based entrepreneurial activities across Europe.

7.3 Assessing Mega Trends

Mega Trends are large-scale, sustained forces of development—such as urbanisation, connectivity and convergence—set to define the future world with their far-reaching impacts on societies, economies, businesses, cultures and finally on individuals' lives. Each Mega Trend is made up of numerous sub-trends that, in turn, are likely to affect particular areas, including the whole space sector, especially in its new entrepreneurial dimension.

The identification and assessment of Mega Trends and the relevant sub-trends is fundamental for any organisation willing to anticipate changes so to adequately face threats and seize opportunities. This is particular the case of ESA, considering the fast evolution of the space sector towards the Space 4.0 era.

Supported by Frost & Sullivan, which has provided access to its assessment of future trends across multiple areas and sectors, the European Space Agency has launched an Agency-wide initiative aiming at:

- **sharing global mega trends** across the organisation and with main stakeholders;
- identifying and analysing opportunities for and threats to the **space ecosystem**; and
- focussing on the **role of ESA** to mitigate or avoid threats and catch opportunities.

A dedicated ESA Team for future perspectives has been set up, with representatives from all the areas of the Agency. Senior managers, technical experts and young staff members have been selected and invited to join the team with the objective of ensuring the largest possible representation of different areas of expertise and levels of experience. Furthermore, in order to enlarge the reflection to other space stakeholders, several representatives of the European aerospace ecosystem have been invited to participate to the discussions.

Trend analysis was performed between October and December 2017 with a tight calendar. Two large workshops took place in ESTEC, the European Space Research and Technology Centre of ESA in Noordwijk, the Netherlands.

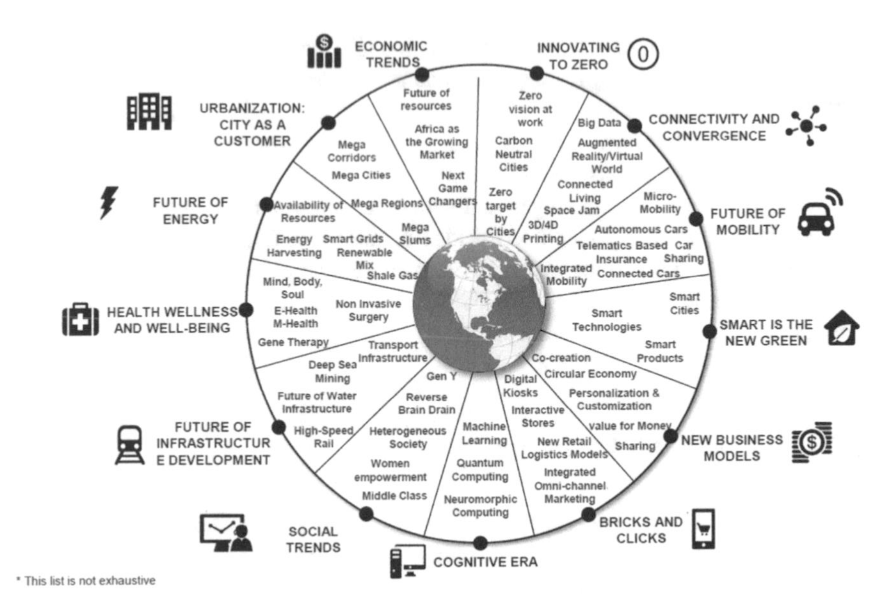

Fig. 7.1 Mega trends. *Source* Frost & Sullivan

These two workshops, both hosted in the Erasmus building, had different objectives and have reached different goals. During the first workshop, a general presentation on Mega Trends was given in the morning and webcast over the internet for the benefit of specialised users and the general public. The presentation addressed the Mega Trends shown in Fig. 7.1 and identified sub-trends potentially affecting the space ecosystem. Fifty-four opportunities and threats were identified during the working session of the afternoon, to which thirty additional ideas were added, obtained through a general consultation of colleagues across different ESA sites and establishments.

Starting from eighty-four opportunities and threats received during the first phase, thirteen themes were identified and used as basis for further work during the second workshop, when reflections on the role of ESA for the thirteen themes were conducted and nine opportunity priorities were selected for the Agency, as depicted in Fig. 7.2.

7.4 Identifying Priority Opportunities for the Agency

Out of the nine areas, nine priorities were identified for initiating and promoting programmes and activities aiming at seizing the opportunities that Space 4.0 has to offer. In parallel, four priorities for transformation processes inside the organisation were also identified, with the objective of easing the preparation of the Agency to face the new challenges. The scheme of the findings is depicted in Fig. 7.3.

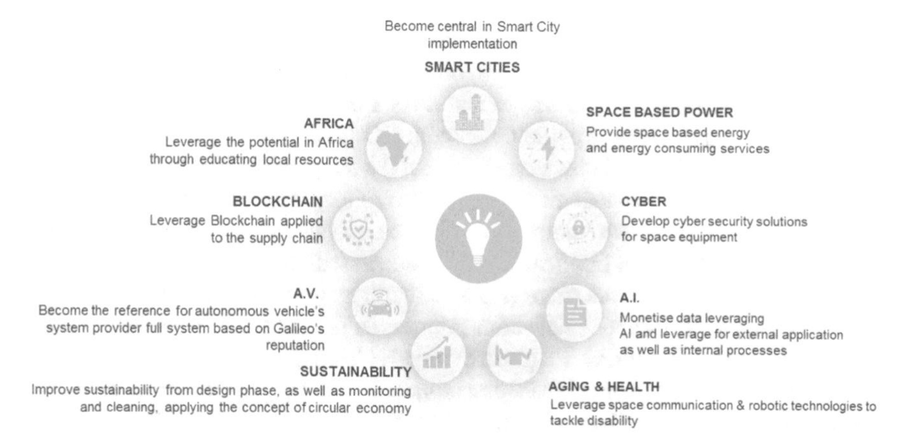

Fig. 7.2 9 Areas of opportunities and threats that were prioritized by the team. *Source* Frost & Sullivan

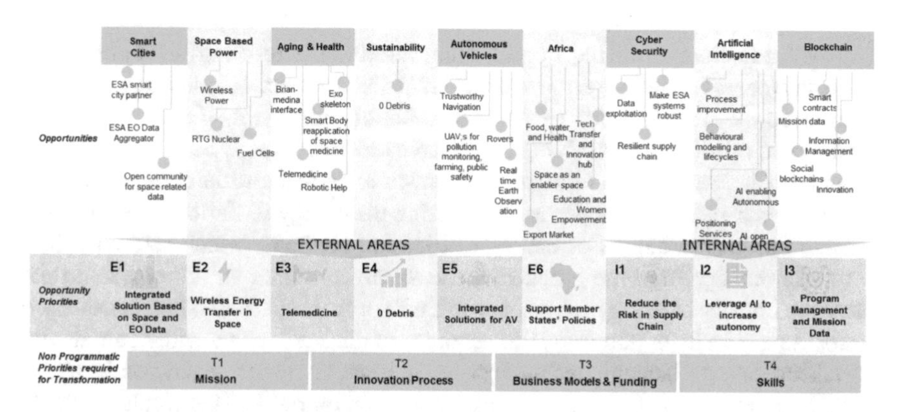

Fig. 7.3 The macro to micro process defined programmatic areas, both internal and external. *Source* Frost & Sullivan

7.4.1 *Programmatic Priorities*

7.4.1.1 Smart Cities

Smarter cities will deliver operational efficiencies and value to end-users through the transition from a traditional city structure, which is currently mostly inefficient, to cities of tomorrow, where sustainability and connectivity will be pursued thanks to digital platforms, data analytics, and the use of artificial intelligence. The current fragmentation of technologies and lack of integrated services will be the major challenges to overcome in the coming years to improve the operational efficiency of cities and deliver value and improved experience to citizens. In particular, the exercise found that the diverse city stakeholder groups, due in part to the lack of coordination and communication between them, do not currently gain sufficient value from **Earth Observation (EO)** and **Space Data.** As a priority opportunity, the solution proposed was for ESA to provide cities with EO and more in general space data solutions, acting as both a knowledge broker and a technical integrator. As a broker, the Agency could ensure that EO data relevant to a specific region is delivered to open data portals to increase both the relevance and value of the data. On the other side, ESA could act as an **integrator of aerospace technologies to provide smart city solutions**, creating a new ecosystem of suppliers providing a single solution to serve many users. ESA's technical competencies in the fields of Earth Observation, Navigation and High Altitude Platform Station (HAPS) are clear assets. In addition, experience in managing space HAPS assets and relevant integration and development skills will deliver clear value to smart city ecosystems.

7.4.1.2 Space-Based Power

Energy production is moving from burning fossils fuels to using low emission or renewable energy to reduce climate change. Whilst the renewable industry has been slow to develop, mainly due to the high price of energy resulting in a need for government subsidies, renewable energy sources are increasingly used. Space has the potential to deliver high quantities of **solar power** to Earth due to the consistently high levels of solar radiation. However, transferring the energy to Earth is not currently feasible due to limitations of energy distribution methods, which rely on cables to transfer energy from the production site to the consumer and of course do not allow the transfer of energy from space to Earth or from space to other space-based assets. As a priority opportunity, ESA could take the lead, with a wider community of stakeholders, to launch a technology programme on power generation in space, to beam power between space locations rather than on Earth. This would be based on wireless energy transfer concepts, to provide power to the Earth, the Moon and artificial satellites.[3]

7.4.1.3 Aging and Health

The world is getting older due to advances in food science, nutrition and medicine. By 2025, the global median age will have risen 2.6 years; Europe will have the highest median age, while APAC's (Asia and Pacific) median age will increase the most, reflecting decreasing birth rates and migration slowdowns. Aging trends will disproportionately affect developed economies, and their consequences are mostly related to Western societies who will require new technology and services to support complex needs of the older segments of the population. In this context, the priority for ESA could be to further strengthen the use of its satellites and build upon its

[3]Man-made satellites launched into orbit.

unique expertise in space communication technologies and applications to alleviate the issues related to aging, through e.g. delivering the space infrastructure required to enable remote consultations (telemedicine). In addition to gathering and transmitting data, ESA's know-how and expertise could be leveraged to advance the use of medicine delivery by drones which would provide a solution to remote communities. ESA could provide secure communications and enable personal health monitoring. Furthermore, the Agency could take advantage of its Human spaceflight missions to develop Integrated Mobility & Health Monitoring solutions on Earth.

7.4.1.4 Sustainability

Today there is a significant amount of space debris and a continuously increasing number of abandoned artificial satellites. In addition to the existing debris, the large number of small satellite constellations—known as Mega constellations—that are currently planned for deployment in the coming years are expected to further crowd the Low-Earth Orbits (LEOs) and increase the operational risks of collisions and damages. This is exacerbated by the lack of global regulations and guidelines, the inhomogeneous application of existing standards, as well as the reduced spacecraft reliability accepted by operators to reduce costs and increase the production rate necessary to deploy massive constellations of thousands of satellites.

Inspired by the circular economy principles based on the 3R (Reduce, Reuse and Recycle), ESA could address the above challenges with the following concepts:

- Application of standards and regulations
- Space servicing Vehicles (Space tug/In-orbit manufacturing)
- Space debris removal
- Satellite recycling
- Satellite tax (e.g. pay per use €5 k/kg per year).

Some of the above-mentioned opportunities have already generated activities within the Agency such as the Clean Space and the Space Debris programmes.

7.4.1.5 Autonomous Vehicles

The development of Autonomous Vehicle (AV) technologies is advancing quickly. Thousands of hours of autonomous car testing on public roads have been performed across the world and we may imagine to see one million fully autonomous cars on the road by 2030. Equally, advances in unmanned aerial systems imply that this technology is being used in a variety of applications, such as agriculture, and new applications are still emerging. Commercial aerospace will also become increasingly autonomous with the near-term goal being a single pilot aboard each aircraft. However, Autonomous Vehicle implementation is currently restrained by technology, test hours and legislation hurdles. There is significant industry fragmentation with a large number of niche providers developing a series of non-standardised solutions.

It's proposed that ESA could become the trusted integrated solutions partner to the European AV industry, especially for the maritime, farming and mining sectors. This would include thought leadership, advisory and integration related to Navigation, Earth Observation, software integration and operations.

7.4.1.6 Africa

Development of Africa's economies, governance structures and standards of living remains slow and varied with the top 10 countries in Africa generating 78% of the continent's growth. Challenges with social development, such as low educational performance and poor healthcare systems, persist and need to be overcome to guarantee a sustainable development pace. Improving broadband connectivity and using consistent Earth Observation data can help alleviate challenges though differences in national policies, wealth and objectives continues to stifle progress whilst there is no space-based agency that aligns the needs of the continent. NGOs and governments in Africa are seeking solutions and ESA could play a key role considering that its Member States have political, economic and social links with many African countries. ESA could therefore support its Member States' policies and open up new opportunities for European downstream services while promoting the development of African industries and operators, promoting local research and development in the following areas:

- **Social**: space for all. Unlock access to all people including remote communities, providing Africa with a broadband system from geostationary orbit
- **Geopolitical**: manage migration patterns to help increase welfare whilst promoting a secure continent through monitoring of criminality and provision of EO data as tool to enable peace. Use satellite data to manage population growth and related urban planning
- **Resource Management**: promote African management of its resources and greater sustainability, from ores to water
- **Governance**: Help create a pan African space agency that can operate a global system.

7.4.1.7 Cyber-Security

The European space ecosystem is increasingly vulnerable to cyber-attacks with threats ranging from enterprise level data loss, to interception of data being transmitted by satellite to terminals, to more malicious acts. A serious data breach to or disruption of a space mission through cyber-attacks would have wide consequences for ESA, including increased programme costs, damage to the Agency's image and reputation as well as impacts on the whole space industry. Cyber-Security was concluded to be a huge threat for the space ecosystem and for the Agency. Actions were identified, to be undertaken by the Agency to mitigate the risk all along the entire supply chain, considering that the interconnectivity of satellite manufacturing to launch, operations and downstream services requires a joint industry approach and common standards. Cyber-Security should be seen as a new discipline to be applied end-to-end to space projects with ESA as a key facilitator of standards, best practice and technology partners. ESA would be ideally positioned to create standards and tools and provide governance to ensure resiliency of the wider space supply chain, evolving them over time to meet future threats. This includes the protection of ground systems, data links to and from space and the satellite supply chain. ESA could also develop a new set of competencies and become an advisor to industry stakeholders which should include standards, regulations, best practice and advisory support. ESA should become a driver for cyber compliance that will reach beyond the European supply base to the global supply chain, thereby becoming the link between cybertechnology vendors and space technology and service providers. ESA would be also well positioned to take a leadership role in the development of standards and advisory services through its holistic end-to-end knowledge of the space value chain, experience of developing cyber secure space systems and intimate knowledge of European Standards.

7.4.1.8 Artificial Intelligence

Space mission operations face several challenges that may generate inefficiencies and make tasks long and complex. Time delay in communication between the spacecraft and the operation centre can significantly extend the length of tasks, for example a rover sending messages back to control and awaiting its next instructions about its direction or action. This can lead to a reduction of scientific opportunities in case the number of samples collected, for example, is reduced due to a lack of rover autonomy. Equally, operational costs of missions are high and could be reduced through applying Artificial Intelligence (AI) to mission control and operations. AI is currently used for docking operations and there are opportunities to extend its application from launch to re-entry operations. As a priority opportunity, the Agency could use AI to achieve a higher degree of autonomy on ESA programmes, leveraging internal capability and the European AI network to build industrial capacity. ESA would drive the application of AI critical enabling technologies for an **augmented autonomy of missions,** so increasing the overall autonomy of space systems, from navigation to landing. Furthermore, through investing in on-board autonomy for both planning and troubleshooting, ESA would be able to improve operational performance.

7.4.1.9 Blockchain

A Blockchain is a list of records, or blocks, that are linked, timestamped and secured using cryptography. Investments in the Blockchain technology have rapidly increased as potential applications for a “distributed ledger” have emerged. Blockchain is a disruptive technology and is likely to impact how information is shared, bypassing third parties and providing greater transparency. Bitcoin is the most known application of Blockchain technology, though applications are emerging across various industries. In air transportation applications include flight operations, borders and immigration, Maintenance Repair and Operations (MRO), airline ticketing and loyalty management. Opportunities in space, apart from mission data, may include smart procurement to optimise the supply chain, or information management where applications may include virtual spacecraft (traceability and configuration consistency) and configuration control (hardware/documentation). However, the diversity of the space value chain in terms of both the number and type of organisations, results in a lack of data standardisation across missions and agencies. Data is not always transparent or readable and requires a technological solution to provide tools and develop standards. As a priority opportunity, ESA could lead the space industry to adopt Blockchain technology for the organisation and the management of scientific data, ensuring that programme management and mission data are encrypted. This could include the development of the tools and standards related to how mission data is generated, transmitted, collected and shared. The ability to capture structured data in a distributed ledger would also provide transparency to all industry stakeholders.

7.4.2 *Non-programmatic Priorities*

Besides the programmatic activities to be initiated to seize the opportunities identified by the analysis of Mega Trends and sub-trends in the different areas, some non-programmatic priorities were identified as necessary for the Agency to play its new role of broker, facilitator and promoter of space-based infrastructures and services (Fig. 7.4).

This would require to **redefine ESA's mission** in line with the identified opportunity areas. ESA should be a global agency open to the world, collaborating with

Fig. 7.4 The internal capabilities that could be enhanced. *Source* Frost & Sullivan

private actors in partnership to reach new opportunity areas and creating awareness by sharing data and information, create visibility and effectively communicate.

It would also require the Agency to implement an **Ecosystem Innovation Process** that would centralise efforts between the programmes and facilitate their coordination, and develop or acquire tools for Open Innovation Platform (O.I.P.), effective Program Management (P.M.), as well as Global platforms for promoting Social Economy Forum and Global Space Economy Workshops.

ESA would also be required to attract **private and public funding** to the space ecosystem, in particular from:

- industry sectors in the scope of the opportunity areas, from which to seek funding from e.g. exploitation of space for new business, tourism, mining, energy, health, …
- new policies where space can help
- the European portfolio of start-ups that ESA supports in getting investment
- the European portfolio of opportunity areas and large projects seeking for investors
- New investors.

Finally, to be able to seize the vast range of opportunities identified, the Agency should also acquire new **technical skills** for opportunity areas as well as new soft skills necessary to develop the required capabilities: innovation, ecosystem and partnership building, financing.

7.5 Conclusion

The Mega Trend initiative has stimulated reflections across the Agency and among stakeholders on threats and opportunities generated by the analysis of future social, economic and technological patterns. The reflection has been focussing on the space ecosystem and on the role of ESA to mitigate threats and seize opportunities.

Though the outcome of the initiative does not represent the ESA positions on the different identified areas, it has been brought to the attention of the Director General and the Executive Board of Directors, leading to a number of actions and initiatives in different areas.

This paper was presented at the 69th International Astronautical Congress, 1–5 October 2018, Bremen, Germany. www.iafastro.org.

Chapter 8
"KiboCUBE"—UNOOSA/JAXA Cooperation Program for Capacity Building by Using the Innovative CubeSat Launch Opportunity from ISS "Kibo"

Fuki Taniguchi, Hiroki Akagi and Kunihiro Matsumoto

Abstract In recent years, a growing number of universities and companies around the world have been developing the Micro/Nano-satellite (mainly CubeSat). These satellites are attracting much attention not only for their short-term and low-cost development, but also for their ability to perform various types of difficult missions, such as earth observation, technology demonstration, and planetary exploration. At the beginning of the Micro/Nano-satellite history, the method of transporting a satellite into orbit was as a piggyback payload carried aboard a launch vehicle, although launch opportunity was limited due to the launches of major satellites. The Japan Aerospace Exploration Agency (JAXA) developed a unique system called the JEM Small Satellite Orbital Deployer (J-SSOD). to deploy and inject satellites into orbit from the Japanese Experiment Module (JEM) known as "Kibo"—one of the International Space Station (ISS) modules—by taking advantage of its one and only function of having both the JEM-Airlock (JEM AL) and the JEM-Remote Manipulator System (JEMRMS), a special kind of robotic arm. In 2012, we successfully deployed five satellites on the first J-SSOD mission, which opened new capabilities for Kibo/ISS utilization and contributed to broadening the possibilities for Micro/Nano-satellites.

8.1 Introduction

In recent years, a growing number of universities and companies around the world have been developing the Micro/Nano-satellite (mainly CubeSat[1]). These satellites

[1]California Polytechnic State University, CubeSat Design Specification rev.12, 2009.

F. Taniguchi (✉)
Japan Aerospace Exploration Agency (JAXA), Paris, France
e-mail: Taniguchi.fuki@jaxa.jp

H. Akagi · K. Matsumoto
Japan Aerospace Exploration Agency (JAXA), Tsukuba-shi, Ibaraki, Japan

S. Ferretti (ed.), *Space Capacity Building in the XXI Century*, Studies in Space Policy 22,
https://doi.org/10.1007/978-3-030-21938-3_8

are attracting much attention not only for their short-term and low-cost development, but also for their ability to perform various types of difficult missions, such as earth observation, technology demonstration, and planetary exploration. At the beginning of the Micro/Nano-satellite history, the method of transporting a satellite into orbit was as a piggyback payload carried aboard a launch vehicle, although launch opportunity was limited due to the launches of major satellites. The Japan Aerospace Exploration Agency (JAXA) developed a unique system called the JEM Small Satellite Orbital Deployer (J-SSOD).[2] to deploy and inject satellites into orbit from the Japanese Experiment Module (JEM) known as "Kibo"—one of the International Space Station (ISS) modules—by taking advantage of its one and only function of having both the JEM-Airlock (JEM AL) and the JEM-Remote Manipulator System (JEMRMS), a special kind of robotic arm. In 2012, we successfully deployed five satellites on the first J-SSOD mission, which opened new capabilities for Kibo/ISS utilization and contributed to broadening the possibilities for Micro/Nano-satellites.

As of March 2018, more than 200 satellites have been deployed from Kibo by the Deployer developed by Japan or the US. Deployment from Kibo has attracted global attention as a new space transport system for satellites. Since 2015, JAXA has collaborated with the United Nations Office for Outer Space Affairs (UNOOSA) for providing CubeSat deployment opportunities from Kibo, in order to facilitate improved space technologies in developing countries. The program is called as "KiboCUBE". This paper introduces an overview of this innovative launch opportunity for Micro/Nano-satellites and our "KiboCUBE" program.

8.2 New Space Transport System for Satellites: "J-SSOD"

8.2.1 What Is "J-SSOD"

Japan has participated to the International Space Station (ISS) program as the only Asian partner. Japan contributed to the ISS through the Japanese Experiment Module (JEM), known as "Kibo" which means "hope" in Japanese, and unmanned vehicle H-II Transfer Vehicle (HTV), known as "KOUNOTORI". Kibo is Japan's first human-rated space facility, and it was designed and developed with a view to conducting scientific research activities on orbit.

Kibo has very unique design in the ISS. First, Kibo has an airlock which is the gateway from inside the ISS to outer space. This airlock can transfer some payloads between the Kibo Pressurized Module (PM) and the Exposed Facility (EF), and it is not designed for egress/ingress of extravehicular activity (EVA) crew members. Secondly, the EF provides a multipurpose platform where experiments can be deployed and operated in the exposed environment in space. The payloads attached to the

[2]Japan Aerospace Exploration Agency (JAXA), JX-ESPC-101133-B, JEM Payload Accommodation Handbook (2015)—Vol. 8-Small Satellite Deployment Interface Control Document, Revision B, January 2015.

EF can be exchanged or retrieved by JEM Remote Manipulator system(JEMRMS) which is Kibo's robotic arm. Using these very unique advantaged facilities, JAXA developed the world's first satellite deployment system "J-SSOD" from Kibo on ISS. This is a new space transport system for satellites.

J-SSOD is a mechanism for deploying Micro/Nano-satellites designed based on the CubeSat design specification (from 1U to 3U, 1U:100 × 100 × 113.5 mm) and 50 kg-size micro-satellite (550 × 350 × 550 mm) from Kibo into space. J-SSOD mainly consists of three components as shown in Fig. 8.1: Satellite Install Case with a spring deployment mechanism, a Separation Mechanism to maintain satellites inside the cases by holding the hinged doors of the Satellite Install Cases, and two Electronics boxes. J-SSOD will be installed on the Multi-Purpose Experiment Platform (MPEP) for back and forth transport through the JEM AL, and for handling the JEMRMS. The JEMRMS will position the platform with J-SSOD towards the aft-nadir direction to assure retrograde deployment. The ballistic number of a satellite shall be less than 100 kg/m^2 for faster orbiting decay of the satellite than that of the ISS. When trigger commands are initiated, the separation mechanism rotates and opens the hinged door of the satellite install case. The spring deployment mechanism in the case pushes the satellite out with spring force, and thus deploys the satellite. The separation mechanism and the electronics box are reusable on-orbit. The satellite install case has no heater, but is covered by Multi-Layer Insulation (MLI) for passive thermal control.

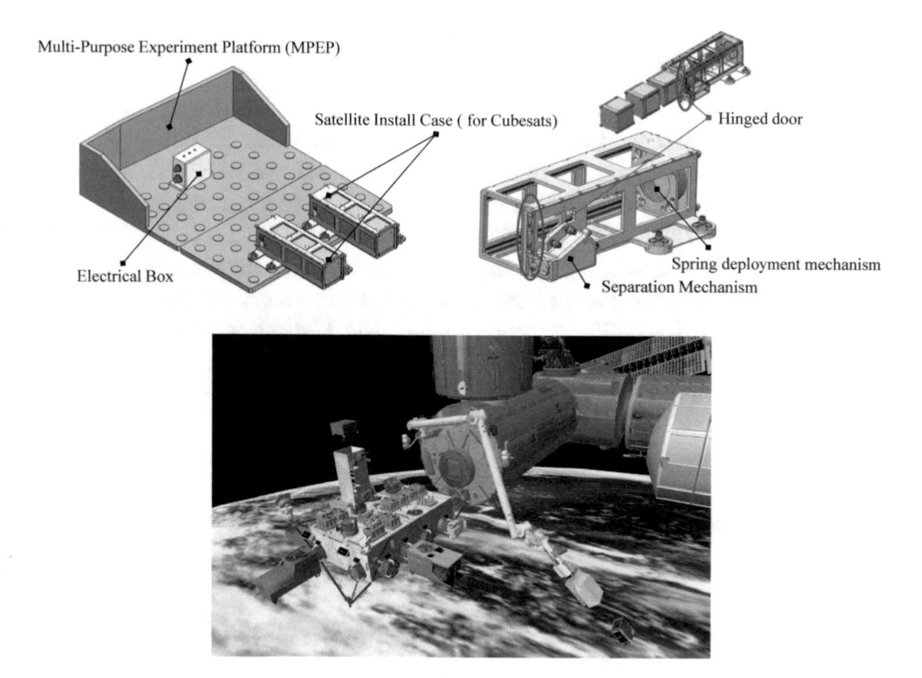

Fig. 8.1 Overview of the J-SSOD system. *Credit* JAXA

8.2.2 Advantage of J-SSOD

At the beginning of the Micro/Nano-satellite history, the method for transporting a satellite into orbit was as a piggyback payload carried aboard a launch vehicle, although launch opportunity was limited due to the launches of major satellites. However, using J-SSOD, the satellites are delivered to Kibo/ISS as part of the spaceship cargo load carried to the ISS by such transfer vehicles as HTV, SpaceX Dragon and Orbital-ATK Cygnus, thereby increasing launch opportunities to resupply the ISS. Therefore more opportunities are ensured for satellite launches. At present, satellite deployers other than J-SSOD that use Kibo include the NanoRacks CubeSat Deployer (NRCSD)[3] and Cyclops (Space Station Integrated Kinetic Launcher for Orbital Payload Systems).[4] JAXA operates the entire system from Tsukuba Space Center in Japan. As of March 2018, more than 200 satellites have been deployed from Kibo and the deployment system has attracted global attention as a new space transport system for satellites.

In addition, the Satellite Install Case is stowed in a Cargo Transfer Bag (CTB), which is a soft-cushion bag filled with cushion foam to hold cargo for the ISS. Therefore, the mechanical environment such as random vibrations during the rocket launch are mitigated, and shock testing with a payload attach fitting (PAF), usually required for a piggyback payload rocket, is not required.

8.2.3 Technical Interface

JAXA defines the technical interface requirements and safety requirements in the Interface Control Document for a satellite to be deployed from J-SSOD. A satellite provider shall show compliance that the satellite meets the requirements defined in the document, and shall attend the following review panels and report on the results of satellite design, manufacture, testing, etc. The interface requirements between J-SSOD and a satellite are defined based on the CubeSat Design Specification published by the California Polytechnic State University with JEM-unique requirements.

8.2.3.1 Safety Review

JAXA is responsible for conducting safety reviews for a selected satellite in the primary design phase (phase 0/I), detailed design phase (phase II), and acceptance

[3]NanoRacks, LLC, NR-SRD-029, NanoRacks CubeSat Deployer (NRCSD) Interface Control Document, 2013.

[4]Newswander, D., Smith, J., Lamb, C. and Ballard, P. "Space Station Integrated Kinetic Launcher for Orbital Payload Systems." Proceedings of the 27th AIAA/USU Conference on Small Satellites, Logan, Utah, August 10–15, 2013.

test phase (phase III). The satellite provider shall submit a Safety Assessment Report (SAR) and necessary support documents for review by JAXA.

8.2.3.2 Compatibility Verification Review

JAXA is responsible for conducting a review to confirm that the satellite verification results comply with the requirements defined in this document before the satellite is delivered to JAXA. The satellite provider shall conduct necessary verifications and submit necessary documents, such as drawings, analysis reports, and test reports for review by JAXA.

8.2.3.3 Confirmation Before Satellite Installation

JAXA is responsible for confirming that all remaining action items, which are identified in the Safety Reviews and Compatibility Verification Reviews, have been closed before a satellite is loaded into the J-SSOD Satellite Install Case. The satellite provider shall close all action items and show that the necessary documentation processes have been completed.

8.2.3.4 Radio Frequency Capability and Authority

A satellite with radio frequency capability shall be certified for space operation in the desired/planned operating frequency bands prior to integration into the launch vehicle. Certification of the satellite is achieved by obtaining an equipment operating license from the National Regulatory Agency. The license, along with the positions of any ground station assets that will be used to communicate with the satellite, shall be submitted to the NASA JSC Frequency Spectrum Manager for notification. A satellite that has intentional RF radiating and/or receiving devices shall be approved and certified by the NASA JSC Frequency Spectrum Manager for the use of a specified frequency band.

8.2.4 The Opportunity for Using J-SSOD

JAXA provides Micro/Nano-satellite deployment opportunities from Kibo, while bearing expenses to help ensure easy and fast launches and operation of Micro/Nano-satellites by private companies and universities, expand the application of space development, foster human resources, and promote a new industry centered on space development. JAXA also provides the opportunities for Asian nations and developing countries as a gateway for sharing the values of Kibo/ISS, and to promote capacity-building to enrol young researchers, engineers and students for utilizing Kibo.

For example, on January 13, 2016, a 50 kg-size Micro-satellite called "DIWATA-1" (meaning "fairy" in Filipino) was handed over to JAXA. This satellite was co-developed by the Department of Science and Technology (DOST) of the Republic of the Philippines, the University of the Philippines, Hokkaido University, and Tohoku University. DIWATA-1 is the first satellite owned by the Philippine government, developed also by Filipino engineers. The satellite is intended to observe Earth, monitor climate change, and develop human resources. DIWATA-1 was launched by a Cygnus developed by Orbital-ATK on March 23, 2016, and then deployed from Kibo on April 27, 2016. It is still operational as of March, 2018.

Other example is BIRDS project. On February 8, 2017, five satellites were handed over to JAXA. Joint Global Multi Nation Birds, also known as the BIRDS project, is a constellation of five 1U CubeSats developed by Japan, Ghana, Mongolia, Nigeria and Bangladesh, and organized by the Kyushu Institute of Technology, Japan. The BIRDS satellites will perform six missions. Three of these missions (i.e. Camera mission, Digi-Singer mission, Single Event Latch-up measurement mission) will be achieved through onboard mission payloads, and three novel missions (i.e. determination of precise satellite location, measurement of atmospheric density, demonstration of ground station network for the CubeSat constellation) will be achieved on the ground using the advantages afforded by the constellation of five CubeSats being operated simultaneously from seven ground stations. This project hopes to provide great leverage to students from developing nations for a hands-on satellite project. These satellites were launched aboard a Dragon spacecraft developed by SpaceX on June 3, 2017, and deployed from Kibo in July 7, 2017 (Fig. 8.2).

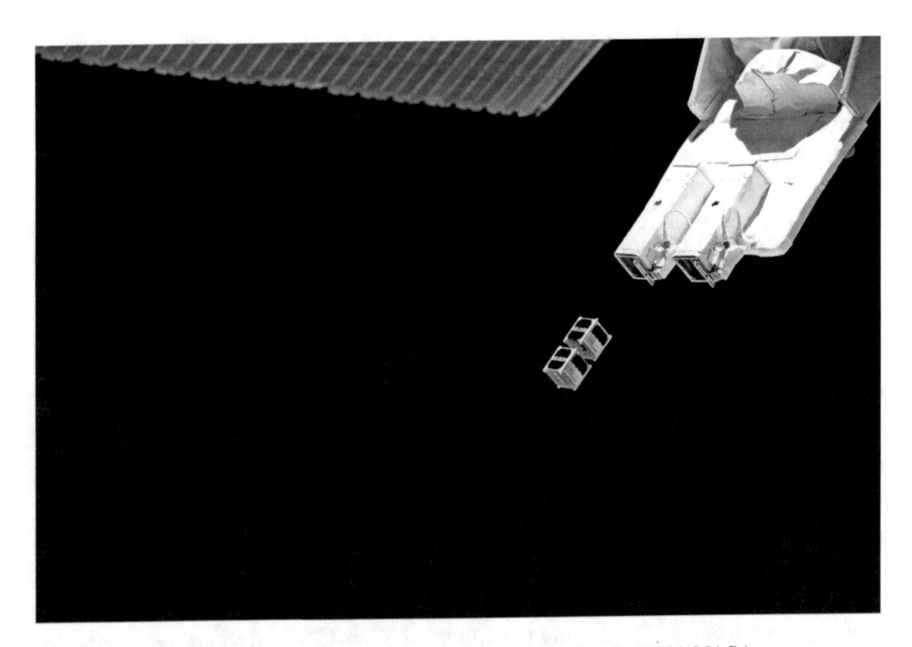

Fig. 8.2 Deployment of BIRDS project satellites from Kibo. *Credit* JAXA/NASA

8.3 KiboCUBE Program

8.3.1 Outline of the Program

Since 2015, JAXA has also collaborated with the UNOOSA for providing CubeSat deployment opportunities from Kibo, in order to facilitate improved space technologies in developing countries.[5] Small satellites can be manufactured at low cost and by utilizing relatively simple technology; therefore, they can be useful for education, communication, disaster mitigation, and human resource development. These two entities launched a three year "KiboCUBE" program in September 2015, which offered developing countries the opportunity to deploy small satellites from Kibo. This program provides the opportunity of 1U size CubeSat deployment at once a year from 2017 to 2019. While taking maximum advantage of the Kibo's strong points and employing the network and know-how with UNOOSA, JAXA will offer opportunities of small satellite utilization from the Kibo to developing countries so that JAXA and UNOOSA contribute to the improvement of their space technology. The outline of procedure for KiboCUBE is:

(a) Announcement of offer from UNOOSA
(b) Application from UN Member States to UNOOSA
(c) Selection by UNOOSA and JAXA
(d) Having an agreement between selected entity and JAXA
(e) Designing and developing CubeSat by selected entity
(f) Technical coordination between selected entity and JAXA
(g) Hand over the CubeSat to JAXA
(h) Launch to ISS and deployment from "Kibo" by JAXA
(i) Operation of CubeSat by selected entity.

8.3.2 Conditions for Application

This opportunity is open to entities located in developing and emerging countries that are Member States of the United Nations. The entities should be heads of research institutes, universities, and other public organizations. Private companies are ineligible. And also, entities located in countries that have the means to transport artificial satellites into space and place them in orbit are not eligible, taking into account the objectives of this Opportunity.

Entities applying for this Opportunity are responsible for the development of their CubeSat including the design, manufacturing, test and verification, as well as its operation and utilization after the deployment. Therefore, to be eligible for this

[5] UNITED NATIONS Office for Outer Space Affairs (UNOOSA), Announcement of Opportunity, KiboCUBE Second Round (2016–2017), September 26, 2016.

Opportunity, applying entities must have sufficient capability; CubeSat development and testing, ability to transport the CubeSat to JAXA, submission of safety assessment reports, coordination of the CubeSat's radio frequency, ability to obtain a license of radio stations, and development of the ground station facility.

The Selection Board consists of members nominated by OOSA and JAXA and will review the incoming applications. These are some criteria, in accordance with Announcement of Opportunity, including enough capability to comply with the technical interface, schedule, and peaceful use of the outer space. In addition, the entities should have clear vision of scientific and technical value of the CubeSat under this program, as determined by either:

(a) The CubeSat's expected contribution to developing human knowledge and capacity to undertake activities in the field of space science and technology in the applying entity's home country or abroad; or
(b) The CubeSat's expected contribution to enhancing research and development through the technological demonstration of deploying and operating the CubeSat in the applying entity's home country or abroad.

8.3.3 *First Round for KiboCUBE*

In August 2016, UNOOSA and JAXA selected a team from the University of Nairobi, Kenya, as the first to benefit from the KiboCUBE program. On January 16, 2018, the team from the University of Nairobi handed over to JAXA the satellite it has developed, known as "1KUNS-PF", or "First Kenyan University Nano Satellite-Precursor Flight". The handover took place at the JAXA Tsukuba Space Center. 1KUNS-PF launched to the ISS in April 2018, and was deployed from Kibo with a robotic arm during the northern hemisphere spring (Figs. 8.3 and 8.4).

8.3.4 *Second Round for KiboCUBE*

In September 2017, UNOOSA and JAXA selected a team from the Universidad del Valle de Guatemala, as the secound to benefit from the KiboCUBE program, and their CubeSat is planned to be launched in FY2020. The team plans to use its Guatemalan CubeSat to test equipment for monitoring the concentration of harmful cyanobacteria over inland bodies of water. The Universidad del Valle de Guatemala and JAXA conducted technical coordination once a month in preparation for the CubeSat launch and deployment from Kibo.

Fig. 8.3 Technical coordination meeting with JAXA Engineer. *Credit* JAXA

Fig. 8.4 Handover ceremony for 1KUNS-PF at Tsukuba Space Center. *Credit* JAXA

8.3.5 Third Round for KiboCUBE

UNOOSA and JAXA announced the Third Round[6] of the opportunity until March 31, 2018. One entity from the Republic of Mauritius was selected from among the proposals received.

[6]The United Nations/Japan Cooperation Programme on CubeSat Deployment from the International Space Station (ISS) Japanese Experiment Module (Kibo) "KiboCU BE". http://www.unoosa.org/oosa/en/ourwork/psa/hsti/kibocube/2018.html.

8.4 Conclusion

KiboCUBE is the first step in UNOOSA's mission to create opportunities for developing countries to participate in space development, collaborating with JAXA, which developed an innovative launch system called "J-SSOD" to deploy and inject satellites into orbit from Kibo/ISS taking advantage of its unique function on Kibo/ISS. Through the experiment of developing a CubeSat, selected entities can go to the next stage for larger and more practical satellite in a relatively shorter step. Obtaining knowledge and technology for satellites means that developing countries also acquire the tools for problem-solving, for example disaster monitoring or improving infrastructure, by themselves. Indeed, the KiboCUBE program could be an effective base for sustainable and stable development, and contribute to the universal spirit of the Sustainable Development Goals (SDGs).

Chapter 9
ICE Cubes—International Commercial Experiment Service for Fast-Track, Simple and Affordable Access to Space for Research—Status and Evolution

Hilde Stenuit and Mauro Ricci

Abstract We are at the dawn of a new commercial spaceflight era. Space is no longer in the domain of agencies and governments only, but is being democratized. Everyone can now access space at reasonable costs and the barriers that limit the access to space are being torn down. Upon the formal signature of the commercial partnership with the European Space Agency (ESA) in June 2017, the International Commercial Experiment Cubes (ICE Cubes) service has been established as the first European service to provide fast, simple and affordable access to the International Space Station (ISS) for research, technology development, capacity building as well as education and inspiration. The ICE Cubes service provides access to space for a wide range of user groups: scientists/research institutes for fundamental and applied sciences; industrial companies for their research and development (R&D) activities; technology providers for in-orbit testing, validation and demonstration of technologies and processes; emerging space-faring nations for capacity building and schools for educational experiments in the STE(A)M areas. Past studies show that the potential of microgravity is high and the interest is significant. However, the level of awareness is low. Unless previously involved with spaceflight, the number of industrial and academic researchers and developers seeking solutions in microgravity is low. Therefore, significant effort is being made to extend the users' network and to establish value propositions for specific R&D areas to take advantage of the space environment. To this purpose, dedicated state-of-the-art diagnostics will be established as part of the ICE Cubes program to further enable the performance of R&D activities in space. At the time of writing, the first ICE Cubes facility has been launched to the ISS and has been successfully commissioned. The ICE Cubes service is therefore as of today fully operative with the first batch of research experiment cubes being operated. The ICE Cubes service for the ISS is the first step of a space access service scenario and capabilities that will potentially include free flyers, external platforms, lunar landers/rovers and post ISS infrastructures. A marketing

H. Stenuit (✉) · M. Ricci
Space Applications Services NV/SA, Zaventem, Belgium
e-mail: Hilde.Stenuit@spaceapplications.com

M. Ricci
e-mail: Mauro.Ricci@spaceapplications.com

S. Ferretti (ed.), *Space Capacity Building in the XXI Century*, Studies in Space Policy 22,
https://doi.org/10.1007/978-3-030-21938-3_9

strategy has been elaborated to take steps forward to further evolve and enhance the ICE Cubes service by expanding from access to the ISS to a range of other space platforms, based on related efforts to increase awareness, arouse inspiration, consolidate and establish partnerships. This chapter provides a status of the ICE Cubes Service for access to ISS and on the establishment of the service for each of the user groups and presents the status and the evolution into a wider space access service and capabilities scenarios.

9.1 Introduction

The International Space Station (ISS) is a great technical achievement, designed as a human permanently inhabited Earth-orbiting laboratory for carrying out long-term scientific research in the unique environment of space.[1] Thanks to its high modularity, the ISS is able to support science in a wide range of disciplines.

Unfortunately, to make use of the ISS has long been rather unappealing to a large number of potential users,[2] due to the burden of complex rules and long procedures associated with developing and operating equipment on board the Station.

The policy of the European Space Agency (ESA) about the utilization of the ISS Columbus module has changed, opening to the possibility of accessing the ISS on a commercial basis.[3] The shift in policy of the various space agencies is allowing for the establishment of new commercial partnerships and services supporting the performance of additional science and technological Research and Development (R&D) activities.

One of these initiatives is the International Commercial Experiment Cubes (ICE Cubes) service. After the launch of the first ICE Cubes facility to the ISS in May 2018, the ICE Cubes service is now providing for the first European commercial access service to the ISS, providing the opportunity to conduct research experiments in different scientific disciplines and perform various technology demonstrations in weightlessness, inside ESA's Columbus laboratory on-board the ISS.[4]

---

[1]ESA. (2014). *User guide to low gravity platforms*. HSO-K/MS/01/14, Issue 3 Rev. 0. December 2014.

[2]ESF.(2012). *Independent evaluation of ESA's Programme for Life and Physical Sciences in Space (ELIPS)*. Final Report. ISBN: 978-2-918428-77-0. 2012 December 13. Available from: http://www.esf.org/fileadmin/Public_documents/Publications/elips_01.pdf.

[3]ESA. (2015). *Call for ideas "space exploration as a driver for growth and competitiveness: opportunities for the private sector"*. March 03, 2015. Available from: http://emits.sso.esa.int/emits-doc/ESTEC/News/ESA_CFI_Space_Exploration.pdf.

[4]ESA. (2017). *ESA signed an agreement with Space applications services for the first commercial European opportunity to conduct research in space*. June 20, 2017. Available from: http://www.esa.int/spaceinimages/Images/2017/06/ESA_signed_an_agreement_with_Space_Applications_Services_for_the_first_commercial_European_opportunity_to_conduct_research_in_space4.

The commercial partnership agreement with ESA allows the ICE Cubes service to ensure committed and recurring launches, which is an important asset for the potential commercial users of the Low Earth Orbit (LEO) market.

The ICE Cubes service is suitable for a number of utilisation areas, including:

- Fundamental and applied research by universities and research centres;
- Industrial and commercial R&D activities;
- In-orbit Testing and validation of technologies and processes;
- STEAM educational experiments, technology demonstrations and inspirational projects;
- Emerging spacefaring nations' capacity building.

Considering the above, the ICE Cubes service is now trailblazing to render access for space scientific and technology R&D fast, simple and cost-effective for both terrestrial and space exploration benefits.

9.2 Current Status of the ICE Cubes Service

The ICE Cubes service has by now established:

- A permanent multipurpose facility (ICE Cubes Facility, ICF) on-board the ISS allowing for the accommodation of scientific and technology R&D projects and experiments in the form of cubes (so-called Experiment Cubes);
- A set of special-purpose or repeated-use multi-unit Experiment Cubes dedicated to specific R&D areas;
- The ground infrastructure for the management of the ICF and the Experiment Cubes allowing for users to have direct and real-time interaction with their space experiment from their own premises;
- The end-to-end commercial service allowing utilization of the ICF and of the aforementioned multi-unit Experiment Cubes;
- Additional ad hoc engineering services (e.g. design, development, assembly, testing), on a commercial basis, as needed.

On May 21st 2018, the Orbital ATK-9 cargo mission launched with on-board the first ICE Cubes facility (Fig. 9.1). A couple of days later, the ICE Cubes facility reached its destination, when the Orbital Cygnus vehicle was captured and berthed to the ISS (Fig. 9.2).

The ICE Cubes facility was then installed on June 5th in Europe's ISS laboratory Columbus (Fig. 9.4). The on-orbit crewmember Alexander Gerst and the ICE Cubes Mission Control Centre at Space Applications Services in Sint-Stevens-Woluwe (Fig. 9.3) cooperated for the smooth and successful installation of the ICF inside the European Physiology Module (EPM) Rack.

Fig. 9.1 Launch of orbital ATK cargo mission 9 with on-board the first ICE Cubes facility. *Source* NASA/Orbital ATK

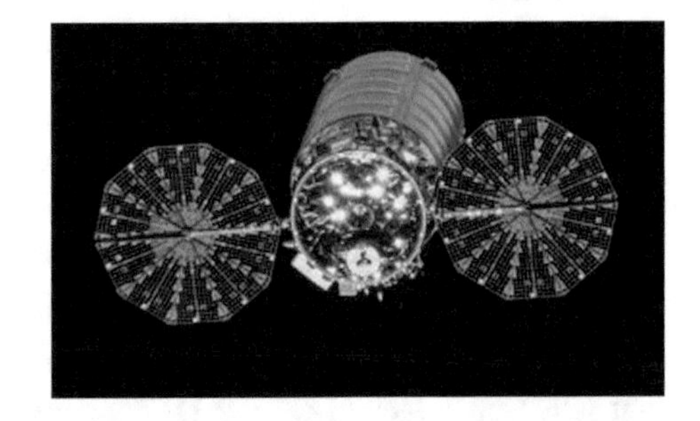

Fig. 9.2 Orbital ATK arriving for berthing at ISS. *Source* NASA/Orbital ATK

9.3 First ICE Cubes Experiments Operational

While waiting for the launch of the ICF, the end-to-end interface testing (Fig. 9.5) was being completed on the ground for the very first ICE Cubes experiments.

The first experiment cubes running inside the ICE Cubes facility are from the International Space University (ISU) and are already demonstrating the versatility of the experiments that can be carried out through the ICE Cubes service. The first cubes launched on SpaceX-15 on June 29th. Once on-board the Space Station, the

Fig. 9.3 ICE Cubes mission control centre in Brussels area during ICE Cubes facility installation. *Source* ICE Cubes

Fig. 9.4 ICE Cubes Facility (ICF) newly installed on-board ISS

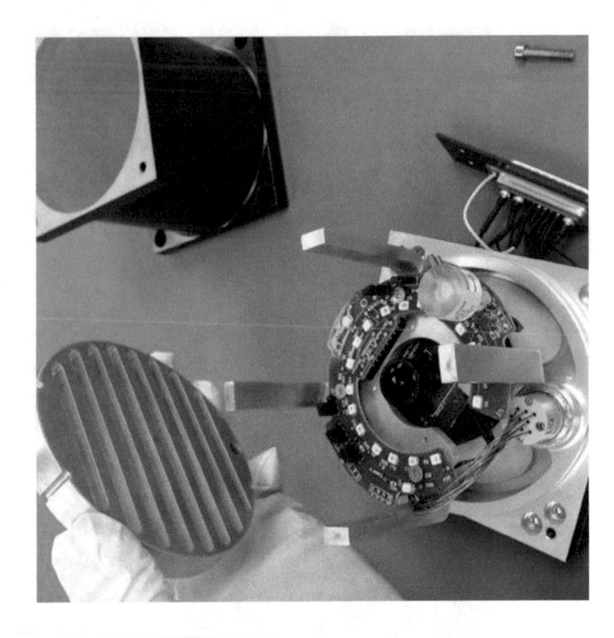

Fig. 9.5 ISU orbital greenhouse experiment planned for SpaceX-16 Nov-2018 during end-to-end interface testing. *Source* ICE Cubes/ISU

plug-and-play cubes were installed into the facility by ESA astronaut Alexander Gerst (Figs. 9.6 and 9.7).

One of the cubes studies microgravity's effect on microbes to evaluate the feasibility of using them on asteroids to produce methane. Eventually, this methane could

Fig. 9.6 ICE Cubes mission control centre during the installation of the first experiment cubes. *Source* ICE Cubes

Fig. 9.7 ESA astronaut Alexander Gerst installing the first ICE Cubes experiments. *Source* ESA/NASA/ICE Cubes

be used as a propellant, one of the most important resources a visiting spacecraft needs.

Another cube from the ISU contains a more artistic and inspiring demonstrator that will be used to have people interact in real time with this project from ground to space. Specifically, a kaleidoscope inside the cube will be actuated from the ground based on people's heartbeats. Live interaction will be possible through the installation of a portable ground station at specific events and venues.

ICE Cubes offers unique 24-h access to experiments via a dedicated mission control centre at Space Applications Services' premises in Sint-Stevens-Woluwe (Brussels area), Belgium. Customers can simply connect at any time to their experiment from their own location over internet to read the data and even directly send commands.

9.4 ICE Cubes Commercial Offer

The ICE Cubes service allows any country, any institute, any company or any entity to directly develop and conduct their experiment on the ISS. The service streamlined the related processes and removed as many barriers as possible related to accessibility, paperwork, planning and budget uncertainties for the users. This allowed for a fast-track route in which a space project can fly in 12 months (for new setups) or down to 6 months (for use of recurrent setups) from contract agreement to launch.

The resulting ICE Cubes service provides a complete "end-to-end service" from Experiment Cube development guidance, interface testing with the ICE Cubes facility, arrangement of the experiment certifications including safety, launch, on-orbit installation and operations support.

Additional services are also available such as engineering support for the hardware development, temperature-conditioned stowage, late access to launch vehicle, additional mass per Experiment Cube, return of the hardware/samples, early access to return vehicle, dedicated crew activities.

In addition to the Experiment Cubes inside the ICE Cubes facility, two other opportunities also exist:

Firstly, a payload which does not comply well with a cube form factor can also be accommodated externally to the facility—still in the internal pressurized Columbus module—wired to the ICF (Fig. 9.8). For example, a payload to test a heat pipe in space which needs to be long and stretched could be accommodated in this way. This approach also allows for easy access to this wired "external" payload. In addition, it could be possible to provide access to vacuum/venting line.

Secondly, the ICE Cubes service can also accommodate payloads which are free-flying inside the European Columbus module connected to the ICF through a private wireless connection for their telemetry/telecommands (TM/TC). This opportunity allows for projects in innovative areas such as: testing guidance and navigation algorithms/strategies; autonomous technologies in formation flying; collision avoidance; proof of concept for miniaturized docking/berthing subsystems; capturing of non-cooperative bodies; environmental monitoring surveys; sloshing; etc. Similar areas have been addressed in the past by the MIT Spheres project (Fig. 9.9).

Fig. 9.8 ICE Cubes with an "external" wired payload

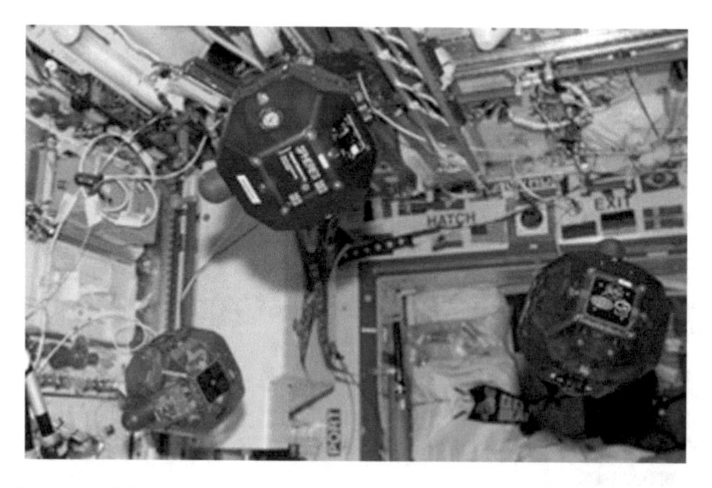

Fig. 9.9 Spheres by MIT as example for free-flying internal payloads. *Source* NASA/MIT

Thanks to an out-of-the-box software suite, users are able to directly operate their own Experiment Cubes. The ICE Cubes Mission Control Centre activates the cubes and opens the tunnels. From then on, customers can operate their space project directly via internet from their company, research institute, lab or school. Standard internet protocols (IP) are used to communicate with the facility, enabling near-real time TM/TC of each single Experiment Cube with data downlink rate up to 4 Mbps and uplink up to 0.5 Mbps.

In terms of IP rights, customers remain the only owners of their Experiment Cubes and of the results of their experiments.

9.5 ICE Cubes Usage Areas

9.5.1 Research by Universities and Research Centres in ICE Cubes

The ISS is a unique laboratory in space which is in a constant state of free fall around our Earth. This means that the station and all that is inside do not experience gravity as we do on Earth. Everything, including any organism and any material or fluid on-board experiences microgravity.

Over the last decade and more, relevant scientific results have been achieved in different research disciplines highlighting all the advantages microgravity offers, showing space research is valuable for the conduction of scientific and technology research that cannot be done in terrestrial laboratories.

All the following scientific areas are extremely relevant to use the ISS as scientific testbed: Thermal and Fluid Science, Foams, Colloids, Particle Physics, Proteins

Crystallization, Materials Science, Radiation Biology and Physics, Cell Biology, Microbiology, Plant Biology, Regenerative Medicine, 3D Tissue engineering, Cancer research and many more.

9.5.2 Industrial and Commercial R&D Activity in ICE Cubes

Important value features are established in the Life Sciences domain. Microgravity allows, for example,

- To establish 3D cell cultures, better mirroring the configuration cells have in the human body;
- Weightlessness accelerates cell growth and
- Ageing processes
- Alters genes expression
- And increases virulence
- Leads to higher tissue fidelity
- And organisms that behave differently

These different value aspects pave the way for great opportunities to life sciences research in, for example, drugs testing and diseases' understanding and therapies. In addition, crystals grow larger, faster and with higher quality in microgravity, enabling improved protein crystallisation studies.

Microgravity hence permits to address very relevant applied R&D in areas such as Pharmaceutical/Drugs R&D, Biotechnology, Organ-on-a-chip, Microbiology/Cell biology, Protein crystallization, 3D Tissue Engineering/3D bioprinting, food and cosmetics R&D, Cancer/Stem cell R&D, Plants/Agriculture/Horticulture, Micro-encapsulation, etc.

Strong advantages are also provided by microgravity for research in the areas of materials, fluids or more in general physical sciences, resulting in the publication of striking results in important scientific journals that are now the building blocks that can be utilized for new and innovative commercial R&D:

- No solute buildup
- No sedimentation
- No convection
- Defect free, Homogeneous, Perfect spherical shape manufacturing
- Controlled, symmetric growth
- Higher resolution, Containerless processing/Free suspensions

In this regard, microgravity allows to address R&D areas such as: Novel & Smart Materials, Soft matter/Fluids/Foams/Emulsions, Coatings/Catalysts, Oil & gas, Cosmetics, Food, Photonics, Optical fibre, Nanoparticles, Additive manufacturing, etc.

9.5.3 *In-Orbit Testing and Validation of Technologies and Processes in ICE Cubes*

The need exists for industries to demonstrate and validate their developed technologies and processes in a relevant space environment, such as on-board the ISS. By performing In-Orbit Demonstration/In-Orbit Validation (IOD/IOV), the related Technology Readiness Level (TRL) can be raised, allowing the technologies to enter the market as well as their use in space missions inside satellites and/or cubesats.

Some example areas for technology IOD/IOV are listed here:

- Radiation hardened electronics development;
- Studies on software techniques to guard against radiation upsets;
- High performance computer systems/miniature space computers;
- Delay-and disruption-tolerant networks testing;
- Heat exchangers/heat pipes;
- Autonomous navigation capabilities/autonomous rendezvous & docking/constellation flying;
- Miniaturized space robotics and servicing.

The ICE Cubes service can also accommodate the space qualification of new sensors, new actuators, Micro Electro Mechanical Systems (MEMS) or other devices such as space-grade pressure transducer or others.

By conducting IOD/IOV through this fast-track and direct route, companies can de-risk innovation and accelerate their development.

In addition, it is acknowledged that the space environment of the ISS may particularly be suited to benefit both state-of-art and innovative manufacturing technologies and processes, such as 3D (bio) printing, coatings, etc.

9.5.4 *Educational STEAM Demonstrations, Capacity Building and Inspirational Projects*

The ISS is one of the most iconic results of humankind's fascination and curiosity for Space which also led to one of the biggest human international collaborations. The ISS has many objectives and here we report some of them particularly applicable to education:

- To promote partnerships between industries, research institutes and educational entities;
- To promote the image of science and engineering, influencing the educational paths chosen by future generations;
- To sustain and reinforce the highly technological aerospace industry;
- To satisfy the age-old human nature of exploration.

ICE Cubes can inspire young generations and contribute to education by supporting teachers and students to learn about the exciting aspects of space research and by stimulating creative and autonomous thinking, cooperation, teamwork and inclusion.

The ICE Cubes service is open to educational STE(A)M projects of different schooling levels (primary, secondary, university), and different STEM curricula. As such, it allows for outstanding intra-school or intra-level collaborations.

Recently, the importance of establishing the "knowledge triangle" and closing the gap between academic, industry and educational entities has increasingly been highlighted, and the service is particularly suited for that.

Additional services can be provided on request such as courses/workshops to familiarise with space environment, space careers and ad hoc support for a target age.

9.6 ICE Cubes Future Enhancement

The ICE Cubes service to the ISS is the first step of an access service scenario that will be extended to other space platforms, orbits and locations.

Steps are in work to enhance the service with:

- Space exposure external accommodation;
- Space viewing capability;
- Earth viewing capability;
- Free flying capability;
- Powered upload/download capability;
- Access to moon landers/rovers;
- In situ diagnostics capabilities;
- Etc.

Potential collaborations are also in work with future commercial platforms providers.

The objective is for the ICE Cubes service to provide fast, simple and direct commercial access for small experiments for all types of space research and technology.

9.7 Results and Discussion

The ICE Cubes service is now fully operational with following key features:

- Allows any entity, any company, any research centre to directly develop and conduct their innovation project on the ISS,
- It is fast: 12 months to launch for new setups and even faster for recurring setups,

- It is simple & direct: processes streamlined and unique point of contact. Service takes care of: certification, safety, manifesting, interface testing, interface with agency & launcher,
- It ensures regular launches: 3 times per year, every ~4 months;
- It provides real-time interaction with space experiments. Users are provided with the tools to create their own control centre. High data rate (downlink up to 4 Mbps)
- A catalogue of space-qualified H/W is being established. Commercial-of-the-shelf H/W & adaptations reduce the risk and make usage (cost-) efficient;
- It can provide for return of experiments/samples from space
- It is cost-effective & adopts a fixed-pricing policy.

9.8 Conclusion

With the ICE Cubes service, an enabling end-to-end service has been established and is now operational to provide fast, simple and direct access for research and technology through regular launches to the ISS. As such, researchers have a flexible and modular accommodation at their disposal with unique real-time interaction capability to investigate in the unique high-quality microgravity & radiation environment of the ISS.

The ICE Cubes service allows for fast-track innovation & disruption in a "NewSpace" dynamic environment for on-orbit validation of technologies and for R&D activities. By testing in a true space environment, researchers and companies can de-risk their innovation and accelerate development.

This paper was presented at the 69th International Astronautical Congress, 1–5 October 2018, Bremen, Germany. www.iafastro.org

Chapter 10
Space Exploration and Innovation: An ESA Perspective

Veronica La Regina and Bernhard Hufenbach

Abstract Within the context of space 4.0 and looking ahead at future space exploration programs, this chapter sets the context in which space innovation and exploration activities are being conducted by the Space Agencies and the private sector in the main spacefaring nations. One instrument analyzed to implement these activities in ESA, is the one of partnerships between actors with complementary knowledge and capabilities. Each ESA Directorate has started implementing initiatives along these strategic guidelines, and the chapter offers concrete examples, focusing mainly on the "Innovation in Action" program set out by the ESA Space Exploration teams. This initiative sources ideas for the upcoming exploration destinations such as LEO, the Moon and Mars, following four key steps: fertilization, seeding, selection and implementation. First outcomes from this activity are presented in terms of new partnerships created.

10.1 External Scenario

Nowadays the world is increasingly interconnected; the distance between continents has shrunk and the vicinity of people relating to each other through business and in other ways is undeniably closer. Due to this proximity, knowledge and ideas about the improvement of life spread rapidly from one side of the planet to the other. To reflect these changes, the most accurate representation of the world is the Dymaxion map (see below). In addition, individual initiative has great potential for propagation and therefore emulation; thus the entrepreneurs known as 'the four space-boys' are highly visible and known for their technological achievements and related business activities. This *converged* world also implies a more rapid transfer of learning from one case to another, thus improving the chances of success and of making an impact.

V. La Regina (✉)
NanoRacks Space Outpost Europe, Turin, Italy
e-mail: vlaregina@nanoracks.com

B. Hufenbach
European Space Agency, Noordwijk, The Netherlands

S. Ferretti (ed.), *Space Capacity Building in the XXI Century*, Studies in Space Policy 22,
https://doi.org/10.1007/978-3-030-21938-3_10

10.1.1 World Map

The Dymaxion map "is the only flat map of the entire surface of the Earth which reveals our planet as one island in one ocean, without any visually obvious distortion of the relative shapes and sizes of the land areas, and without splitting any continents[1]" (Fig. 10.1).

In the context of an interconnected world, the dynamics of the space exploration ecosystem have greater influence and they are more quickly received by the diversified stakeholders, e.g. industrial players, visionary entrepreneurs, space agencies, policy makers and users. For instance, the Google Lunar XPRIZE went on for a decade and raised global attention with the formation of almost 16 teams until July 2015, which now have ownership of potential lunar capacity and/or capability. Later in early 2018 the 5 finalist teams were considered not in the position to meet the challenge goal to reach the Moon by the 31st of March 2018 (Table 10.1).

While Google's challenge remained unmet, the competition was a catalyst for the more effective use of space agency budgets. It has incentivized space entrepreneurs to bring forth a new era of affordable access to the Moon and beyond by inspiring the next generation of scientists, engineers, space explorers, and adventurers. Information released by the press reports[2] a cumulative spending value—as at July 2015—of almost US$60.9 Million and an equivalent of 945 person-years (Fig. 10.2).

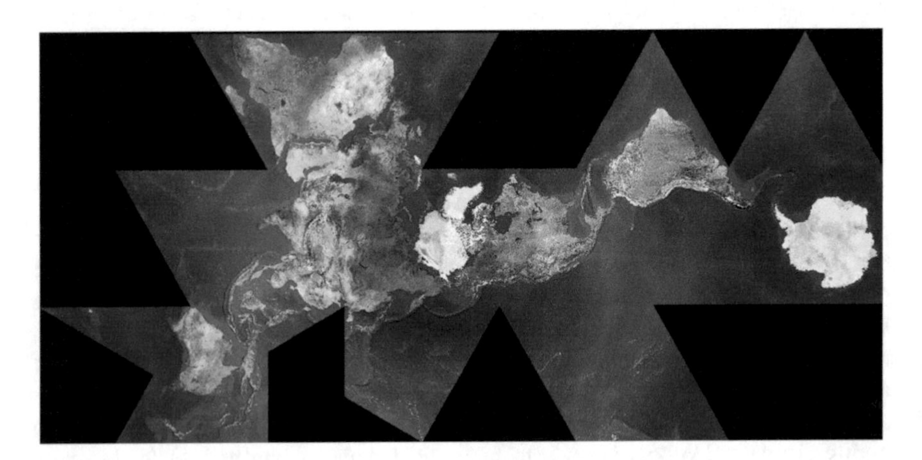

Fig. 10.1 Dymaxion map. *Source* Buckminster Fuller Institute

[1]Life Presents R. Buckminster Fuller's Dymaxion World (1943), LIFE: 1 March, pp. 41–55.

[2]The information was displayed during the EIRO Forum 2016 in Manchester (UK) on July 2016 by Alan Wells, Emeritus Professor of Space Technology and Co-Chair Judging Panel, Google Lunar X-Prize University of Leicester.

Table 10.1 List of teams from Google Lunar XPRIZE

Team	HQ country	Finalist
SpaceMETA	Brazil	
Plan B	Canada	
Angelicum Chile	Chile	
Part Time Scientists	Germany	
PULI	Hungary	
Indus	India	✓
SpaceIL	Israel	✓
Team Italia	Italy	
Hakuto	Japan	✓
Independence-X	Malaysia	
Euroluna	Denmark	
Astrobotic	USA	
Moon Express	USA	✓
Omega Envoy	USA	
STELLAR	USA	
Synergy Moon	USA	✓

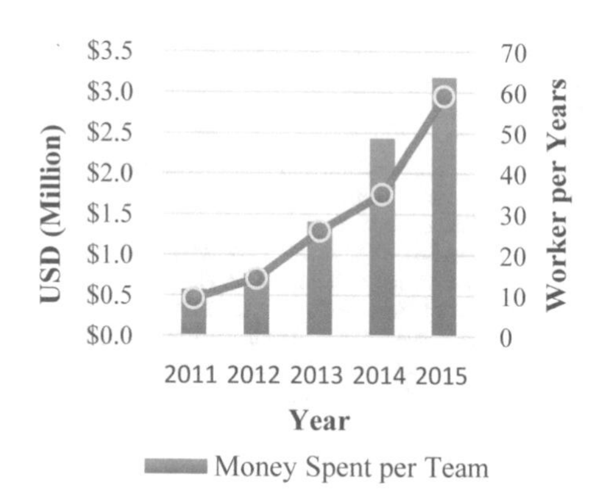

Fig. 10.2 Money spent per year by Google Lunar XPRIZE teams

10.1.2 The Four Space-Boys

In the international arena of 'NewSpace', the scenario is visually dominated by four entrepreneurs: Elon Musk, Jeff Bezos, Paul Allen and Richard Branson. Each of them has been primarily inspired by an almost philanthropic ambition to improve the world (Table 10.2).

The table above presents the main business performance of the well-known big four of the US NewSpace scenario in terms of vision, business footprint, private investment and NASA procurement contract value along with the current commercial offer, e.g. access to infrastructure, development of technology and procurement of

Table 10.2 NewSpace entrepreneurs and their business performance

Entrepreneur	Vision	Business footprint	Private investment	Space agency's procurement	Space exploration offer
Elon Musk	Build a greenhouse on Mars to assure food provision to remote and disadvantaged areas	• PayPal—*initial endeavour*—(billing/finance) • Tesla (automotive) • Solar farm (renewable energy) • SpaceX (launcher) • Starlink (Broadband services) • DeepMind (AI) • The boring company (tunnelling)	1.8 Billion US$[a]	~5 Billion US$[b]	Falcon 9, Falcon Heavy, Dragon
Paul Allen	Enable everyone to be connected	• Microsoft (IT) • Stratolaunch (launcher)	300 Million US$[c]	5.1 Million US$[d]	Stratolaunch
Richard Branson	Assure survivability to everyone	• Virgin Media (info transport) • Virgin Galactic (sub-orbital transport)	680 Million US$[e]	~4.5 Million US$[f]	Sub-orbital human spaceflight
Jeff Bezos	Satisfy the wish of having everything in your hand	• Amazon (delivery service) • Blue Origin (space transportation)	500 Million US$ and disinvesting 1 Billion US$ from Amazon per year[g]	~25 Million US$[h]	BE-1, -2, -3, -4, Glenn, Shepard

[a] *Source* https://uk.rs-online.com/web/generalDisplay.html?id=elon-musk-empire
[b] Atlantic Council (2014-06-04), Discussion with Gwynne Shotwell, President and COO, SpaceX,; NASA selects SpaceX's Falcon 9 booster and Dragon spacecraft for cargo resupply services to the International Space Station (Press release) SpaceX. December 23, 2008; NASA announces next steps in effort to launch Americans from U.S. Soil—NASA. August 3, 2012; NASA chooses american companies to transport U.S. Astronauts to International Space Station". NASA. September 16, 2014
[c] *Source* The Space Review: Stratolaunch: SpaceShipThree or Space Goose?". www.thespacereview.com
[d] *Source* http://spacenews.com/nasa-agreement-sign-of-stratolaunch-engine-development-program/
[e] Schreck, Adam, Abu Dhabi partners with Virgin Galactic spaceship firm, ABC News.; Abu Dhabi's Aabar boosts Virgin Galactic stake". Market Watch. 2011-10-19; Kleinman, Mark (2014-06-12). "Google In Talks To Take Virgin Galactic Stake". SkyNews
[f] "NASA Buys Flights on Virgin Galactic's Private Spaceship". Fox News. 2011
[g] Foust, Jeff (July 18, 2014). "Bezos Investment in Blue Origin Exceeds $500 Million"; St. Fleur, Nicholas (2017-04-05). "Jeff Bezos Says He Is Selling $1 Billion a Year in Amazon Stock to Finance Race to Space". New York Times, 2017
[h] Blue Origin Space Act Agreement with NASA, 2013; Morring, Frank, Jr. (April 22, 2011). "Five Vehicles Vie To Succeed Space Shuttle". Aviation Week (2011) The CCDev-2 awards, … went to Blue Origin, Boeing, Sierra Nevada Corp. and Space Exploration Technologies Inc. (SpaceX)."

services. The business stories of each of these entrepreneurs have four primary points in common:

1. The philanthropic mission of their business vision to improve aspects of life for individuals on Earth, where Space plays the role of an enabler. Most of these visions are simple dreams expressible by everyone;

2. The tenacity shown in the pursuit of their mission, endorsed by their own investment to kick-start the business case;
3. The ability to mobilize investors and/or trustors into the business with a high value of private annual equity; and
4. The prompt access to R&D via NASA and other sources, e.g. the Defense Advanced Research Projects Agency (DARPA), etc.

10.1.3 Transfer of Learning

Interconnectivity and interdisciplinary are two features of the current era; this implies also the cross-fertilization of best practices and guidelines that prove to be effective from one field of application to another. This learning transfer is an additional common element of the four entrepreneurs' stories.

Transfer of learning is the dependency of human conduct, learning, or performance on prior experience. The notion was originally introduced as transfer of practice by Edward Thorndike and Robert S. Woodworth,[3] who explored how individuals would transfer learning in one context to another, similar context—or how "improvement in one mental function" could influence a related one. Their theory implied that transfer of learning depends on how similar the learning task and transfer tasks are, or where "identical elements are concerned in the influencing and influenced function", now known as the identical element theory.

Today, transfer of learning is usually described as the process and the effective extent to which past experiences (also referred to as the transfer source) affect learning and performance in a new situation (the transfer target).[4] However, controversy remains as to how transfer of learning should be conceptualized and explained, what its prevalence is, what its relation is to learning in general, and whether it exists at all.[5] There are a wide variety of viewpoints and theoretical frameworks apparent in the literature, which can be categorized as:

- a taxonomical approach that categorizes transfer into different types;
- an application domain-driven approach that focuses on the developments and contributions of different disciplines;
- the examination of the psychological functions or faculties transfer models invoke; and
- a concept-driven evaluation, which reveals, compares and contrasts theoretical and empirical traditions.

---

[3]Thorndike, E. L. and Woodworth, R. S. (1901) "The influence of improvement in one mental function upon the efficiency of other functions", Psychological Review 8: Part I, pp. 247–261 https://doi.org/10.1037/h0074898 Part II, pp. 384–395 https://doi.org/10.1037/h0071280 Part III, pp. 553–564 https://doi.org/10.1037/h0071363.

[4]Ellis, H. C. (1965). The Transfer of Learning. New York: The Macmillan Company.

[5]Helfenstein, S. (2005). Transfer: review, reconstruction, and resolution. Thesis, University of Jyväskylä. ISBN 951-39-2386-X.

Knowledge transfer involves the application of previously learned knowledge while completing tasks or solving problems.

10.2 Needs of Change for Space

Since 1980 there has been a revolution in work every decade, through the introduction of new and/or disruptive tools and technologies. The figure below shows a comparison between the evolution of Information Communication Technology (ICT) and the Space sector where the periods of stability are longer, i.e. two or three decades between revolutions instead of one (Fig. 10.3).

In the 1980s job tasks were delivered via paper documents, in the 1990s via software running on PCs and similar devices. Later, in the 2000s, digital technologies paved the way for the Web and SIMship of information and exchange became dominant. Ten years later the need for sharing—cloud networks—became central to improving performance. Into the 2020s and beyond we see the growth of *convergence,*[6] implying the demand for personalization of needs and related solutions. It can be observed that the same smartphone is not the same for each user, because each user downloads different apps and the related functions according to her/his requirements. This analysis reveals that nowadays technology solutions are increasingly personalized to individual needs. This assures the business success and sustainability of private initiatives and the endeavours behind them.

In the case of the Space sector, Space 4.0 is facing the challenge of making space accessible to everyone after almost six decades of having only a few space-faring nations and related entities in academia and industry being active in, and attaining benefits from, the sector. The European Space Agency (ESA), at the 16th Ministerial

...	1980	1990	2000	2010	2020
	Boom!	Localization	Globalization	Integration	Convergence
	Documents	Software	SIMship	Integrated enterprises systems	
	Papers	Digital	WEB	Social Shared	Personalized
Space 1.0	Space 2.0		Space 3.0		Space 4.0
Astronomy	Space Race		ISS		?

Fig. 10.3 Periods of change in ICT and in the Space sector, *Source* ESA

[6]The entire phenomenon is named convergence and it can be classified into four categories: convergence of services, convergence of transmission lines, convergence of terminals and convergence of providers. The driver of this phenomenon is user mobility and the desire to know increasingly about where the user is, where other users are and what is available in the vicinity. This situation implies a continuous "local awareness" of people who travel and move very often. V. La Regina, SatCom Policy in Europe, ESPI Report no. 32, 2011, p. 11.

Table 10.3 Benchmark of different frames for working together

	Procurement	Collaboration	Investment	Partnership
Outcome	Existing solutions	Specific solutions	Scalable solutions	Innovative solutions
Feature	Purchasing agreement	Development agreement	Control agreement	Sharing agreement
Right	Property exchange	Co-property	Property	Co-property
Knowledge	No new	Shared	No new	Augmented
Parties	Buyer-to-seller	Problem owner to solution developer	Investor to investee	Stakeholder-to-stakeholder
Pay-off	Low price/low quality	Satisfaction	Profit/survive	Win-win

Council in December 2016, approved a strategy document named Space 4.0[7] aiming to enlarge the engagement of new actors in order to increase the return of benefits to society. This document illustrated the different paths taken by each iteration of Space strategy: i.e. Space 1.0 as astronomy dominated, Space 2.0 labelled as the Space Race of the Cold War, Space 3.0 opening up from the Cold War with the International Space Station (ISS), and the current Space 4.0 with the goal of overcoming the Manichean distinction between the space and non-space sectors.

In order to follow this avenue of innovation, the best way of working together with new actors is via partnership agreements. Partnerships open up ways of working together among diversified players. The following table offers a benchmark of different methods of working together (Table 10.3).

For collaborations aiming to pursue innovation and introduce disruptive technologies, partnerships between diversified stakeholders are the most promising in terms of outcomes, knowledge creation and sustainable pay-offs.

10.2.1 What ESA is Doing for Innovation

The ESA Space 4.0 is a policy concept from the ESA Director General (DG) outlining the new era for Space. It has been framed as an action plan with four 'Is' describing the way ESA is to perform its role in the current scenario; these are:

- Inform through the reinforcement of the link with large public and user communities;
- Inspire through the launch of new initiatives and programmes, both current and future generations;

[7]ESA Space 4.0 https://www.esa.int/About_Us/Ministerial_Council_2016/What_is_space_4.0.

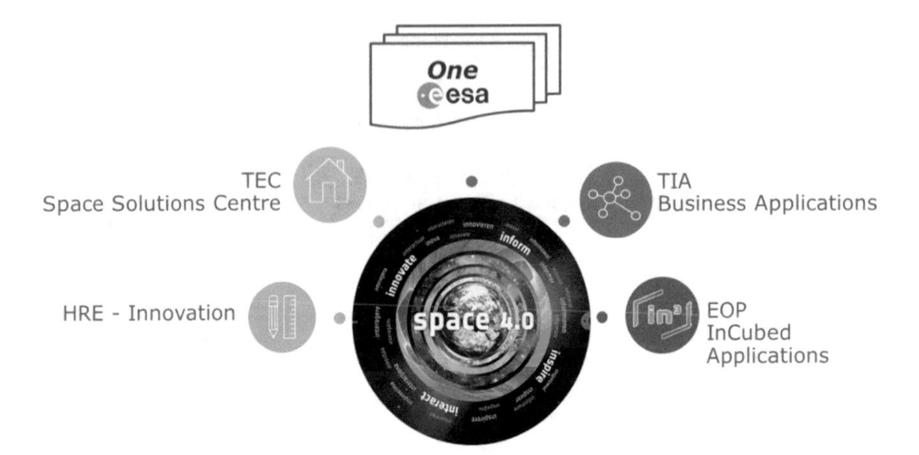

Fig. 10.4 OneESA and innovation offer across directorates, *Source* ESA

- Interact through enhanced partnerships with Member States, European institutions, international players and industrial partners, and
- Innovate through more disruptive and risk-taking technologies.

Each of the ESA directorates is adopting a series of new policies and programmatic activities in line with the 4Is paradigm.

The innovation offer from ESA is also delivered in a coordinated manner via different working groups, evaluation panels and inter-agency groups. For the directorate devoted to technology development (TEC), in Fig. 10.4, the ESA Space Solutions Centre is conceived to be an outpost of the ESA at regional and/or national level with the mission of fostering space business through the creation of start-ups, exploiting space technologies into other businesses via technology transfer and supporting value creation from ESA Intellectual Property, e.g. ESA's patents.

The TIA directorate, responsible for the development of space utilization via integrated applications, opened a brand—ESA Business Applications—in order to foster a commercially-driven approach to space projects and related development.

The Earth Observation (EOP) directorate, which plays a crucial role in the big data domain, recently launched a new framework to work together with new coming partners. This is labelled *Investing in Industrial Innovation in Earth Observation*—also known as '*InCubed*'.

The InCubed program aims to accompany the European economic operators in the Earth Observation market's rapid evolution currently witnessed in Europe and overseas. InCubed offers a new process in ESA Earth Observation programmes to strengthen and stimulate the realization of endeavours near-to-the-market with Economic Operators. Activities to be supported include the development and/or demonstration of space assets (missions, instruments, prototype elements of EO constellations), innovative mission management solutions, including for future EO constellations, the development of application platforms, etc. The actual scope of the proposed activities is defined by the economic operators themselves in accordance with the InCubed objectives. The proposals to be supported by InCubed will have to be co-funded by the economic operators (partnership agreements), as a sign of their engagement and belief in their own business case.

10.2.2 Focus on ESA Space Exploration's Innovation Action

The ESA Space Exploration Strategy outlines the long-term planning for Europe's participation in space exploration. The strategy provides ESA's conclusions on the reasons to explore space, and focuses on three common mission goals for the next decades:

- Exploitation of human-tended infrastructures in Low Earth Orbit (LEO) beyond 2020 for advancing research and enabling human exploration of deep space;
- Returning samples from the Moon and Mars;
- Extending human presence to the Moon and Mars in a step-wise approach.

The "Resolution on Europe's Space Exploration Strategy" has been developed by ESA and adopted by its Member States' Ministers at the Council meeting at Ministerial level in 2014. The Resolution reaffirms ESA's exploration destinations (LEO, Moon and Mars) and objectives for space exploration. The Strategy promotes four strategic goals, which are science, economics, global cooperation and inspiration. The innovation activities put in place are in line with the 2014 strategy and they are endorsing the goals and related benefits.

The innovation is conceived as a simple way of putting in place new things and for this to happen, a defined step-approach has been developed as shown in (Fig. 10.5).

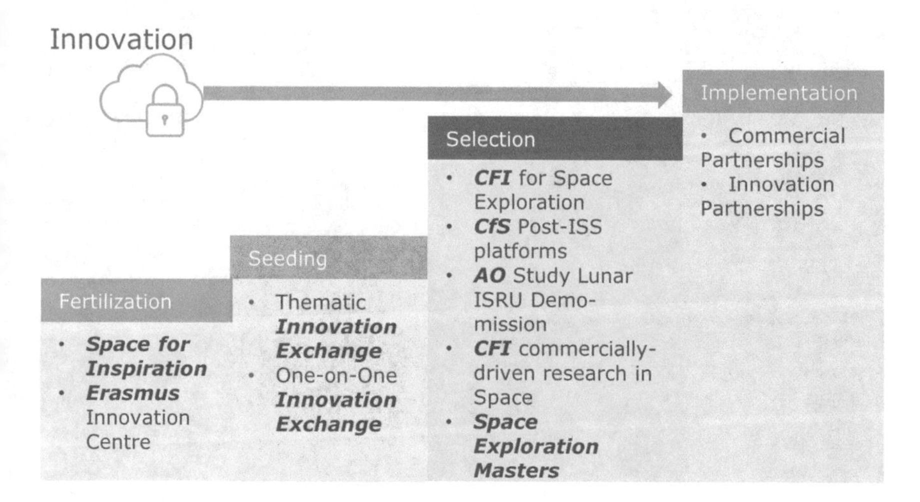

Fig. 10.5 Innovation approach at the ESA Directorate of Human and Robotic Exploration, *Source* ESA

The innovation is a phenomenon demanding a preparation—or fertilization-step, consequently a kick-starting opportunity is provided for the pooling of different sources of ideas from which to select the most promising ones according different criteria, in order to implement the ones with higher expectations and related support from the different stakeholders.

10.2.2.1 Fertilization Stage

The fertilization stage to pursuing innovation requires physical space where people can exchange on their issues and concerns and identify potential further collaborations. The directorate offers the ESA ERASMUS Innovation centre based in Noordwijk (NL) within the ESA Technology Centre—ESTEC—as a multi-dimensional space to inspire space exploration development from restricted meetings hosted in the MML room to interactive events in the Auditorium and/or 3D theatre and to wide networking opportunities in the High-bay displaying several space exploration related facilities for the development and prototyping of missions (Fig. 10.6).

Another opportunity to facilitate interaction for innovation is 'Space for Inspiration'—the bi-annual conference providing food for thought to the overall space

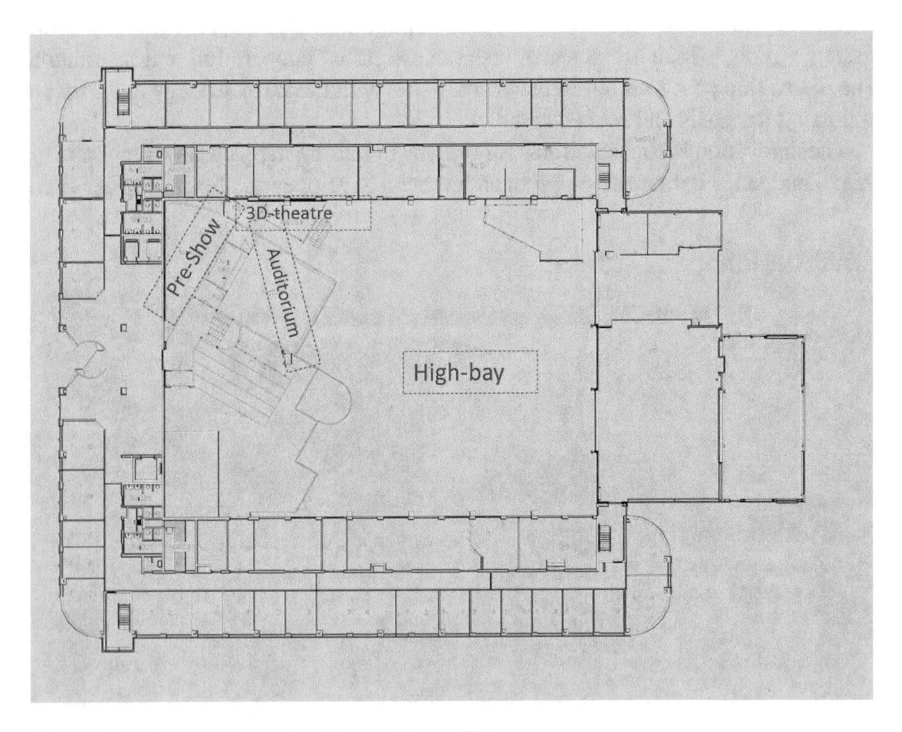

Fig. 10.6 ERASMUS Innovation Centre, *Source* ESA

exploration community by celebrating achievements, raising concerns, embracing common challenges and sharing visions for innovative advancements.

Interactions that take place in these two settings can evolve into narrower groups of engaged actors who may be encouraged into a seeding phase for a partnership agreement.

10.2.2.2 Seeding Stage

The seeding stage aims to unlock the broad potential of synergies between ESA and emerging partners towards the identification of a common goal. This happens through ESA Innovation Exchange—dedicated working-together days hosted on a thematic basis or with single entities willing to share needs and solutions with ESA for space exploration purposes. Primarily these exchanges are intended to bridge the space and non-space sectors to deliver greater benefits to society from space exploration investment and effort.

The Innovation Exchange is a single day working together on a single topic or with a single stakeholder in order to:

- Increase awareness among internal and external stakeholders;
- Disrupt barriers to unlock resources and broaden ESA's user-base;
- Explore opportunities for partnerships;
- Address strategies for the benefit of humankind from the cultural and inspirational aspects; and
- Strengthen European collaborations to serve global challenges.

10.2.2.3 Selection Stage

The ideas stimulated by the fertilization and seeding phases arrive at a selection process in order to proceed through the ESA programmes. For this the ESA offers the following venues (Table 10.4).

10.2.2.4 Implementation Phase

After making it through a careful selection process with a diversified evaluation board composed according to the type of required expertise, the ideas mature into a partnership agreement approved and/or validated by ESA participating states as owners of the resources involved. The type of agreement is shaped in ad-hoc fashion among the partners and it usually ends in a cooperative agreement without exchange of funds. The established partnership usually enables the commercial partners to open up new business with a broader community of users. This unlocks the opportunity for ESA to work indirectly with new entities outside of its geographical footprint and within industrial sectors where the Agency does not have direct influence and relationships.

Table 10.4 ESA Selection phase for innovative ideas

Venue	Status	Description	Remarks
Call for ideas: space exploration as driver for growth and competitiveness	Permanent open call	A unique opportunity for the private sector to shape and engage as a strategic partner in the future global space exploration undertaking. ESA is looking into novel ways to partner with the private sector and facilitate the realisation of European exploration ambitions	3 commercial partnerships have been established with ESA
Co-funded studies on platforms and facilities in low earth orbit	Closed	As the operational life of the International Space Station comes to an end in the next decade, new partners are welcome to work on orbital infrastructures and transportation for affordable research and novel applications	4 proposals have been submitted
Lunar ISRU demonstration mission definition study	Study on going with two teams per work-package; lunar transportation, ISRU payload and communication and navigation capability	In Situ Resource Utilisation (ISRU) involves the extraction and processing of local resources into useful products such as water and oxygen. ESA is looking into lunar mission concepts to break explorers' reliance on Earth supplies. ESA wants to leverage on existing private sector investment	35 proposals have been submitted

(continued)

Table 10.4 (continued)

Venue	Status	Description	Remarks
Call for ideas commercially-driven research in Space	Closed	ESA is inviting public and commercial organisations from ESA participating states to propose ideas for commercial and application-driven utilisation of ISS, and the other space environment facilities. The aim is to generate ideas leading to new products and/or services that will create businesses and generate commercial opportunities. The call is organised under ESA's business applications together with the European Exploration Envelope Programme (E3P) of the agency	N/A
Space exploration masters	2017 and 2018 editions closed	Innovation ideas competition to meet ESA and partners' challenges for the advancement of space exploration and related benefits to society	2017 Edition collected 135 business cases from 34 nations

10.3 Conclusion

In conclusion, a concern with innovation is a definite and growing feature of the Space sector. In response, the Agency has renewed its role in order to enable the European Space community to be competitive and play a leadership role at the global level. At the same time, innovation also creates some inefficiency in the early fertilization and seeding phases, thus there is often a set of concerns relating to the optimal allocation of resources. This, however, should not lead the Agency to close avenues, indeed it should continuously promote opportunities to do more in this direction with the ambition of improving life on Earth.

Chapter 11
The Dream Chaser Spacecraft: Changing the Way the World Goes to Space

Luciano Saccani

Abstract As one of the most innovative U.S. companies in space, Sierra Nevada Corporation's (SNC) Space Systems business area designs and manufactures advanced spacecraft and satellite solutions, space habitats and environmental systems, propulsion systems, precision space mechanisms and subsystems, and SNC's celebrated Dream Chaser. In the first of its kind mission, the United Nations Office for Outer Space Affairs (UNOOSA) is partnering with SNC to offer United Nations Member States the opportunity to participate in an orbital space mission using SNC's Dream Chaser spacecraft. This mission would support non-space faring countries and institutions so that the Dream Chaser mission can really enable inclusive access to space.

11.1 Dream Chaser® Spacecraft Overview

As one of the most innovative U.S. companies in space, Sierra Nevada Corporation's (SNC) Space Systems business area designs and manufactures advanced spacecraft and satellite solutions, space habitats and environmental systems, propulsion systems, precision space mechanisms and subsystems, and SNC's celebrated Dream Chaser.

With decades of space heritage working with the U.S. government, commercial customers, and the international market, SNC has participated in more than 450 successful space missions and delivered 4000 + systems, subsystems and components around the world.

Owned and operated by Sierra Nevada Corporation (SNC), the Dream Chaser spacecraft is a reusable, multi-mission space utility vehicle. It is capable of transportation services to and from low-Earth orbit and is the only commercial, lifting-body vehicle capable of a gentle runway landing, which offers better protection for sensitive science.

L. Saccani (✉)
Sierra Nevada Corporation, Louisville, USA
e-mail: luciano.saccani@SNCorp.com

S. Ferretti (ed.), *Space Capacity Building in the XXI Century*, Studies in Space Policy 22,
https://doi.org/10.1007/978-3-030-21938-3_11

The vehicle is capable of landing at a regular commercial airport, anywhere in the world. The uncrewed cargo version of the spacecraft does not require a pilot and is completely autonomous. The Dream Chaser cargo system is designed to deliver up to 5500 kg of pressurized and unpressurized cargo.

Another feature is the less toxic propulsion used for launch abort, orbital translations, attitude control and deorbit. The less toxic environment enables access to the cargo immediately after landing without any needs for decontamination or quarantine. This important feature allows quick retrieval of critical and sensitive specimen so they can be immediately transported to a lab or to a conditioned environment.

The configuration and architecture of the Dream Chaser is very versatile. The uncrewed version consists of two key components: the Dream Chaser vehicle and the cargo module which attaches to the back. The vehicle is reusable, but the cargo module is unique and can be modified, developed and manufactured for each mission. There is no limit to options and flexibility for use of the vehicle. Since the cargo module has to be adapted and manufactured consistently with the specific mission requirements, it could easily be developed and manufactured by any entities that want to heavily contribute.

The multi-mission vehicle can be used in a variety of ways:

- **Science**—Microgravity laboratory for on-orbit science and technology validation
- **Observation**—Remote sensing for Earth and space imaging and observation
- **Manufacturing**—Unique microgravity manufacturing and production
- **Servicing**—Satellite servicing, deployment, assembly, retrieval, deorbit and propellant transfer
- **LEO Transportation**—Carries and returns astronauts and cargo to other LEO destinations.

The Dream Chaser Cargo System was selected by NASA to provide cargo delivery and disposal services to the International Space Station under the Commercial Resupply Services 2 (CRS-2) contract. The spacecraft will provide a minimum of six cargo missions to and from the space station starting in fall 2021. However, the spacecraft can be used 15 times.

Dream Chaser was originally designed as a crewed spacecraft, capable of carrying up to seven astronauts to and from the space station and other low-Earth orbit (LEO) destinations. The crewed Dream Chaser was developed in part under NASA's Commercial Crew Program, and has environmental control and life support systems, windows for crew visibility, an integral main propulsion system for abort capability and major orbital maneuvers.

Dream Chaser can be customized for both domestic and international customers through vehicle configuration, launch site, destination, orbit altitude and inclination, landing site, duration and a host of other variables. SNC entered into agreements with multiple international space agencies and together we are developing technologies, applications and missions for Dream Chaser-based space systems.

11.2 International Presence

In the first of its kind mission, The United Nations Office for Outer Space Affairs (UNOOSA) is partnering with SNC to offer United Nations Member States the opportunity to participate in an orbital space mission utilizing SNC's Dream Chaser spacecraft. This mission would support non-space faring countries and institutions so that the Dream Chaser mission can really enable inclusive access to space. The partnership is designed to provide nations without their own space programs, including developing countries, access to space.

A Memorandum of Understanding was signed in 2016 between the United Nations, represented by UNOOSA, and SNC to coordinate a space mission for multiple United Nations Member States utilizing the Dream Chaser vehicle. Under the agreement, UNOOSA and SNC pledge to work jointly with United Nations Member States to define a multi-country mission opportunity that will provide Member States, and especially non-space-faring ones, affordable access to space to develop and fly payloads for missions related to experiments in microgravity science, remote sensing or space hardware qualification.

United Nations Member States are invited to provide payloads or experiments to be flown in LEO, with the requirement that they advance one or more of the Sustainable Development Goals. The initial concept envisions a two-week free-flight mission with multiple payloads.

The mission would amplify, support and be in line with UNOOSA's 17 Sustainable Development Goals: no poverty, zero hunger, good health and well-being, quality education, gender equality, clean water and sanitation, affordable and clean energy, decent work and economic growth, industry, innovation and infrastructure, reduced inequalities, sustainable cities and communities, responsible consumption and production, climate action, life below water, life on land, peace, justice and strong institutions and partnerships for the goals.

For the UNOOSA mission, the Dream Chaser would orbit Earth for two-three weeks and then return on a runway. Launch site, launch vehicle and landing site have not been identified yet.

In September 2017, UNOOSA jointly with the Sierra Nevada Corporation launched a Call for Interest (CFI) during the 68th International Astronautical Congress held in Adelaide, Australia. Developing countries were invited to propose experiments to be flown on board the Dream Chaser vehicle that address any of UNOOSA's Sustainable Development Goals.

The aim of CFI was to ascertain the level of interest from Member States in this initiative, and to determine the level of support each interested party might require. The CFI was open for less than two months and its initial deadline was postponed due to high volume of interest received.

The CFI received more than 150 applications from 75 countries. No outreach campaign was done prior to the CFI being released and there was a relatively short time-frame in which applications needed to be submitted. Most applications were from Mexico, followed by Nigeria and the United States.

The massive quantity and quality of the responses to the CFI show the appetite that both emerging and more developed space countries have for flight opportunities. Space transportation is often the bottleneck that does not allow scientists and researchers to finalize their work. UNOOSA and SNC are working to help them to get affordable and easier access to space.

On January 10, 2018, as a follow-up to the successful Call for Interest, UNOOSA, together with SNC, held a technical briefing at the United Nations in Vienna to elaborate the technical capabilities and other details of the Dream Chaser spacecraft to allow more in-depth understanding enabling interested parties to be better equipped for the future Announcement of Opportunity.

SNC representatives presented the company and its vision, mission overview, payload types supported on the orbital mission, draft of Payload User's Guide and the technical specifications of the Dream Chaser spacecraft to 13 participants in person and 124 participants connected through a webinar service.

11.3 Conclusion

As SNC advances Dream Chaser for CRS-2 and other missions, we are inspired by the legacy of the space shuttle and the 135 missions that came before us. We are grateful for the opportunity to work with NASA centers nationwide, leveraging 40 years of space shuttle heritage.

The Dream Chaser has been developed with knowledge, experience and lessons learned from the space shuttle era. The spacecraft has been under development since the 1980s, when NASA developed the HL-20 vehicle that SpaceDev, now SNC, later took over for additional development. Several important tests have already been successfully performed. With the help of NASA's Armstrong Flight Research Center, SNC was able to execute the approach and landing test at Edwards Air Force Base in November 2017. This heritage and partnership allows us to benefit from NASA's expertise in the development of the Dream Chaser.

In addition to NASA, SNC partners with dozens of companies and universities on various components of the vehicle. We are proud to work with international organizations to advance space for the world and look forward to being a world leader in spaceflight (Figs. 11.1, 11.2 and 11.3).

Fig. 11.1 Dream Chaser configuration for United Nations Mission. *Photo Credit* Sierra Nevada Corporation

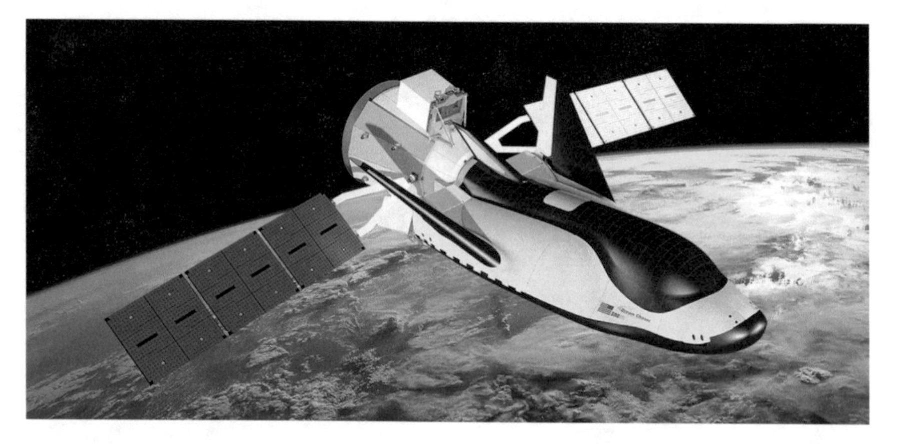

Fig. 11.2 Dream Chaser with cargo module and solar arrays, the cargo module burns up at re-entry in the atmosphere disposing of the cargo that does need to be brought back to Earth. *Photo Credit* Sierra Nevada Corporation

Fig. 11.3 Dream Chaser without cargo module and solar arrays. *Photo Credit* Sierra Nevada Corporation

Part II
Space for Global Health

Chapter 12
Japan's Space Activities for Global Health

Chiaki Mukai, Yoko Kagiwada and Nanoko Ueda

Abstract Space technology is indispensable for every aspect of our lives. It is the basis of every nation's infrastructure, varying from disaster prevention, agriculture, communication, education and more. Japan has been actively engaged in outer space activities, which improve the lives of humankind. Recognizing that space technology contributes to the creation of a sustainable society, Japan has been developing the ways and means of space applications for sustainability. Technology has greatly advanced in the recent years, and astronauts will be starting their exploration to the moon in the near future. However even today, many developing countries suffer from limited health care. Space technology could be one of the effective tools for solving health issues on the ground, and the promotion of international cooperation for further enhancing the utilization of space technology in this area is essential. Japan's space activities for global health introduced in this paper are some of Japan's efforts towards a sustainable society.

12.1 Introduction

Space technology is indispensable for every aspect of our lives. It is the basis of every nation's infrastructure, varying from disaster prevention, agriculture, communication, education and more. Space technology allows countries to take the bold and transformative steps, which are urgently needed to shift the world onto a sustainable and resilient path. The world has embarked on a collective journey, and improved access to space is the key for the benefits of all humankind from space. The Seven Thematic Priorities, endorsed by the United Nations Committee on Peaceful Uses of Outer Space (COPUOS) in 2016 as a milestone opportunity to address the overarching long-term development concerns of humankind, with concrete deliverables pertaining to space for development, reflect how space technology contributes to

C. Mukai (✉) · Y. Kagiwada
Japan Aerospace Exploration Agency, Tokyo, Japan
e-mail: mukai.chiaki@jaxa.jp

N. Ueda
Ministry of Foreign Affairs of Japan, Tokyo, Japan

S. Ferretti (ed.), *Space Capacity Building in the XXI Century*, Studies in Space Policy 22,
https://doi.org/10.1007/978-3-030-21938-3_12

solving various problems. One of the Seven Thematic Priorities is "strengthened space cooperation for global health." The objective is to improve the use of space technologies and space-based information and systems in the global health domain; to promote enhanced cooperation and sharing of information in emergencies, epidemics and early warning events, as well as on environmental parameters; to enhance capability in integrating health data in disaster management plans and to strengthen capacity building in advancing space technologies in global health efforts, while identifying governance and cooperation mechanisms to support this objective.

Japan has been actively engaged in outer space activities, which improve the lives of humankind. The Basic Space Law, enacted in 2008, states that space development and use shall be carried out in order to improve the lives of the citizens; to ensure a safe and secure society; and to mitigate disasters, poverty and various other threats to the survival and lives of humankind. Recognizing that space technology contributes to the creation of a sustainable society, Japan has been developing the ways and means of space applications for sustainability. Japan's space activities for global health introduced in this paper are some of Japan's effort towards a sustainable society.

12.2 Utilization of International Space Station (ISS) and Japanese Experiment Module "Kibo" for Global Health

12.2.1 Contribution of Astronaut's Health Research to Ground

Astronauts experience several health risks aboard the International Space Station (ISS) due to the unique environment of space. The harsh environment of space could be characterized by microgravity, confined spaces, and cosmic radiation. These characteristics affect astronauts' health, and many studies are conducted in the field of space medicine to prevent or reduce such health issues (Fig. 12.1).

One of the health issues caused by a prolonged stay in space in a microgravity environment, is the weakening of bones and muscles. The loss of bone density in space is 10 times greater compared to the loss on the ground. Menopausal women living on earth usually lose bone density at a rate of only about 1–2% a year, whereas the average loss of bone density in astronauts living in space for 6 months is 9%. The muscle atrophy of astronauts in space is 2 times greater compared to a bedridden patient on the ground.

The second health issue is related to psychological stress. Aboard the ISS, crew members from different nationalities live and perform various experiments together in a closed, confined environment for six months. They must cooperate under a tight schedule while overcoming differences in cultural background. The environmental difference between space and ground causes severe mental stress to astronauts.

Fig. 12.1 Japanese astronaut Chiaki Mukai aboard the ISS. *Source* JAXA

Another health issue is due to the great cosmic radiation in space. Astronauts are exposed to ten times the amount of radiation compared to the amount on the ground. This increases the risk of cancer and may damage the immune system of astronauts.

Space medicine plays a key role in alleviating or preventing such problems. Astronauts themselves have become the subjects of experiments to investigate mechanisms for maintaining their health and validating training equipment.

One research that JAXA has conducted with NASA is the investigation of effectivity of bisphosphonate as a countermeasure for bone loss in space. Bisphosphonate is a drug used to treat osteoporosis on Earth. In addition to their daily exercise, several astronauts took the bisphosphonate drug. The findings from this study indicate that the astronauts who exercised and took bisphosphonate exhibited a smaller decrease in bone density compared to astronauts who performed exercise only.

Another research that JAXA is conducting is stress research in a closed environment. Using the isolation chamber in Tsukuba Space Center, JAXA has been conducting a closed environment experiment for establishing a stress assessment method to be applied for astronauts. Since astronauts cannot receive a face-to-face assessment on their mental and psychological health aboard the ISS, such biomarker for stress assessment is important for astronauts to easily check their stress level in space. For each experiment, eight participants stay in the isolation chamber for two weeks, and total of five experiments have been conducted so far. JAXA is now investigating to define the stress marker based on the results of these experiments.

By conducting space medicine research aboard the ISS, it could contribute to overcome health problems on the ground. The health issues that affect astronauts can be assimilated to an accelerated model of aging-like physiological changes on the ground. For example, bone loss and muscle atrophy are common issues for astronauts and the elderly alike. Therefore, knowledge obtained from space medicine research could be applied to medical research on the ground.

One idea for the application of space medicine technology to ground is the on-orbit remote medical care. Flight doctors look after astronauts from the ground using monitors, and this method is called tele-medicine. Such a method could be applied to areas where doctors cannot easily get access to their patients, such as in developing countries where the number of doctors is limited. If patients could consult their doctors through monitors or e-mail, this would increase the survival rate of patients.

12.2.2 Contribution of Microgravity Experiments Aboard Kibo to Ground

The Japanese Experiment Module, "Kibo" is Japan's first human-rated space facility on the ISS. Using the microgravity environment of space, various experiments related to material science, life science, and medical science are being conducted aboard Kibo. The outcomes of these experiments cannot be recreated on Earth and are expected to solve many problems on the ground (Fig. 12.2).

Fig. 12.2 Japanese module KIBO of the International Space Station. *Source* JAXA

Fig. 12.3 Protein crystals grown in Space. *Source* JAXA

The areas of experiments in Kibo consists of the Pressurized Module which is filled with air at a pressure of 1 atmosphere, and the Exposed Facility where exposed experiments (such as evaluation of space materials in exposed environment or earth observation) are conducted. Experiments related to medical science are conducted inside the Pressurized Module.

One of the experiments related to medical science is the high-quality protein crystal growth experiment. The microgravity environment of space makes it possible to crystallize high-quality protein, which is necessary to analyze the protein crystal precisely and to unravel molecular structures that affect diseases. Over one hundred different proteins provided by universities and companies have been crystallized on Kibo. These are expected to contribute to the design of drugs for diseases such as influenza, cancer, and muscular dystrophy.

In 2016, JAXA has succeeded in the synthesis and structural analysis of artificial blood for dogs, in cooperation with the national university. This study contributes to overcome the existing challenge in blood transfusion therapy for pets. This research is expected to be applied for human blood transfusion technology (Fig. 12.3).

12.3 Remote Sensing Satellites for Global Health

Remote sensing enables the analysis of the environment and characteristics of the given object from distance. Earth observation using a remote sensing satellite collects environmental information of a wide area regularly. Since the 1970s, it is used for global health and environmental epidemiology and its advantage has been recognized to spot the habitats of living organisms that carry an infectious disease. In recent years, greater amount and improved quality of observation data is collected thanks to the growing number of earth observation satellites and the remarkable progress of technology. This is significant especially in developing countries, where environment data is missing.

Japan Aerospace Exploration Agency (JAXA) and the World Health Organization (WHO) worked on a research project, which aimed to contribute to Polio eradication. Polio remains a threat to those who live with underdeveloped hygienic sewage systems. Environmental surveillance supplements the surveillance of acute flaccid paralysis by monitoring wastewater for poliovirus circulation.[1] Its propagation, reinfection, and efficiency of countermeasures can be monitored by detecting the Polio virus from water samples in sewage. Since epidemic areas are widely distributed and located in remote areas, the WHO needs to narrow down their locations to collect water samples. To help the work of the WHO, JAXA provides a 3-D view of the Earth using both a digital elevation model and satellite images, which enables us to estimate the source of pollution. JAXA and WHO analyzed wastewater flow to optimize selection and placement of sampling sites with higher digital surface model (DSM) resolution.[2] In the research, the newly developed 5 m mesh DSM from the panchromatic, remote-sensing instruments for stereo mapping on-board the Japanese advanced land observing satellite was used to estimate catchment areas and flow of sewage water based on terrain topography. The data assisted the identification of optimal sampling sites for environmental surveillance to maximize sensitivity to poliovirus circulation, and population data were overlaid to priorities selection of catchment areas with dense populations. This technology has been tested in Nigeria catchment areas, flow paths and pour points are investigated as possible sampling locations. The research concluded that "the analysis demonstrated the feasibility of using DSMs to estimate catchment areas and population size for program planning and outbreak response with respect to polio".[3] The WHO considers the method useful in flat areas and complicated landform. The research demonstrated that the satellite application has a significant potential to contribute to Sustainable Development Goal 3 "Good health and well-being."

[1] Asghar, H., Diop, O. M., Weldegebriel, G., Malik, F., Shetty, S., El Bassioni, L., et al. (2014) Environmental surveillance for polioviruses in the Global Polio Eradication Initiative. *National Center for Biotechnology Information*. https://www.ncbi.nlm.nih.gov/pubmed/25316848.

[2] Takane, M., Yabe, S., Tateshita, Y., Kobayashi, Y., Hino, A., Isono, K., et al. (2016). Satellite imagery technology in public health: analysis of site catchment areas for assessment of poliovirus circulation in Nigeria and Niger. *Geospatial Health*. https://www.geospatialhealth.net/index.php/gh/article/view/462/512#citations.

[3] See Footnote 2.

12.4 Conclusion: Future of Global Health

Technology has greatly advanced in the recent years, and astronauts will be starting their exploration to the moon in the near future. However even today, many developing countries suffer from limited health care. Space technology could be one of the effective tools for solving health issues on the ground, and the promotion of international cooperation for further enhancing the utilization of space technology in this area is essential.[4,5,6,7,8]

---

[4](JAXA HP). Kibo module. http://iss.jaxa.jp/en/kibo/.

[5](JAXA HP). Space medicine. http://iss.jaxa.jp/med/index_e.html.

[6](NASA HP). Bisphosphonates as a countermeasure to space flight induced bone loss. https://www.nasa.gov/mission_pages/station/research/experiments/explorer/Investigation.html?#id=232.

[7](JAXA HP in Japanese). Stress research in Tsukuba facility. http://iss.jaxa.jp/med/studies/heisa.html.

[8](JAXA HP). Success of synthesis and structural analysis of artificial blood for dogs. http://iss.jaxa.jp/en/kiboexp/news/20161111_pcg.html.

Chapter 13
The Use of Space Assets for Global Health: Tele-epidemiology, which Contribution for Earth Observation Satellite Data? CNES Activities in Tele-epidemiology

Cécile Vignolles

Abstract CNES and its partners have developed a conceptual approach called tele-epidemiology which consists in studying the links between the environment, ecosystems and etiological agents responsible for diseases in human, animal and plant populations, based on space products truly adapted to the needs of health actors. This concept has been applied with success for the Rift Valley Fever (RVF) in the Ferlo region in Senegal. Hereafter results are presented.

13.1 Context

Half the world's population is subject to the risk of emerging or re-emerging infectious diseases. The latter are responsible for 14 million deaths every year.[1] According to the WHO, they are a leading cause of global mortality as they account for nearly one third of deaths in low-income countries.[2]

The context of this situation is a world in transition where rapid environmental changes (climate change, population growth, deforestation and urbanization, agricultural intensification, globalization and increased trade, etc.) fosters pathogens and their dispersal, thereby contributing to endemic and emerging diseases in humans, wild or domestic animals and plants.

Emerging/re-emerging infectious diseases with high epidemiological potential risks, lead public health managers to adapt their policies. Adaptation includes early knowledge of risks. The latter requires new tools to prevent re-emerging risks.

This situation brings us to rethink and improve our knowledge of the relationship between the environment, climate and health. Faced with these social challenges, an

[1]Debré, P., & Gonzalez, J.-P. (2013). *Vie et morts des épidemies* (Vol. 216). Odile Jacib Edition.
[2]WHO. (2014). The top 10 causes of death—Fact sheet n°310.

C. Vignolles (✉)
Centre National d'Etudes Spatiales (CNES), Toulouse, France
e-mail: cecile.vignolles@cnes.fr

S. Ferretti (ed.), *Space Capacity Building in the XXI Century*, Studies in Space Policy 22,
https://doi.org/10.1007/978-3-030-21938-3_13

integrated multidisciplinary research is growing, especially around the concepts of "One Health" and "EcoHealth".

In this context what can be the contribution of satellite Earth observation data?

13.2 The Concept of Tele-epidemiology

The Centre National d'Etudes Spatiales (CNES), the French Space Agency, has developed with its partners a conceptual approach, known as tele-epidemiology, based on the study of the climate-environment-health relationships and an original appropriate space offer.[3] The primary mission is to provide to public health actors additional tools/services helping them in diseases surveillance and in the implementation of strategies for diseases control. The overall objective is to attempt predicting and mitigating public health impacts from epidemics.

This multidisciplinary approach, that combines the physical, biological, and social sciences aims to understand the key mechanisms involved and to identify the factors that affect the surge and spread of these pathologies. These factors can be environmental, climatic, demographic, socio-economic and/or behavioural. Some can be identified from space, which requires the development of effective methods to use remote sensing for risk factor characterization, mapping and monitoring. Data from Earth observation satellites do not directly concern the pathogens causing the disease, but their environment—they will therefore be used to measure these favourable factors. Analysis of those processes is a key step in the development of new and original risk mapping using Earth observation satellite data.

CNES and its partners have worked with the countries' health authorities and with scientists to develop tools to compile entomological risk maps for some infectious vector-borne diseases (presence/absence of water points, presence/absence of larvae breeding sites, larval densities and adult mosquito densities) with high spatial and temporal resolution. The effectiveness of risk prevention could be improved by providing health authorities with these maps predicting "when and where" there will be a risk of emergence of the disease vectors and the risk level. If regularly updated, risk maps could provide useful data to optimize vector control measures. This approach was successfully implemented for the Rift Valley Fever in Senegal and for urban malaria in Dakar. It was also tested for rural malaria in Burkina Faso. Today, it is being developed for dengue in Martinique and Guyana.

The integrated and multidisciplinary approach of tele-epidemiology includes three steps (Fig. 13.1).

1. Monitoring and assembling multidisciplinary in situ datasets to identify physical and biological mechanisms at stake.

[3]Marechal, F., Ribeiro, N., Lafaye, M., & Guell A. (2008). Satellite imaging and vector-borne diseases: The approach of the French National Space Agency (CNES). *Geospatial Health, 3*(1), 1–5.

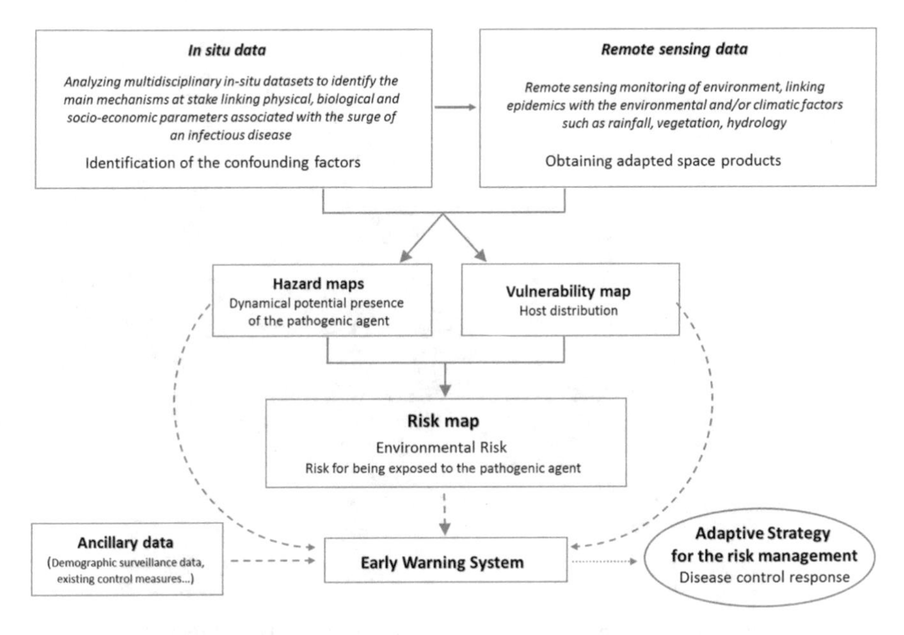

Fig. 13.1 The conceptual approach of tele-epidemiology for vector-borne diseases

2. Remote-sensing monitoring of an environment, linking epidemics with confounding factors such as rainfall, vegetation, hydrology, etc.
3. Dedicated modelling for environmental risk mapping. This last step consists in building predictive models by combining in situ data and remote sensing products derived from Earth Observation satellites, geographic data and meteorological data to produce dynamic high spatio-temporal resolution risk maps.

13.3 The Example of the Rift Valley Fever in Senegal

13.3.1 The Rift Valley Fever

The Rift Valley Fever (RVF) has become one of the most important arbovirosis emerging and/or re-emerging. It is considered a major public health issue with a strong socio-economic impact in breeding areas where it occurs. Though it mainly affects animals, humans can also be infected with the disease which results in a severe pathology. The disease also results in significant economic losses due to death and abortion among RVF-infected livestock.

This viral disease is transmitted by the bite of mosquitoes such as *Aedes vexans* and *Culex poicilipes*, the main vectors in the Ferlo region of Senegal which is our study area (Fig. 13.2).

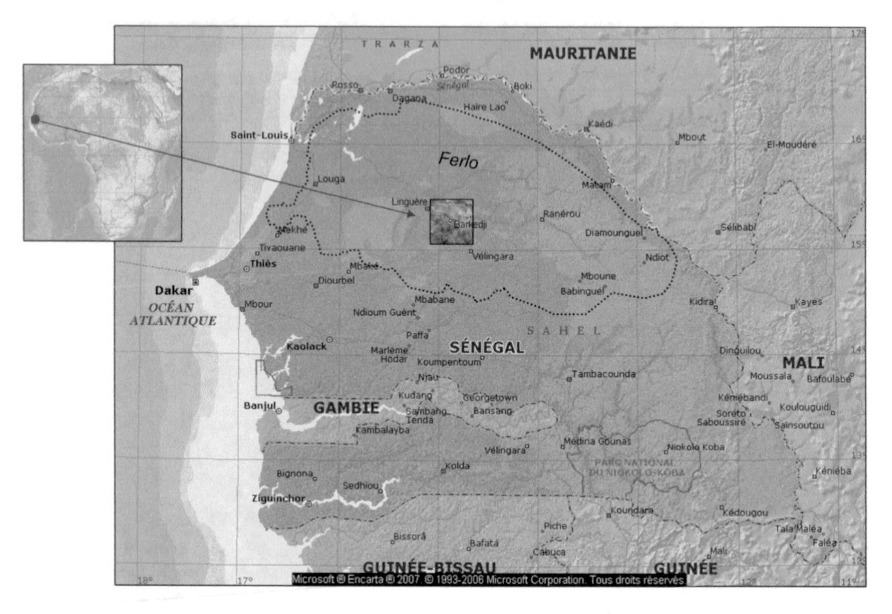

Fig. 13.2 Map of Senegal. The site of the study area around the village of Barkedji in the Ferlo region is within the black rectangle. Map adapted from Encarta 2007 (©), 1993–2006 Microsoft Corporation

RVF is on the top of the list of epidemics of importance according to the Senegalese "Système National de Surveillance des Épidémies (SNSE)", the national epidemics monitoring system. In particular, it is a key public health issue in the Ferlo region of Senegal where socioeconomical impacts are similar to those in Kenya.[4] RVF emergence corresponds to the conjunction of three main factors in time and space:[5]

1. Vector proliferation which depends upon rainfall and environmental variability, among other things,
2. environmental and ecological conditions facilitating virus diffusion, and
3. potential contacts between aggressive infected vectors and hosts (essentially parked livestock).

[4]Rich, K. M., & Wanyoike, F. (2010). An assessment of the regional and national socio-economic impacts of the 2007 Rift Valley fever outbreak in Kenya. *The American Journal of Tropical Medicine and Hygiene, 83*(2 Suppl.), 52–57.

[5]Lafaye, M., Sall, B., Ndiaye, Y., Vignolles, C., Tourre, Y. M., Borchi, F., et al. (2013). Rift Valley fever dynamics in Senegal: A project for pro-active adaptation and improvement of livestock raising management. *Geospatial Health, 8*(1), 279–288.

Entomological studies,[6,7,8] have shown that the abundance of the main RVF vectors in the Ferlo region (*Aedes vexans* and *Culex poicilipes*) is directly linked to ponds' dynamics. The latter is associated with the spatiotemporal variability of rainfall events. Rainfall distribution and its spatial heterogeneity, is thus a key parameter for the emergence of the main vectors of the RVF.

13.3.2 The AdaptFVR Project

AdaptRVF was a French-Senegalese partnership project between the Dakar Ecological Monitoring Center, the Pasteur Institute of Dakar, the Directorate of Veterinary Services of Senegal, Météo-France and CNES. It aimed to apply the conceptual approach of tele-epidemiology to RVF in the Sahel region of Ferlo (Senegal), which is regularly affected by the disease. It was funded by the French Ministry of Ecology through its Management and Impacts of Climate Change Program (Gestion et Impacts du Changement Climatique GICC in French).

The goal of this applied project was to use specific Geographical Information System (GIS) tools and remote-sensing (RS) images and data to detect potential ponds as breeding sites, and evaluate the risk of cattle being exposed to vectors' bites. Subsequently a risk for mosquito's emergence has been modelled and validated using in situ entomological measurement campaigns. It should be acknowledged that a risk is a result of hazard and vulnerability. If hazard is represented by the mosquitoes presence (entomological risk called also vector hazard), vulnerability is represented by parked animals and migrating livestock.

Three steps have been necessary to achieve the goal:

1. Set-up brand-new index for the detection of small and temporary ponds based on SPOT-5 satellite images at 10 m spatial resolution.[9] Satellite remote sensing analysis has provided a global view of the nearly 1300 ponds identified in the study area (see Fig. 13.3).
2. Dynamic modelling of the Zones Potentially Occupied by Mosquitoes (ZPOMs) combining mechanisms linking rainfall variability, dynamic of ponds and density of aggressive vectors (see Fig. 13.4). Remotely-sensed environmental data (i.e. presence/absence of ponds) and meteorological information (rainfall from in situ

---

[6]Chevalier, V., Lancelot, R., Thiongane, Y., Sall, B., Diaïté, A., & Mondet, B. (2004). Rift Valley fever in small ruminants. *Emerging Infectious Diseases, 11,* 1693–1700.

[7]Bâ, Y., Diallo, D., Kebe, C. M. F., Dia, I., & Diallo, M. (2005). Aspects of bio-ecology of two Rift Valley fever virus vectors in Senegal (West Africa): *Aedes vexans* and *Culex poicilipes* (Diptera: Culicidae). *Journal of Medical Entomology, 42,* 739–750.

[8]Mondet, B., Diaïté, A., Ndione, J.-A., Fall, A. G., Chevalier, V., Lancelot, R., et al. (2005). Rainfall patterns and population dynamics of *Aedes (Aedimorphus) vexans* arabiensis, Patton 1905 (Diptera: Culicidae), a potential vector of Rift Valley fever virus in Senegal. *J Vector Ecol, 30,* 102–106.

[9]Lacaux, J.-P., Tourre, Y. M., Vignolles, C., Ndione, J.-A., & Lafaye, M. (2007). Classification of ponds from high-spatial resolution remote sensing: Application to Rift Valley fever epidemics in Senegal. *Remote Sensing of Environment, 106,* 66–74.

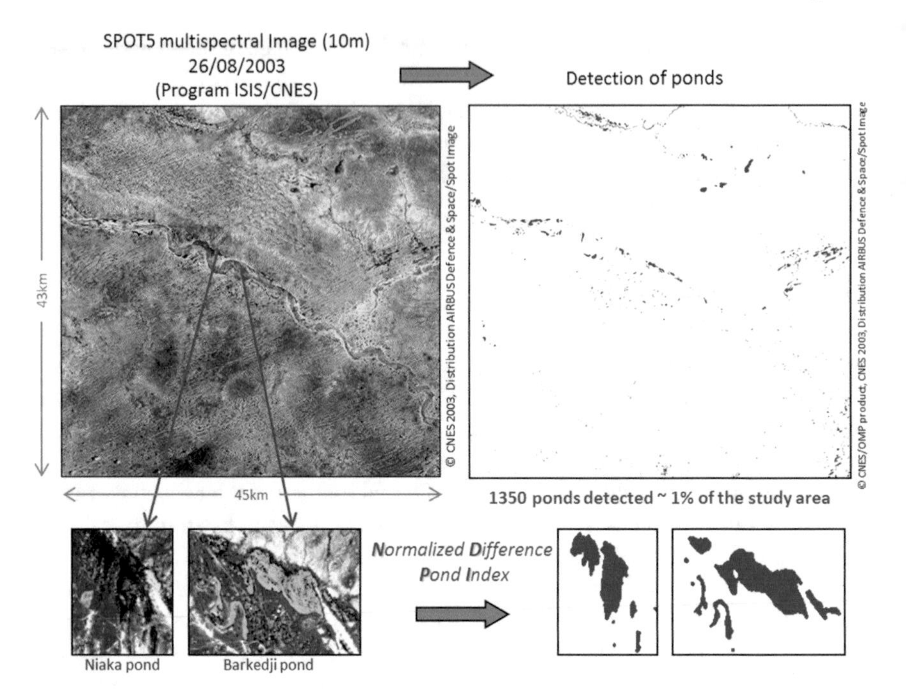

Fig. 13.3 Pond detection based on the Normalized Difference Pond Index (NDPI) index calculated from SPOT-5 images (See Footnote 9)

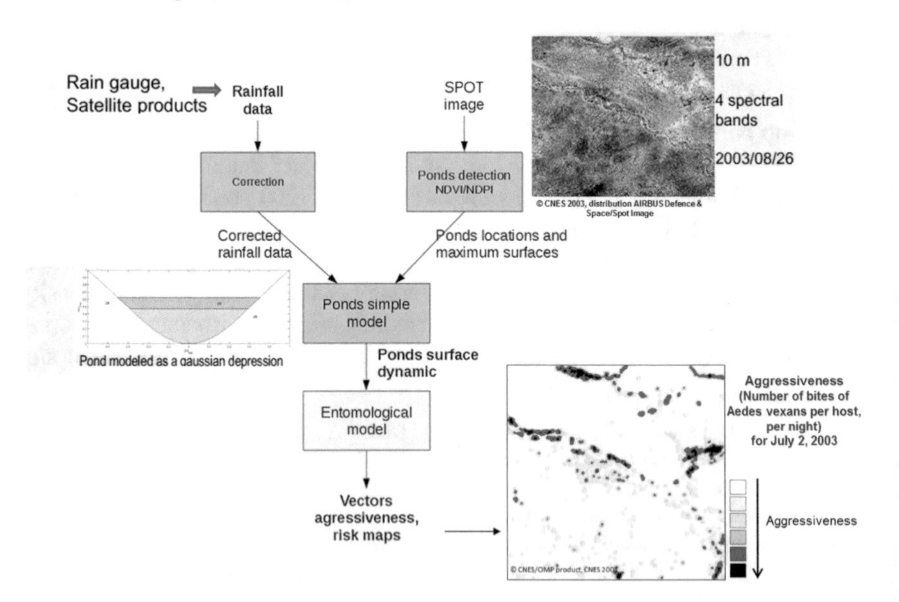

Fig. 13.4 The Rift Valley fever entomological risk modelling

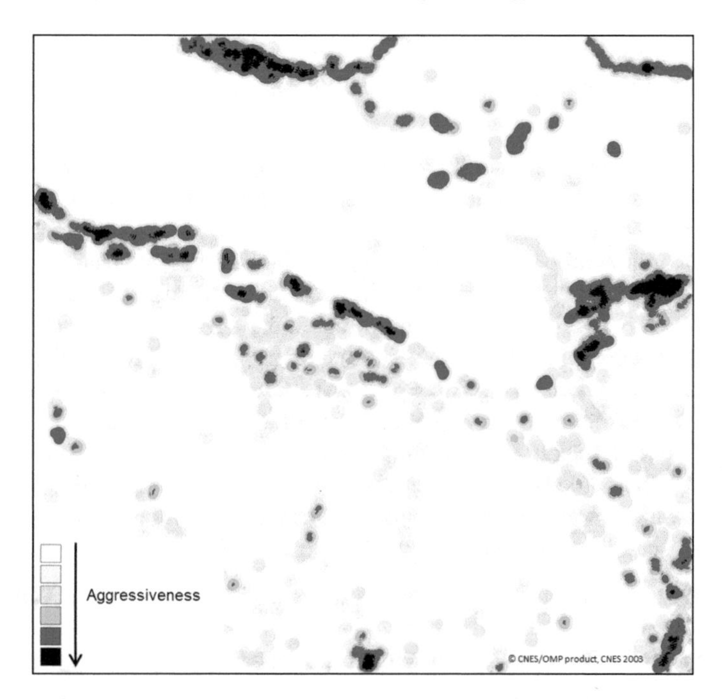

Fig. 13.5 Number of bites of *Aedes vexans* per host, per day—Result of ZPOMs modelling for July 2, 2003

and satellite data including TRMM, GSMaP, RFE CMORPH, PERSIANN) were used to fit a model with hydrological and entomological components, in order to produce dynamic high resolution maps (10-m spatial resolution, daily temporal resolution) to predict the entomological risk (see Fig. 13.5) i.e. the aggressiveness risk (number of mosquitoes' bites par host and per day).[10]

3. Crossing dynamic ZPOMs (vector hazard) with parked livestock positioning (vulnerability of hosts) to assess the environmental risk, i.e., the risk of being exposed to vector bites).[11] The integration of dynamic modelling on mosquitoes proliferation and the positioning of the livestock parks into a geographic information system, provides the Directorate of veterinary services of Senegal with forecast bulletins of the risk zones for the following 10 days, updated on a weekly basis.

[10]Guilloteau, C., Gosset, M., Vignolles, C., Alcoba, M., Tourre, Y. M., & Lacaux, J.-P. (2014). Impacts of satellite-based rainfall products on predicting spatial patterns of rift valley fever vectors. *Journal of Hydrometeorology, 15,* 1624–1635.

[11]Vignolles, C., Lacaux, J.-P., Tourre, Y. M., Bigeard, G., Ndione, J.-A., & Lafaye, M. (2009). Rift Valley fever in a zone potentially occupied by *Aedes vexans* in Senegal: dynamics and risk mapping. *Geospat Health, 3,* 211–220.

The Directorate of veterinary services will be then able to integrate this information into its adaptation strategy of animal health management. This strategy includes:

- park livestock away from zones at risks: warnings in local language have been installed near the ponds to inform breeders to park their animals at least 500 m away from the ponds
- organize anti-larval control: with these bulletins, the Pasteur Institute of Dakar should be able to organize efficient larval and vector control actions
- organize RVF vaccination: with these bulletins, the Directorate of veterinary services of Senegal could optimize vaccination campaigns in the most risky zones
- organize the communication strategy: by integrating the forecasted risks bulletins in the National Information System of Surveillance of Epidemics that feeds the Ministry of livestock in Senegal, the headquarters of the Directorate of veterinary services of Senegal and its local representatives in rural districts. It is planned to broadcast advertising messages in local language through local radio stations to facilitate comprehension and acceptance levels.

This tele-epidemiology concept has been successfully applied to Rift Valley fever in the Ferlo region of Senegal leading to the development of a dynamic mapping of Zones Potentially Occupied by Mosquitoes. Based on space technology, it has been possible to design for local users such as health authorities, an additional decision-aid tool for a better management of animal health with the aim of a better adaptive strategy.

The transfer of all the technologies to the Centre de Suivi Ecologique (service provider), for an operational use, is foreseen, and will include the coupled bio-mathematical model and its main output: predicting the emergence of RVF vectors as a function of heterogeneous rainfall amounts.

Chapter 14
EO Data Supporting MSF's Medical Interventions

Edith Rogenhofer, Sylvie de Laborderie, Jorieke Vyncke, Andreas Braun and Gina Maricela Schwendemann

Abstract Carrying out complex humanitarian and medical operations in emergencies is impossible without accurate and up-to-date information on the area. Most of the time there are no maps or only insufficient geographical information available in areas where MSF and other humanitarians are working, hindering relief efforts.

14.1 Introduction

Carrying out complex humanitarian and medical operations in emergencies is impossible without accurate and up-to-date information on the area. Most of the time there are no maps or only insufficient geographical information available in areas where MSF and other humanitarians are working, hindering relief efforts.

Médecins Sans Frontières Switzerland started creating a specialized Geographic Information System (GIS) unit in 2013, providing GIS support to the entire movement. The GIS unit supports the projects and operations by providing and producing maps, tailor-made tools and GeoApps, trainings and deployment of dedicated GIS officers to projects. Although MSF staff used and considered maps and GPS as basic tools since many years, dedicated GIS officers were deployed to projects for the first time during the Ebola outbreak in 2014.

Three case studies are used to illustrate how MSF's GIS unit is using EO data to support MSF's operations and other NGOs and health professionals working in the same area. The first case study shows how remote sensing and crowd sourced

E. Rogenhofer (✉) · S. de Laborderie
Médecins Sans Frontières, Vienna, Austria
e-mail: Edith.ROGENHOFER@vienna.msf.org

J. Vyncke
Médecins Sans Frontières, London, UK

A. Braun
Universität Tübingen, Tübingen, Germany

G. M. Schwendemann
University of Salzburg, Salzburg, Austria

S. Ferretti (ed.), *Space Capacity Building in the XXI Century*, Studies in Space Policy 22,
https://doi.org/10.1007/978-3-030-21938-3_14

mapping enabled GIS specialists to create maps in Guinee, supporting an Ebola intervention in previously unmapped areas. The second case study shows how crowd sourced mapping and information from GIS teams in Kinshasa resulted in mapping of large parts of Kinshasa that were previously unmapped. These maps enabled MSF teams to plan and carry out a Yellow fever vaccination campaign, vaccinating 760,000 people. The third case study illustrates how remote sensing was used to support Rohingya refugees in Bangladesh camps. By end of 2017, more than 700,000 Rohingya arrived in already existing camps and joined some 200,000 Rohingya who had fled to Bangladesh previously. EO-derived information on camp size and number of shelters supported the implementation of all necessary humanitarian activities in the camps.

14.2 Case Studies

14.2.1 EO Support for Ebola Intervention

After the first cases of Ebola haemorrhagic fever in Guinea were observed in March 2014, medical and epidemiological teams were deployed. The first GIS officer arrived in Guinea in April 2014 to support the teams. With Ebola outbreaks in the neighboring countries, GIS officers were also deployed to Liberia and Sierra Leone.

Basic geographical information such as roads and villages were lacking in all countries and the few maps existing were of very poor quality. Using satellite images, and with the support of the HumanitarianOpenStreetMap Team (HOT) community, the teams on the ground produced 800 maps and related information products from March 2014 to May 2015. Village locations with both official and colloquial names and roads were the most urgent information sought for, but also maps of major cities were urgently needed (Fig. 14.1).

VHR Satellite images were purchased and turned into digital maps. In urban areas, satellite images were used as background information and provided very useful information for the teams on the ground. The International Charter on Space and Major Disasters was activated at a later stage and helped greatly by providing quick and free access to satellite images to MSF and other NGOs.

A Copernicus activation was done to assist identifying areas where palm trees are growing. Oil Palm trees are providing a habitat to bats, the main vectors of the virus. Identifying areas with oil palm cultivation was a priority for epidemiologists for case detection.

VHR satellite images were also used by the volunteers of the HumanitarianOpen-StreetMap Community (HOT).With the help of HOT, the first three priority cities were mapped within three days, and 90,000 buildings were mapped within five days by 244 volunteers.

In addition, GPS coordinates from teams on the ground were collected and uploaded onto OSM. The combined efforts and different resources resulted very quickly in detailed maps, with buildings and roads derived from Very High Resolution (VHR) images and local information on buildings and villages added by teams

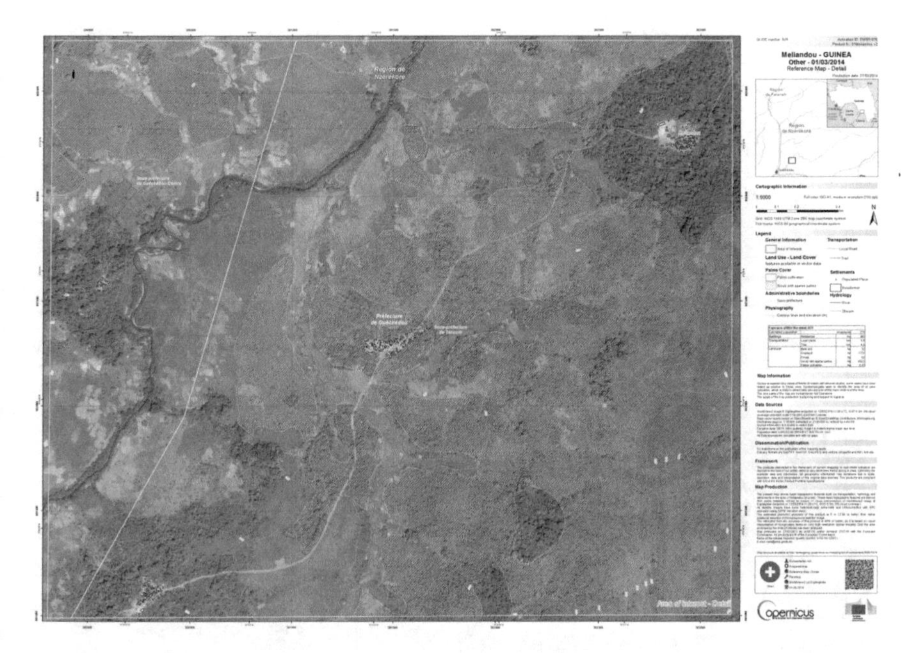

Fig. 14.1 Land cover analysis done via Copernicus activation to identify oil palm trees

on the ground. Over the duration of one year, 3300 volunteers added or edited 16 million objects such as buildings and roads to assist Ebola interventions. Without the help of the HOT community, the area could have not been mapped as accurately and with this speed.

The maps created were vital for the teams involved in patient tracing, enabling the teams to reach people in these villages faster. Weekly updated maps on confirmed and suspected cases visualized epidemiological data. Maps produced contained no patient sensitive data and were shared with health professionals and other NGOs. This helped to understand the level of the epidemics in an easy way for all teams. A case study on the experience using EO-based data to support the medical intervention of staff in the projects and in headquarters has been prepared (Figs. 14.2 and 14.3).[1,2,3]

[1]Füreder, P., Lang, S., Hagenlocher, M., Tiede, D., Wendt, L. & Rogenhofer, E. (2015). Earth observation and GIS to support humanitarian operations in refugee/IDP camps. In: B. Palen, & H. Comes, (Eds.), *Geospatial data and geographical information science—proceedings of the ISCRAM 2015 conference, Kristiansand.*

[2]Lang, S., Füreder, P., & Rogenhofer, E. (2018). Earth observation for humanitarian operations. In C. Al-Ekabi & S. Ferretti (Eds.), *Yearbook on space policy 2016*. Springer, Cham: Yearbook on Space Policy.

[3]Lessard-Fontaine, A. Soupart, M., & de Laborderie, S. (2015). Supporting ebola combat with satellite images: The MSF perspective, GI_Forum 2015—Geospatial Minds for Society, p. 445.

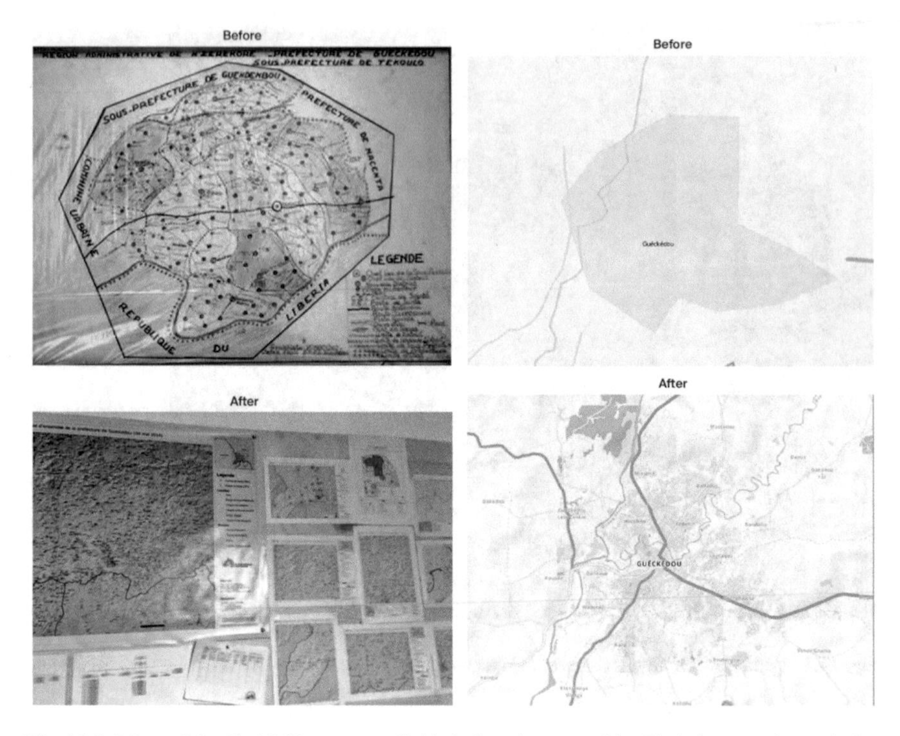

Fig. 14.2 Maps of the Guéckédou area available before the start of the Ebola intervention and after the intervention and mapping

14.2.2 EO Support for Kinshasa Yellow Fewer Vaccination Campaign

A widespread outbreak of Yellow fever occurred in Angola in late 2015 and early 2016, with the disease spreading into neighbouring Democratic Republic of Congo (DRC) in 2016. In March 2016, the first cases of Yellow fewer were confirmed in the border area Angola/DRC. The virus was detected in Kinshasa, the capital of DRC in June 2016. Kinshasa is Africa's third–most populous city with a population over 11 Million. Kinshasa is located beside the Congo River. Its location as well as the city´s slums with their insufficient water supply and sanitation provides an optimum breeding ground for the Aedes aegypti mosquitoes transmitting the yellow fever virus. To reduce the risk of a yellow fever outbreak in Kinshasa town, a massive yellow fever prevention campaign was conducted in cooperation with the world health organization and the DRC Ministry of Health. MSF's operational centre Brussels led the yellow fever prevention efforts in 3 of Kinshasa's 35 health zones.

In order to vaccinate the population living in the 3 health zones MSF was covering, a massive effort was undertaken to organize sufficient staff and logistics. 100 teams comprising of 16 people per team worked in the 3 health zones with the support from 188 staff such as drivers, hygiene promoters and logisticians. In total 188

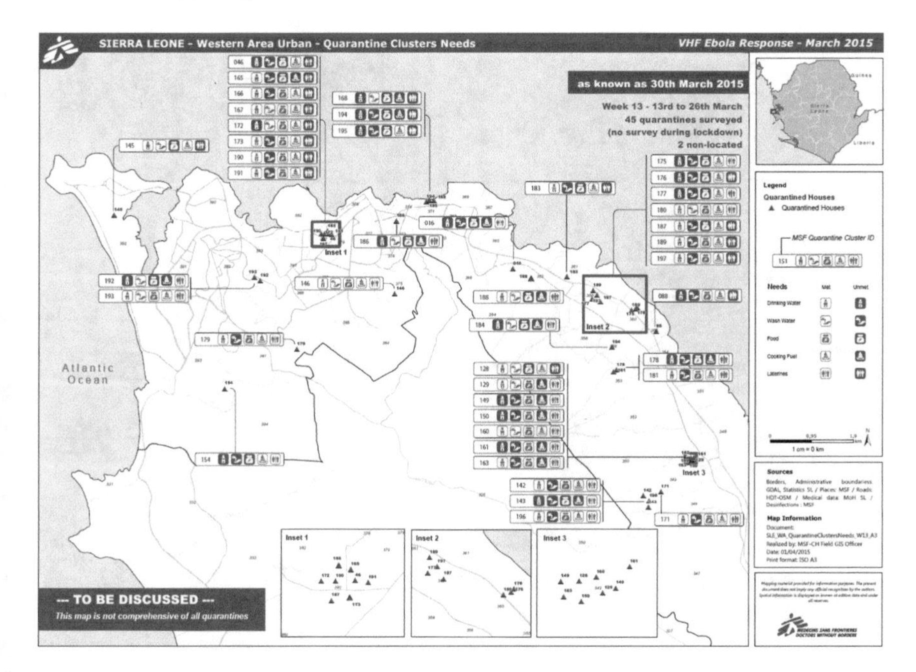

Fig. 14.3 Quarantine clusters needed in Sierra Leone as known by 30 March 2015

Main GIS products and activities to support the yellow fever vaccination campaign			
Cross cutting	Vaccination	Vector control	epidemiology
Base mapping	Supported site selection by mapping distances between vaccination centres	Maps to support vector control activities	Developed and implemented a GIS-based sampling methodology to evaluate coverage following the vaccination campaign
MapKits: sets of common maps for staff	Bespoke maps for different teams (logs, waste management, cold chain, etc)	Developed and implemented a GIS based sampling methodology for a survey	GPS training for survey teams
Coordination and security maps	Daily progress and coverage maps	Analysis maps (via remote support)	Data collection and verification of locations
Installation and management of maps on smartphones	Improvement of the vaccination database		

Fig. 14.4 Main products and activities to support the vaccination campaign

international and national staff supported the vaccination teams by providing transportation, logistics and medical supplies needed, informing the population of the vaccination campaign and carrying out vector control and preventive measures. The population living in the 3 health zones was estimated to be 760,000 people or 10%

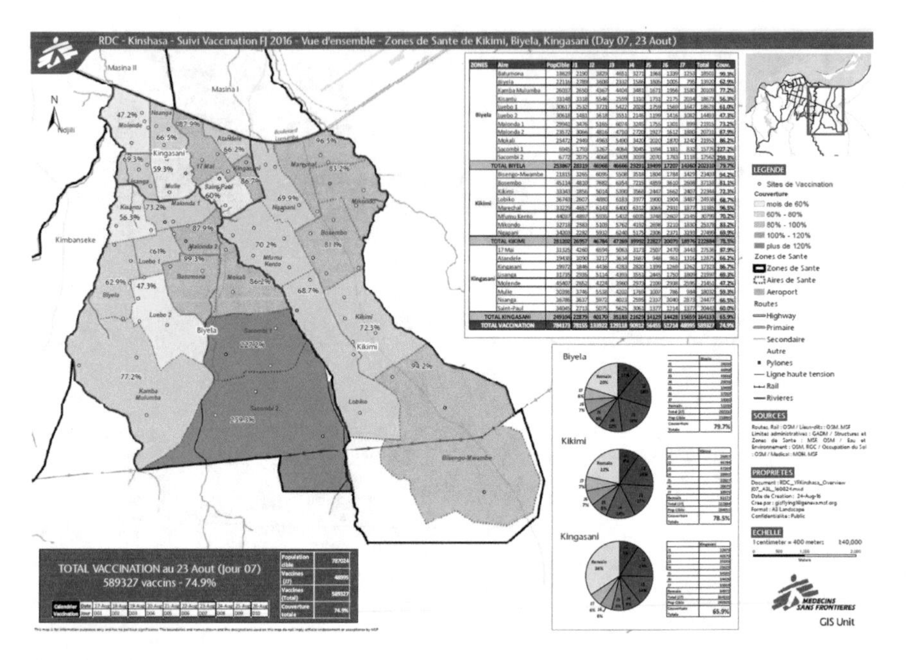

Fig. 14.5 Vaccination follow up as per August 23rd 2016

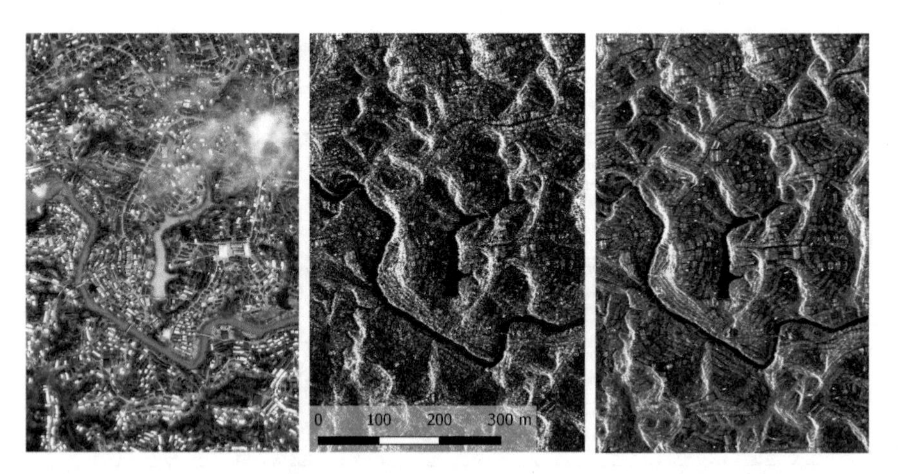

Fig. 14.6 Comparison of optical image (left), TerraSAR-X image from 30.09.2017 (middle), and mean image of all five images (right)

of Kinshasa's 7.6 million person vaccination target. Planning such a massive vaccination campaign provides numerous challenges, such as managing the 65 vehicle fleet in densely populated areas, providing timely and sufficient medical supplies and especially ensuring a perfect cold chain. To achieve a smoothly running cold chain, 4000 ice packs needed to be renewed on a daily basis in different locations.

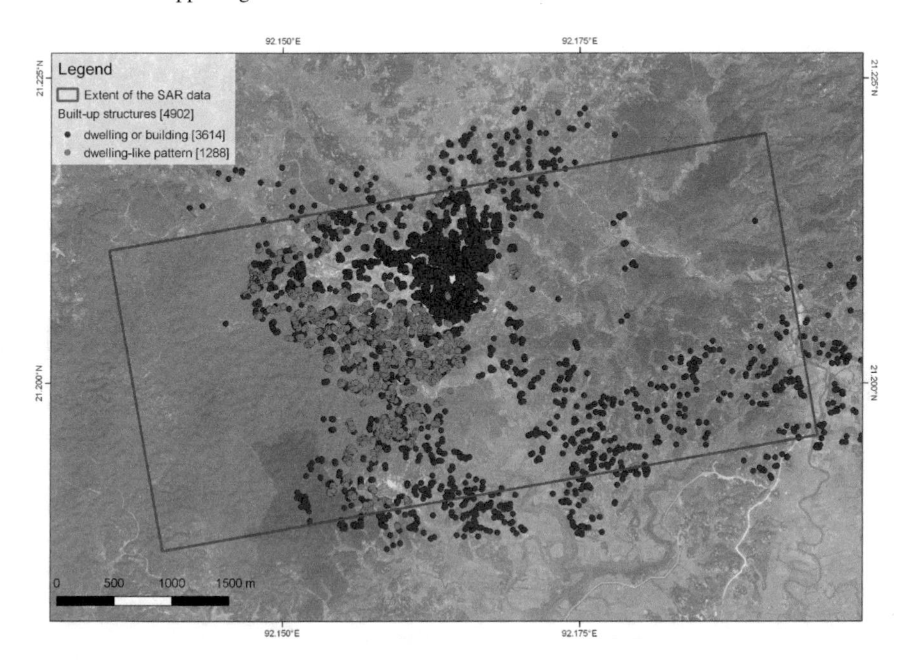

Fig. 14.7 Dwellings identified based on visual interpretation of the TerraSAR-X images from 30.09. and 11.10.2017

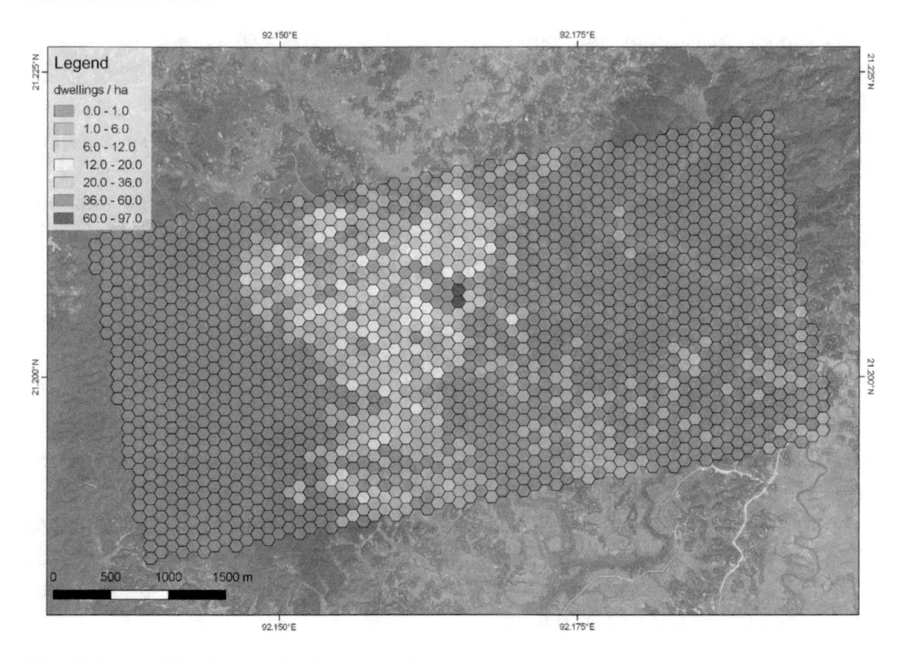

Fig. 14.8 Dwelling density for the area of Kutupalong

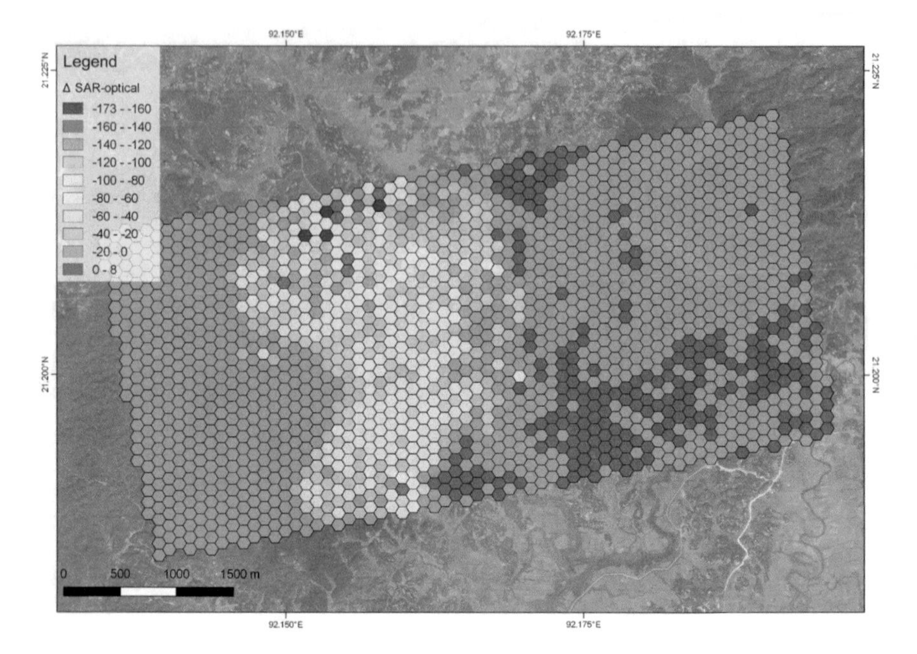

Fig. 14.9 Dwelling density derived from VHR optical data compared to results based on SAR data

To support the vaccination campaign, a dedicated GIS officer was deployed to Kinshasa supported by locally hired GIS staff. From 26 July to 15 September a total of 240 GIS products were produced and a detailed map of the town was produced with the help of the HumanitarianOpenStreetmap Team. Different types of maps and products were produced to support the entire teams and all professions (Figs. 14.4 and 14.5). A case study, showing the roles that GIS, specialists on the ground, and remote sensing played in the success of the campaign, has been published in 2015[4]

14.2.3 EO Support to Assist Rohingya Refugees in Bangladesh Camps

In August 2017, over 671,000 Rohingyas were forced to flee Myanmar, one of the largest displacements of people in recent memory in such a short period of time.

Over the previous years, some hundreds of thousands of refugees settled down in the 3 makeshift camps Kutupalong, Balukhali and Leda. The camps were already overcrowded before the new influx of Rohingya refugees arrived, but still absorbed many of the newly arriving refugees.

[4]Lüge, T. (2015). GIS support for the MSF Ebola response in Liberia, Guinea and Sierra Leone Case Study, 2nd edn. MSF-CH GIS unit GIS unit, MSF 2015. https://reliefweb.int/sites/reliefweb.int/files/resources/GIS%20Support%20Ebola%202015_EN.pdf.

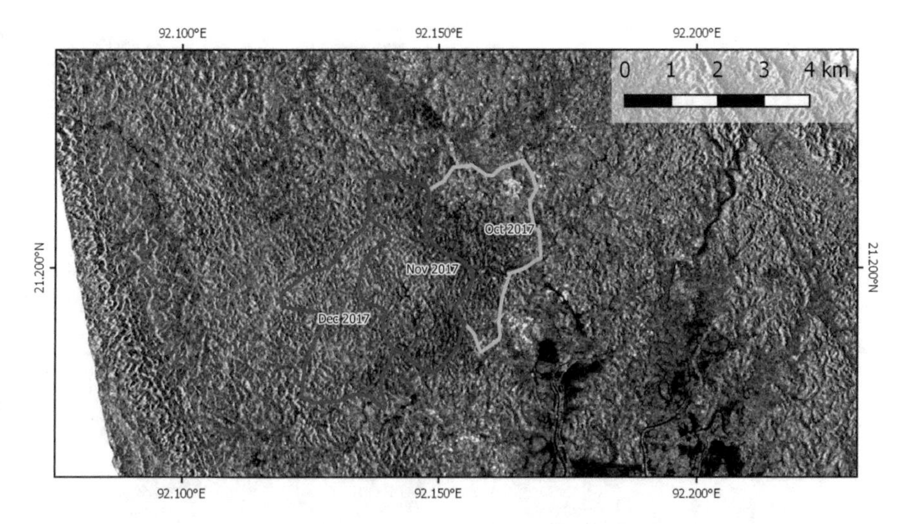

Fig. 14.10 Expansion of camp Kutupalong between October and December 2017

Around 546, 000 new arrivals settled in informal shelters in the Kutupalong/Balukhali expansion site, around 236,000 refugees settled in spontaneous settlements, and around 46,000 refugees could settle within host communities. By early 2018, more than 900.000 Rohingya refugees are living in Bangladesh.

MSF teams were present in the area since some years and were quickly reinforced to cope with the influx of refugees. Dedicated GIS officers were deployed amongst other professionals from the beginning, to support the teams and to work closely together with the epidemiologists and logistics staff.

A threefold approach was chosen to assist the teams in Bangladesh: GIS officers on the ground, support of HOT volunteers, and analysis of VHR data by remote sensing specialists from the Department of Geoinformatics—Z_GIS, University of Salzburg, who are regularly providing EO-based information on displaced populations to MSF.[5]

There were no maps of the overcrowded and rapidly growing camps existing, so one of the urgent tasks was to collect information on the ground to produce the first base maps. Due to the rainy season and the cloud cover, the acquisition of VHR images was extremely difficult and it took some time before the first VHR images were available.

Synthetic Aperture Radar (SAR) images were purchased at the beginning of the crisis as optical data were not available. Images from SAR satellites provide information on surface characteristics and can be acquired independently from sunlight or cloud cover. They therefore serve as a valuable source of information for time-critical tasks. The working group of Geoinformatics at the department of geography at the University of Tübingen, focuses on the derivation of information for humanitarian operations. For Kutupalong, 5 scenes of TerraSAR-X, a German radar satellite, have been ordered in Staring Spotlight mode, allowing a spatial resolution of about 0.6

[5]See for example Footnote 2 or 1.

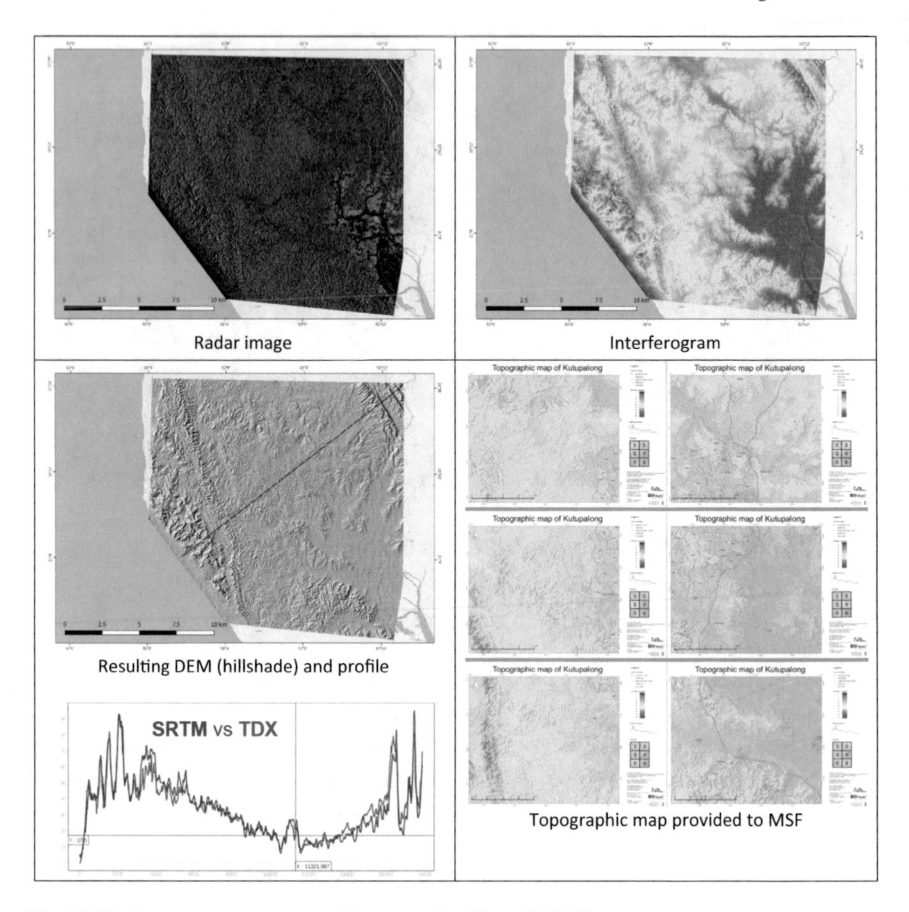

Fig. 14.11 Steps to create topographic map using Terra SAR-X scenes

meters. The images were taken on 30.09.2017, 11.10.2017, 27.12.2017, 07.01.2018 and 18.01.2018. Having multiple images allows computing one mean image with reduced effect of speckle, a noise-like pattern which is present in all radar acquisitions because of signal interference of different surfaces. Furthermore, temporal developments of the camp can be visualized based on the time-series.

Figure 14.6 shows a comparison of the image content of optical and SAR images and illustrates the potential of mean images of multiple acquisitions for obtaining higher contrasts. It also shows that not all dwellings which are visible in the optical image can also be identified in the radar image despite their comparable spatial resolution. This is because radar signals are sensitive to roughness, moisture and orientation of surfaces. This means that houses made from hard materials, such as metals or corrugated sheets cause a higher backscattered signal than wooden roofs or surfaces covered by leaves, for example. The figure furthermore shows that slopes facing to the South-West (looking direction of the sensor) are generally brighter than their surroundings. This topographic effect can be reduced with the integration of

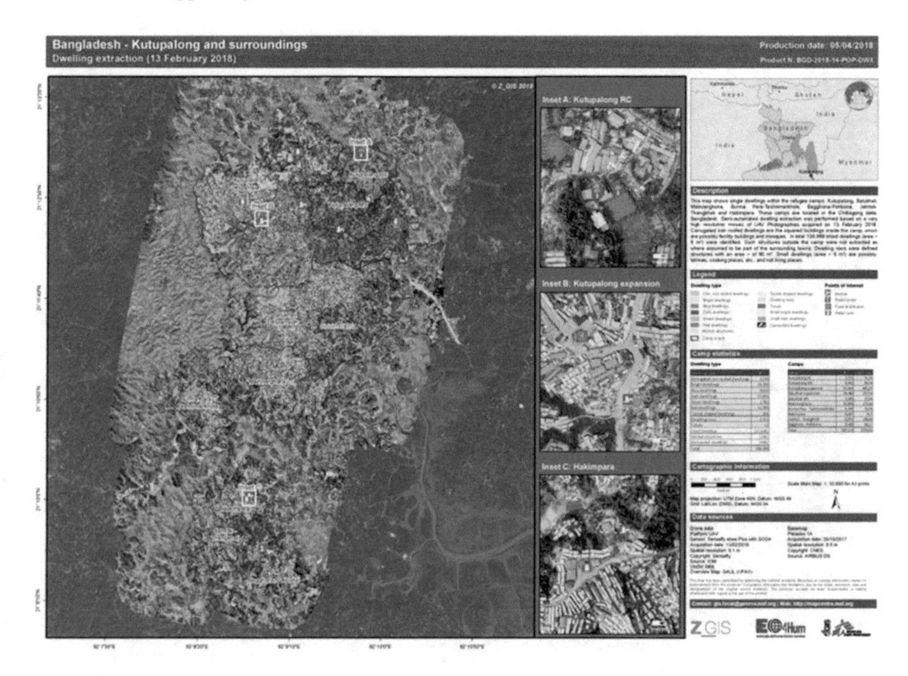

Fig. 14.12 Dwelling extraction in Kutupalong and surrounding camps as of 24.12.2017. Twelve different dwelling types were distinguished according to their roof type, colour and size. In total almost 110,000 dwellings were identified

digital elevation models (DEMs) of sufficient spatial resolution. In this case, the DEM was not good enough to compensate for all topographic influences on radar backscatter.

However, in cases of emergency, the time required to take multiple images is often not available. In order to provide fast results, manual identification of dwellings has been performed based on a mean image of the first two acquisitions. In the case of TerraSAR-X, the repeat cycle between two observations is 11 days (30.09.2017 and 11.10.2017). It can be furthermore reduced to 4.5 days for images of different flight directions and looking angles.

Manual digitization of visually identifiable dwellings was conducted within short time after image delivery and resulted in a total number of 4902 built-up structures, of which 3614 were labeled as "dwelling or building" and 1288 as "dwelling-like structures" (see Fig. 14.7). Both turned out to be dwellings after consultation of local experts.

In order to estimate the dwelling density, hexagons of the size of one hectare (10.000 m^2) were created and number of dwellings per hexagon was calculated. The following image shows the results. Most of the areas show numbers of smaller than 6 dwellings per hectare but they reach 97 dwellings in the center of Kutupalong (Fig. 14.8).

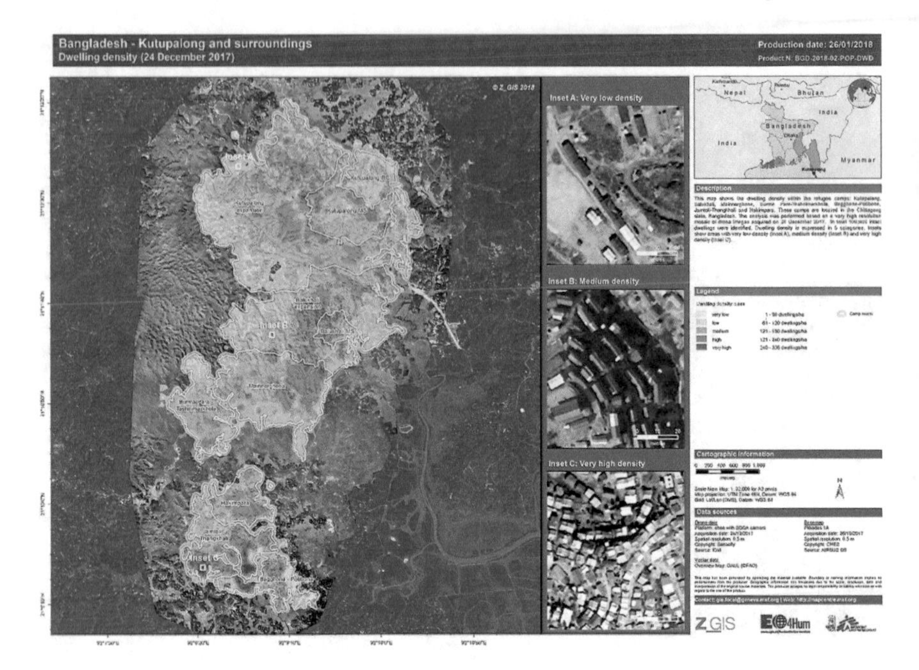

Fig. 14.13 Dwelling density of Kutupalong and surrounding camps as of 24.12.2017. Density is expressed as number of dwellings per hectare (red = very high density with 240–336 dwellings/ha; yellow = very low density with 1–60 dwellings/ha)

Statistical comparison with the dwellings identified based on VHR optical images showed a clear underestimation of dwellings, mostly between 0 and 50 buildings per hectare, but reaching a difference of 175 buildings per hectare in the centre of the camp. This can be explained by the reasons mentioned above: Some dwellings are not visible in the radar image because of their material (wooden and other natural material), their orientation in relation to the looking direction of the sensor and the impact of topography on the signal. Still, SAR can serve as a quick first indication for the spatial distribution (Fig. 14.9).

The following image shows the changes of radar backscatter intensity in the region around Kutupalong. It is a colour composite from three scenes of the Sentinel-1 satellite from 04.10.2017, 09.11.2017 and 03.12.2017. The images are provided at no cost within the Copernicus programme and have a spatial resolution of 10 m. By combining information of three different acquisitions, temporal developments can be shown. In this case, the rapidly growing camp area is shown in red and yellow colours. The camp area roughly grew from 5 km^2 (October) to 9.75 km^2 (November) and 14.1 km^2 (December) (Fig. 14.10).

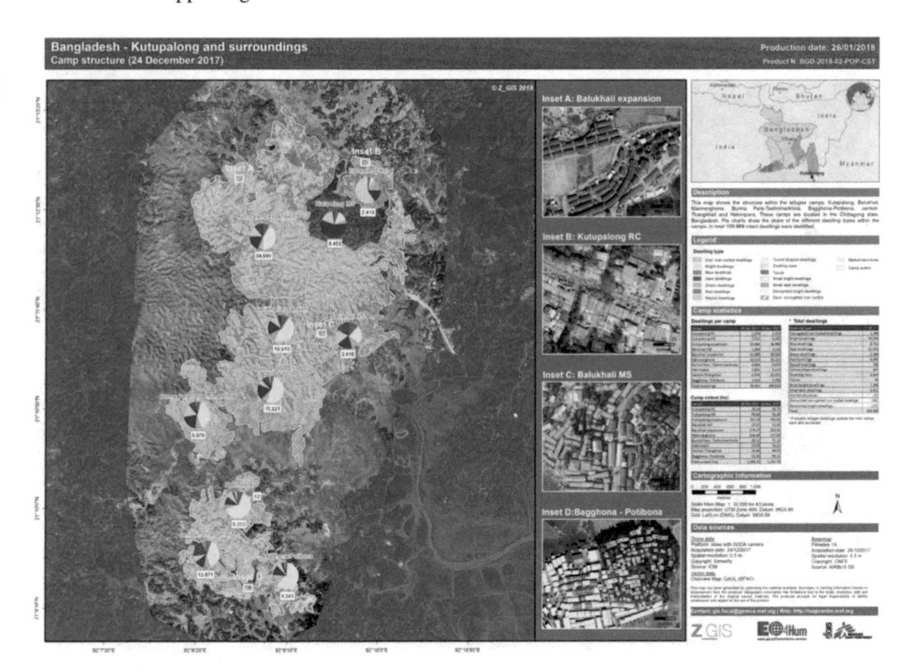

Fig. 14.14 Camp structure analysis of Kutupalong and surrounding camps as of 24.12.2017. Pie charts show the distribution of dwelling types and total number of dwellings per camp

In order to assist the planning of the camp's expansion, a topographic map was created based on a digital elevation model. Elevations were derived from two TerraSAR-X scenes at a spatial resolution of 7.31 provided by the German Aerospace Center (DLR). The generation of a DEM is based on radar interferometry as shown in the image below. The interferogram is generated as a product of the image pair and contains topographic information. It is converted to absolute heights based on the acquisition geometries of the satellites in a second step. The topographic map was divided into 6 parts and enriched by generalized elevation contour lines, local peaks, the main road network and further known settlements (obtained from OpenStreetMap) as well as the current outlines of the camp (Fig. 14.11).

In September, the first optical data was available and HOT volunteers started mapping. The HOT volunteers focused on mapping the roads, footpaths and streets within and around the camps, providing essential information to logistics. The teams on the ground contributed local information and the first dwelling counts were carried out by the team of Z_GIS. Different dwelling types were distinguished according to their roof type, roof colour and size (see Fig. 14.12). In addition a camp extent and dwelling density calculations for an easy-to-grasp overview on the spatial distribution of dwellings (see Fig. 14.13) were provided. The camps around Kutupalong were monitored various times to provide information on changing dwelling numbers, dwelling types and change patterns within the camps (see Fig. 14.14). Overall the camps were analysed five times (25.09.2017, 06.10.2017, 26.10.2017,

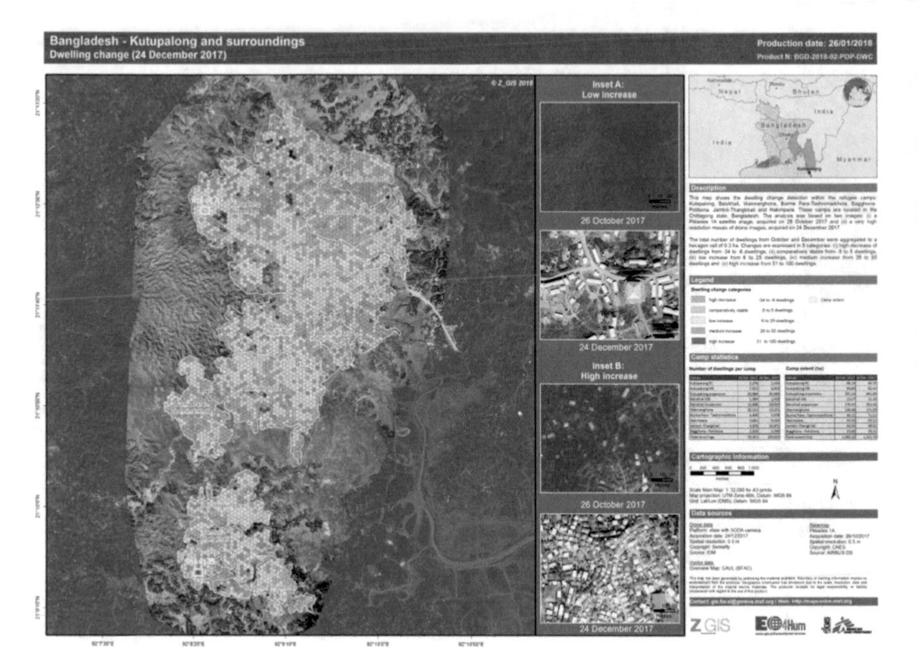

Fig. 14.15 Dwelling change detection in Kutupalong and surrounding camps between October and December 2017. The number of dwellings increased from ~77,000 to ~110,000 and the camp extent increased from ~1000 to ~1330 ha. (number of dwellings aggregated to hexagonal cells; red/orange/yellow = increase of dwellings; blue = decrease of dwellings)

24.12.2017, 13.02.2018) using VHR optical images of different sensors (Pléiades 1A, WorldView-2/3 with 50 cm spatial resolution and UAV with 10–30 cm spatial resolution). Between October and December 2017 the number of dwellings increased tremendously from around 77,000 to almost 110,000. At the same time the camp extent increased from ~1,000 to ~1330 ha. The last analysis in February 2018 showed that the camps are currently quite stable. The camp extent slightly increased in the western part, but at the same time the number of dwellings slightly decreased.

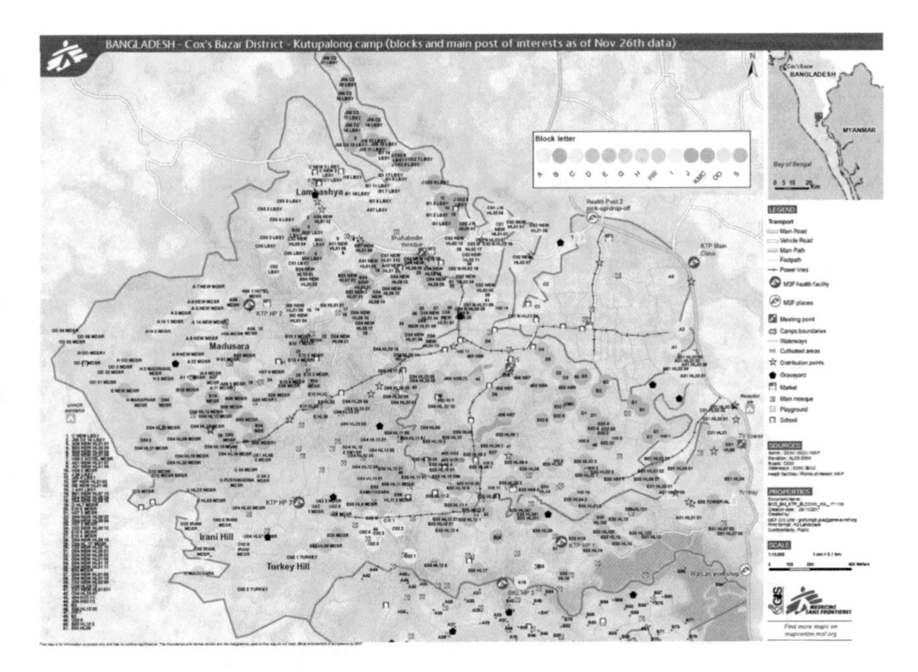

Fig. 14.16 Kutupalong camp, blocks and points of interest, Nov 2016

All information gained by teams on the ground, via HOT and remote sensing experts, was used to produce maps and analysis of the area, supporting the medical and logistics teams working in the camp. Within the first month 50 maps were produced by the GIS officer on the ground (update on the camp, thematic maps such as information on WatSan and maps on context, activities). Z_GIS produced 20 maps related to population monitoring using VHR data. The maps produced were shared with other NGOs and relevant parties on the ground (Figs. 14.15, 14.16 and 14.17).

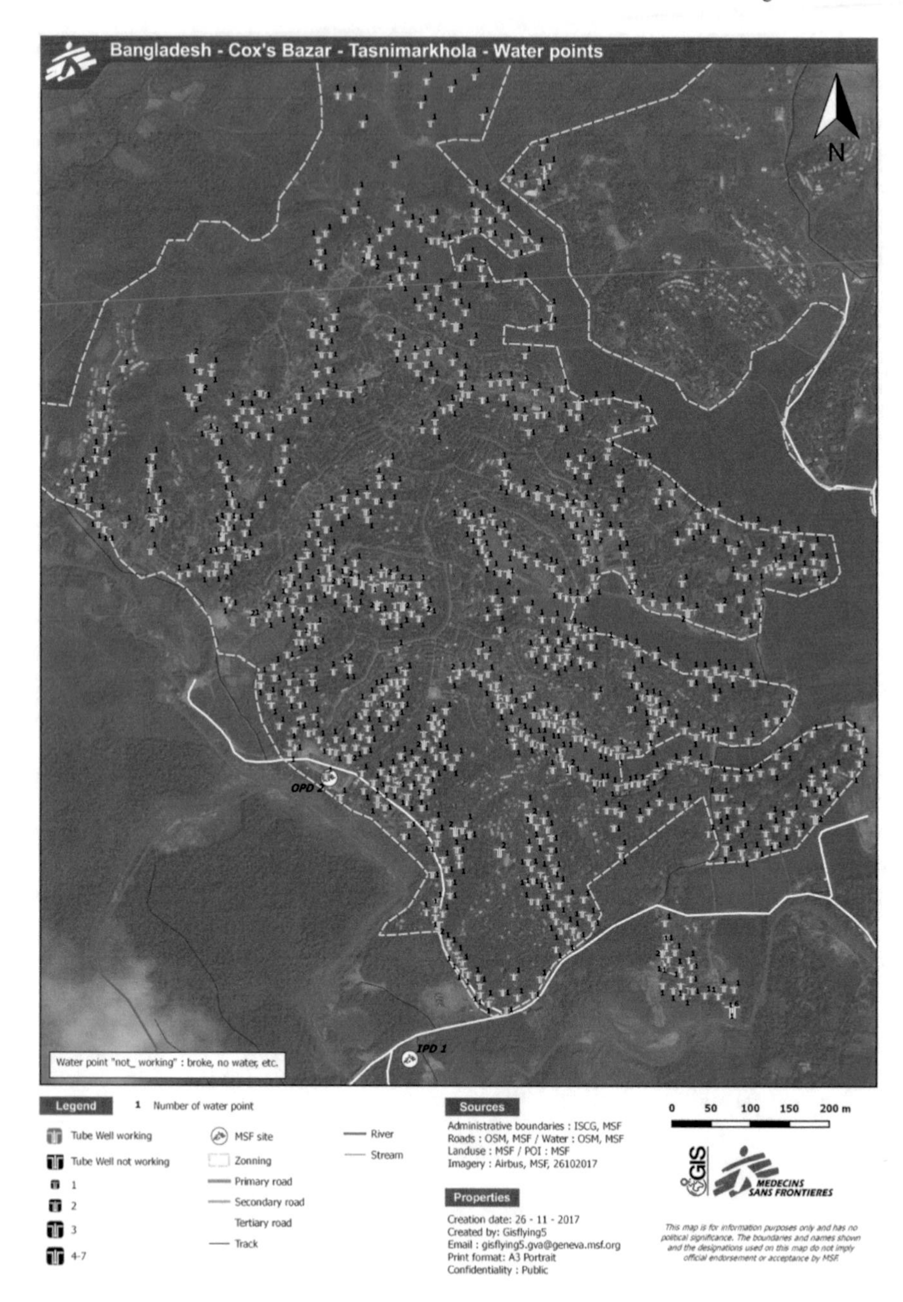

Fig. 14.17 Waterpoints of Tasnirmakhola camp, 26.11.2017

14.3 Conclusion

Information derived from EO data provided very much needed information for teams on the ground as well as for teams and decision makers at the headquarters. [6,7,8]

Mapping entire cities and areas would be impossible for humanitarian organisations alone but is possible by using EO data, the support from remote sensing specialists, the HOT community and teams on the ground. The activation of the international charter and the analysis provided by Copernicus was very valuable to the epidemiological teams on the ground.

The case studies clearly showed the benefits of the collaboration between different actors such as Copernicus, Image providers, the HOT communities, remote sensing specialists and teams on the ground.

---

[6]See Footnote 4

[7]Lüge, T. (2014). GIS support for the MSF Ebola response in Guinea in 2014 Case study. *MSF-CH's GIS unit (CartONG)*. http://osmstories.org/assets/MSF_Ebola.pdf.

[8]Lüge, T: GIS support for the 2016 MSF yellow fever vaccination campaign in Kinshasa, 1st edn. *MSF-CH GIS unit, MSF 2017*. https://reliefweb.int/sites/reliefweb.int/files/resources/GIS%20Support%20Yellow%20Fever%202017_MSF_EN.pdf.

Chapter 15
White Space: Applications of Research and Development Derived from Space Flight Analogues for Developing Solutions for Global Health Worldwide. Concordia, Antarctica Case Study

Beth Healey

Abstract This chapter provides an account of the scientific research conducted at the Concordia Antarctic research station also known as "White Mars". Many technologies tested at Concordia can both enable future manned missions on planetary surfaces, but also support technology transfer activities on Earth. For example innovative Life Support Systems are today providing clean water in Morocco, in a similar fashion to the recycling of water on the International Space Station and at Concordia. Similarly the medical and procedural know-how acquired during space analogue missions, could be shared to provide telemedicine solutions in developing countries. Finally this contribution highlights the challenges of living and working in a remote and isolated continent, in a small international team, providing critical lessons learned for future space exploration ventures.

15.1 Introduction

Antarctica is a continent covering an area larger than Europe and is home to one of the most hostile environments on this planet. It is governed through an international treaty safeguarding its vast, frigid deserts for the aims of peace and science.

Polar exploration is particularly challenging, as illustrated by historical Antarctic expeditions to the South Pole by pioneers like Scott and Amundsen. Today, life in Antarctica is safer and more accessible compared with those early days of polar exploration, not least thanks to technological developments, but there are still places on this southernmost continent that remain unvisited by anyone. As a fellow of the Royal Geographical Society, which supports and promotes geography and related research, education, fieldwork and expeditions, I have always been interested in remote environments how the human body and mind respond to these extreme conditions (Fig. 15.1).

B. Healey (✉)
European Space Agency, Concordia Station, Antarctica
e-mail: bethahealey@gmail.com

S. Ferretti (ed.), *Space Capacity Building in the XXI Century*, Studies in Space Policy 22,
https://doi.org/10.1007/978-3-030-21938-3_15

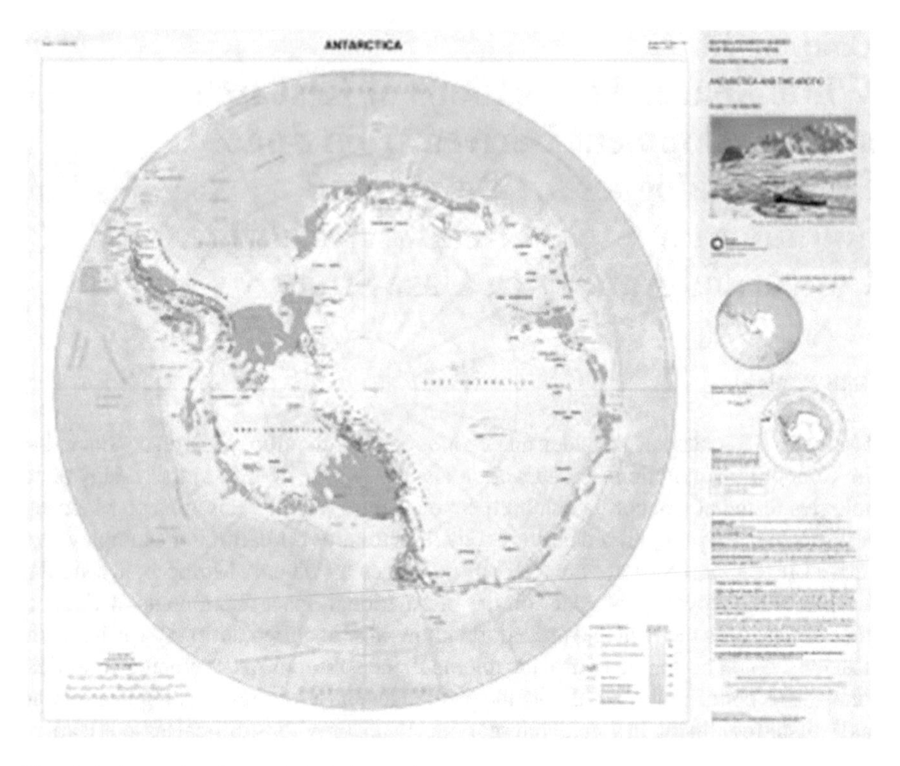

Fig. 15.1 Antarctica. *Source* British Antarctic survey

In 2015 I was appointed the resident European Space Agency (ESA) Medical Doctor at Concordia research station in Antarctica, where I collected data for a number of physiological and psychological experiments on the overwinter crew. On this platform, ESA studies the effects of living in an extreme environment, as the conditions are similar to deep space missions, where the crew need "to be prepared for anything".

15.2 Space Flight Analogues for Developing Solutions for Global Health Worldwide

Space agencies currently simulate different elements of space missions using 'spaceflight analogue' platforms. These terrestrial platforms are used for research and development as well as astronaut training. Each platform in the analogue portfolio has a different research focus, which include the effects of microgravity, isolation, confinement and crew dynamics. A range of often extreme environments are utilised to simulate these different parameters—from cave systems and remote deserts to underwater platforms and bed rest studies (Fig. 15.2).

Fig. 15.2 Space analogue missions (Underwater, Bed rest and Isolation MARS500), *Source* ESA, BBC, ESA

The Global Exploration Roadmap demonstrated the increasing focus by space agencies to once again leaving low earth orbit, and recommencing longer duration exploratory missions, for example to the Moon or Mars.

Concordia is a Franco-Italian Antarctic station situated on the Dome Charlie plateau and is an ESA spaceflight analogue. In terms of both distance and travel time, Concordia can be considered more remote than even the International Space Station. The risks of living at a station as remote as Concordia are real and did require some consideration. It is one of only three inland Antarctic stations situated 1200 km inland, sitting 3320 m high up on an ice mountain plateau on the Antarctic continent. It is one of the few truly isolated stations where evacuation is impossible for nine winter months even in case of emergency, as temperatures (ambient temperatures on the plateau drops to as low as −80 °C during the winter months) are too low for planes or helicopters to function and thus for anyone to reach the station. As if that was not enough, the crew experience 105 days without sunlight and the air is so thin that breathing is a challenge on its own. In such an environment, even apparently minor problems can escalate quickly.

These unique conditions make Concordia an interesting platform to research the physiological and psychological challenges that future astronauts on long duration missions are likely to face, as well as developing the medical models that they may require. This is why Concordia is often referred to as 'White Mars' (Figs. 15.3 and 15.4).

Antarctica is an emblematic location for understanding our own planet's environment and climate which means a research station like Concordia is used for many different fields of study, from seismology to astronomy. Alongside these terrestrial investigations, at Concordia, ESA is able to carry out research experiments in preparation of future human spaceflight, enabling astronauts to work and live for longer periods on the Moon or even on Mars. Space exploration is already proven as the driving force behind research and technological innovation and the goal of travelling deeper into space will benefit not only astronauts in space, but all people here on Earth (Fig. 15.5).

The Concordia platform has now been running for several years and the results of the research is publicly available in scientific journals. This book chapter outlines some examples of the potential applications of the research conducted by the 11th overwinter crew (DC11), particularly the increasing focus on the development of

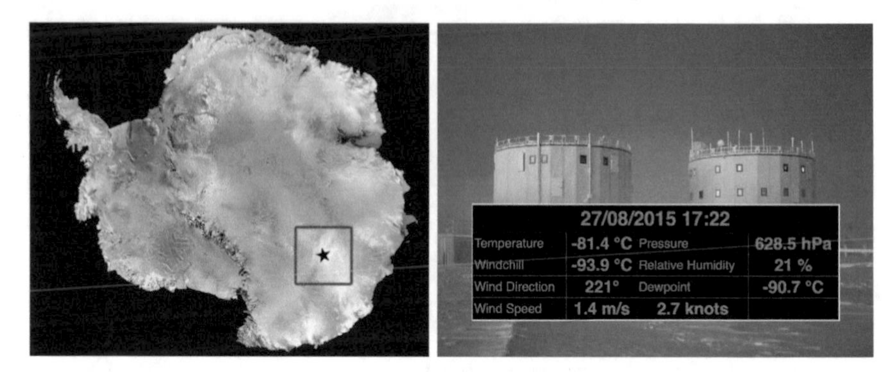

Fig. 15.3 Concordia station in Antarctica, *Source* ESA

Fig. 15.4 The International Space Station and Concordia station: in terms of both distance and travel time even more remote than the International Space Station, *Source* ESA

Fig. 15.5 Enabling astronauts to work and live for longer periods on the Moon or even on Mars is one goal of ESA's research activities in Antarctica, *Source* ESA

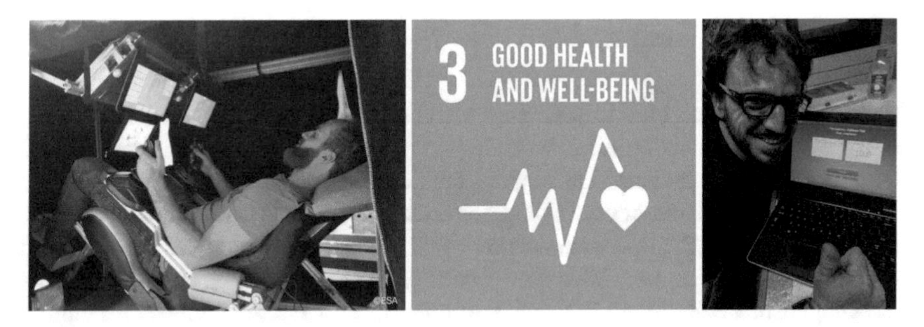

Fig. 15.6 Optimising human performance and addressing SDG3 (United Nations), *Source* ESA and United Nations

remote medical care, optimising human performance and life support systems which offer increasingly relevant and exciting possibilities for solutions to global health challenges and addresses the third sustainable development goal of good health and wellbeing (Fig. 15.6).

15.3 Preparing for the Mission to Concordia Station

Before taking up duty as ESA Medical Doctor in Concordia, I trained alongside other medical doctors going to work in a clinical capacity for the Institute Polaire Francaise in Chamonix in the French Alps. We learnt about the management of altitude sickness due to lack of oxygen, patient retrieval, frostbite, mountain medicine and hypothermia (i.e. your body getting dangerously cold). At the European Astronaut Centre in Cologne, I had the opportunity to meet all the "Principal Investigators", i.e. the research group leaders for each ESA experiment that would be carried out at Concordia, as well as learning how to use the various equipment I would be using in the field (e.g. 24-h blood pressure monitors, saliva samples and urine collection systems, blood tests, functional magnetic resonance imaging scans, cognition tests, etc.).

The whole Concordia overwinter crew were later invited to the European Astronaut Centre (EAC) for baseline data collection and for a special Human Behaviour Performance training. This kind of training is similar to the training astronauts receive before going to live on the International Space Station. We were a group of thirteen people composed of ten men and three women, that learned about how to work together as a team and some of the challenges we might face living in such a small group for a long period of time. We also started to identify our own personality traits and those of team members to help us identify potential problems early. In addition, at EAC it was possible to explore the Space Station training modules and the neutral buoyancy facility where the astronauts train underwater.

15.4 Remote Medical Care

Medical models used on Antarctic platforms have already been used to help inform astronaut field medical training. With limited resources at Concordia, there is a real requirement for remote medical support, for example telemedicine (Fig. 15.7).

Since the first telemedicine link from space in the 1990s as part of a 10 day Skylab mission where, for the first time doctors could study images of an astronaut's heart from mission control, there have been significant advances in both telemedicine and tele-diagnostics technology both on Earth and in space. This includes the ability to diagnose and ablate renal calculi using a small lightweight ultrasound machine (Fig. 15.8).

Fig. 15.7 Medical care on the International Space Station and at Concordia station, *Source* NASA and ESA

Fig. 15.8 Remote medical support in Space and in developing countries, *Source* NASA and ESA

Fig. 15.9 World Health Organisation and Aerospace community knowledge exchange, *Source* WHO and ESA

Concordia is an exciting platform to validate and develop this technology as it provides a real test bed for its evaluation. Telemedicine offers exciting potential for developing healthcare systems, where access to medical specialists is limited.

15.5 Human Performance

Every day in operating theatres worldwide surgeons use the World Health Organisation (WHO) surgical checklist to improve patient safety. This checklist was developed through collaboration with the aviation industry, and like aviation there is also a lot we can learn about patient safety from spaceflight (Fig. 15.9).

With limited access to training, inevitable on long duration missions, a spaceflight simulator has recently been installed at Concordia to consider how training schedules can be optimised remotely. This translates to the training of medical teams and operators of medical equipment operating in remote environments. Prior to departure, all overwintering crew received Human Behavior and Performance training at the European Astronaut Centre. Similar training could also be used to optimise performance within emergency response and medical teams. A cognition battery to monitor Astronaut Performance was also evaluated at Concordia. With doctors working long unsociable hours, often at night, there could also be potential applications of such tools to monitor performance for patient safety.

15.6 Crew Physiology and Psychology

A number of protocols considered the effects of this extreme environment on the physiology of the crew, for example monitoring crew dynamics and the measurement of the response of biochemical markers to stress caused by confinement, isolation and low light levels. All of this research has clinical applications, for example hypoxia experienced at altitude corresponds to that experienced by hypoxic patients in intensive care and likewise, the effect on our eyesight of low light levels and artificial lighting is relevant anyone spending long periods inside, for example factory workers.

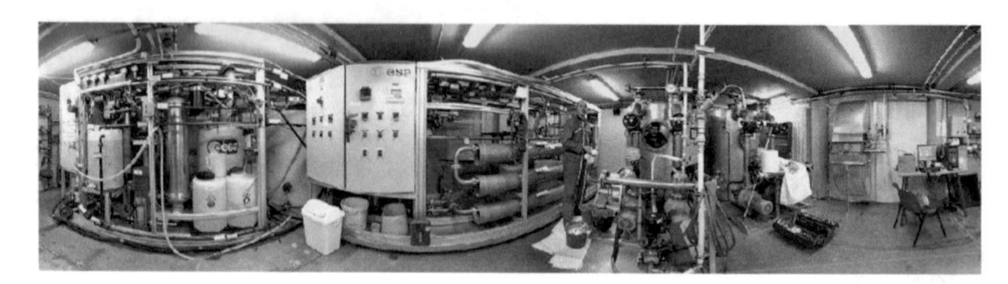

Fig. 15.10 The ESA MELiSSA life support system installed at Concordia station, *Source* ESA

Fig. 15.11 The ESA MELiSSA life support system installed in Morocco, *Source* ESA

15.7 Life Support Systems

The ESA MELiSSA program has developed a number of life support systems at analogue platforms. Concordia currently recycles approximately 95% of its 'grey' water. Likewise, the Eden ISS facility installed at Neumayer station researches innovative farming techniques. Life support systems such as these offer interesting solutions to some of the challenges faced in the developing world (Fig. 15.10).

The water recycling technology developed at Concordia has since been installed also into a remote village in Morocco to provide clean, fresh drinking water (Fig. 15.11).

This summary only touches on some of the extensive application of the research and development being trialled at terrestrial analogue platforms and how this research can translate into innovative and effective solutions to challenges faced in the developing healthcare systems and remote medical settings worldwide.

Chapter 16
Optimizing the Interaction Between Drones and Space Infrastructures

Stefano Ferretti and Alfredo Roma

Abstract Space systems are proving to be particularly effective in delivering solutions when combined with airborne systems (e.g. drones, balloons), especially where regional or local needs cannot be fulfilled by an existing ground infrastructure or when they are out of service. The first point to be addressed is how to best define a clean interface between these systems, in order to ensure full interoperability and integration of data streams in real time. This is a prerequisite for allowing the end users to receive the information in a format which can be directly received and used in the field also by mobile equipment. An example of how international suppliers of space data are working together is given by the ICG of the United Nations in the field of GNSS, and by the GEOSS in the field of Earth Observation. Whether such a coordination mechanisms could be expanded also to other (new) types of platforms worldwide is further discussed in the chapter. We need comprehensive information systems allowing better analysis and assessment of causes of hazard events or events affecting security. It is then important to analyse how this valuable information can be placed at the service of the International Charter of Space and major disasters. In such a case the well-established process of disaster response coordinated with the Charter data, could be enriched with data provided by airborne platforms that have the added value of flexibility, modularity, accuracy and targeted response in a localized area. The clear advantage is that where space systems can provide a fast feedback on a certain situation on the ground, allowing for initial planning and deployment of resources, once the situation evolves it then becomes relevant to have more detailed and localized information that airborne dedicated systems can provide: images, videos, samples of volcanic or nuclear clouds. These processes should nevertheless be completely transparent to the end users, who are interested in receiving information which can be readily used in situ, irrespectively of its source. It is therefore argued that an end to end scenario should be designed according to the needs and requirements of the end users, and proper coordination mechanisms should be put in place in order to acquire and processes the data from space and airborne sources in a fully integrated system. Finally, information must be secured in a data bank system to be consulted for preventing measures.

S. Ferretti (✉) · A. Roma
European Space Policy Institute (ESPI), Vienna, Austria
e-mail: stefano.ferretti@esa.int

S. Ferretti (ed.), *Space Capacity Building in the XXI Century*, Studies in Space Policy 22,
https://doi.org/10.1007/978-3-030-21938-3_16

16.1 Introduction

The idea of receiving and relaying communications signals around the planet was first introduced by Guglielmo Marconi who received the Nobel Prize for his studies in the telecommunications domain.

The possibility of using satellites in geostationary orbit was initially described by Arthur Clarke in 1945.[1] This idea became a reality in 1957 when the first satellite, Sputnik 1, orbited the Earth.

Geostationary satellites keep a constant position over the equator, easing the communication link between them and the ground terminals, providing different functions and services, such as telecommunications, global navigation, broadcasting.

Geosynchronous satellites orbit over fixed points above Earth but need to be tracked by ground equipment, still with the advantage to relay signals faster. In between the two there are several types of satellites, employing a wide range of technologies and operating in various types of orbits. In this context the next step is the development of high-altitude drones to eventually cover some satellite functions. This option would open up new avenues for service delivery, possibly integrating the existing systems in new ways. The question is whether we are now, at the edge of a new revolution, a time when high-altitude sustainable, efficient and reliable drones, with dedicated payloads, will be flying up easily and cheaply, while flying down for upgrades and maintenance as necessary.

The same way the radio was not replaced by satellites, rather the two communications systems were integrated to provide new and higher quality services, it is expected that the new drone systems will complement the options offered today by satellites.

The technology is almost there, moving from the research and development efforts of NASA's Armstrong Flight Research Centre in the past decade, towards the full scale implementation of operational systems made by companies that are successfully testing and flying Unmanned Aerial Systems.

In 2016, European and American efforts have proven quite successful in this domain, with China and other emerging powers running at a fast pace as well: the European company Airbus flew Zephyr T, a drone powered only by solar panels, likewise Facebook successfully tested the Aquila drone, that aims at delivering free internet service around the globe; in the meantime, the Chinese CH-4 UCAV conducted remote split operations, demonstrating their capability to control the vehicle via satellite link for the first time. It is therefore timely and of highest importance that not only roadmaps include these new technologies and options, but also that policies and regulatory frameworks be further developed and refined (Fig. 16.1).

[1]Clarke, A. (1945). Extra-terrestrial relays: Can rocket stations give worldwide radio coverage.

Fig. 16.1 Zephyr high-altitude pseudo satellite (HAPS) *Credits* Airbus defence and space

16.2 Drones

Space systems are proving to be particularly effective in delivering solutions when combined with airborne systems (e.g. drones, balloons), especially where regional or local needs cannot be fulfilled by an existing ground infrastructure or when this infrastructure is out of service.

Unmanned Aircraft Systems (UAS), or more commonly drones, can vary from light, simple, short-range vehicles, radio controlled in the field of view of the operator (Visual Line of Sight), to large scale aircraft, tele-operated overseas via satellite link. All these UAS usually complete missions which are bound by their refueling requirements.

The most recent trends include also new types of vehicles, high-altitude drone concepts which are solar powered and could in principle stay aloft for months or even years, while delivering services as planned.

The traditional larger drones are equipped with jet engines and satellite links, allowing them to fly in a wide range of altitudes and distances, while carrying heavy and complex payloads, such as high resolution cameras, signal intelligence systems (ELINT/COMINT), thermal sensors and ground-surveillance radars.

Smaller and lighter UAS normally fly at lower altitudes in shorter ranges, carrying simple payloads, such as cameras and sensors, usually downlinking data in the field of view of a ground station.

The mission depends on the payload and the ground station's capacities to collect, process and disseminate data for the mission's purposes, drones piloting and the ATM information system. In all cases the satellite links for the larger drones piloting and the ground stations data distribution (satellite based) links for the smaller drones are becoming of growing relevance.

Therefore the key role of Space in drones operations is essential and the on-going technological developments are coherent with the full exploitation of its potential.

However, the elaboration of a regulatory framework that allows drones to fly safely in non-segregated areas is still required, for which the European Union has taken the initial steps.[2,3,4]

Unmanned Aircraft Systems (UAS) are commonly used for high-risk civilian missions, improving the efficiency of civil protection forces but also open up wide market opportunities for state-of-the-art technologies.

16.3 Integrated Applications Using Drones and Space Assets

The clear advantage is that where space systems can provide a fast feedback on a certain situation on the ground, allowing for initial planning and deployment of resources, once the situation evolves it then becomes relevant to have more detailed and localized information that airborne dedicated systems can provide: images, videos, samples of volcanic or nuclear clouds.

Drones can be used for various purposes and in different scenarios, with potential benefits for the society at large, as also recently outlined by the European Commission.

In particular, their use in the field of civil protection can be useful to monitor, prevent, alert and manage post-crisis situations caused by natural disasters.

In the security field, their increasing capacities can be employed in surveillance of coasts or critical infrastructures such as ports, airports and power plants.

Drones can also be placed at service of sustainable development, by monitoring and protecting natural environments. At European level the issues related to civil protection, security and environment are dealt within treaties and specific directives.

The overarching approach is to ensure that appropriate means are in place to monitor, support and enforce the response of the respective authorities, with an increasing role given to information acquisition, interpretation and analysis.

---

[2]European RPAS Steering Group. (2013). *Roadmap for the integration of civil remotely piloted aircraft systems into the European Aviation System*. Final Report from the European RPAS Steering Group, June 2013. Accessed on November 8, 2013. Available at http://ec.europa.eu/enterprise/sectors/aerospace/files/rpas-roadmap_en.pdf.

[3]Roma, A. (2014). *Remotely piloted aircraft system: privacy and data protection implications—Aviation and space Journal*, University of Bologna, January/March 2014.

[4]7th FP: *Assessment of the potential insertion of unmanned aerial system in the air transport system*. Final Report 2013.

Civil protection agencies have amongst their responsibilities to intervene when a disaster occurs, alerting the people to abandon their places and organising active countermeasures to avoid or at least limit the effects of certain disasters.

Considering that the economical and social cost of such disasters is extremely high (e.g. in the order of hundreds billions of Euro), it is of the outmost importance that effective and innovative technologies, including drones and satellites, are harmonized to provide the best possible support in these challenging situations.

For example, following a disaster, one major problem could be to ensure communications with continuous coverage for the entire mission duration, which should be ensured 24 hours per day, keeping active the line-of sight communications among all users. In case of earthquake or tsunami drones can rapidly offer an assessment of damages caused by the disaster.

Drones linked to satellites can provide connectivity in the affected areas, providing valuable assets to both the rescue teams and the population. Another good example is the cooperative effort of drones and earth observation satellites, which can be particularly effective in the case of floods, due to the fast evolution of these phenomena.

In fact, floods can be caused by various conditions, like dam beaks, fast snowmelt, or following heavy uninterrupted rain lasting for some days, the terrain can no longer retain the excess of water, causing the ground to collapse in a flow of water, mud and debris.

All the above conditions have a specific tipping point, triggering the flood, therefore knowing both the global and local evolution in real time is critical to limit the damages. Proper monitoring during the initial phases could help experts in understanding the level of risk, augmenting the probability to evacuate people on the basis of a correctly identified and real danger.

Drones can also support, in real time, teams fighting forest fires which have been detected by satellites. The range of damages induced by forest fires is usually not limited to the loss of human lives and to the cost of damaged property, but usually can last for years, ranging from a constant damage of the soil to water quality, tourism and also climate.

It is already recognised that forest fire emergency can be raised in advance with proper monitoring, and adequate support can be offered to the fire fighters using dedicated payloads mounted on drones to monitor the scene.

Another case in which drones and satellites can prove particularly helpful in the future is earthquakes. The European Commission provides emergency services via Copernicus, but despite the extremely relevant importance in terms of human losses and damages that earthquakes usually cause, Europe does not have at the moment a specific policy.

Mitigation measures are usually demanded to local authorities but prevention is limited to specifications for anti-seismic building construction. In addition to this, ground assessments, using drones in combination with satellites, could effectively support preventive risk analyses of constructions in areas that are considered vulnerable.

The development of new technologies may open up new applications for drones as well. In case of nuclear accidents, the short term and long term contamination of terrain and water needs to be assessed. The loss of human lives is typically limited during the initial emergency but could continue in the following years due to pollution.

While prompt evacuation of the population is essential, the rescue teams need to limit their permanence in the area depending on the level of contamination. In such cases the use of dedicated payloads on the drones can evaluate such levels, by sampling air, soil and water, providing precise information as necessary without the need to expose humans to unnecessary radiation.

Another case where the development of dedicated payloads may open up new possibilities for an integrated use of drones and satellites is the one of volcanic eruptions monitoring. This natural disaster has typically a local impact, usually with casualties in the case of pyroclastic clouds, and generally with damages to properties dependant on the extension of the phenomena.

However, the eruption of Eyjafjallajökull in 2010 created, together with problems to the civil population living close to the volcano, also considerable issues to airlines and passengers for several days, inducing very high losses in the airline business.

Consequently, in the International Civil Aviation Organization (ICAO) Volcanic Ash Contingency, three ash contamination levels have been defined: low (2×10^{-4} g/m^3 < ash concentration < 2×10^{-3} g/m^3), medium (2×10^{-3} g/m^3 < ash concentration < 4×10^{-3} g/m^3), and high (ash concentration > 4×10^{-3} g/m^3) contamination. Radar Systems able to detect ash and its concentration are under development, to be ideally embarked on a number of airlines, but alternative means of detecting ash, such as drone payloads should be considered as well.

In addition, the aerial and space components of security programs are recognised as particularly relevant for the high value content and the frequency of the information provided to the security networks for further comprehensive analyses.

For example, coastal monitoring, both from an environmental standpoint and for borders control, can be enhanced combining satellite data with drones, because of their operational flexibility and persistence. Critical infrastructures, water and natural resources reserves, can all benefit from similar advantages as well.[5]

16.4 The Interface Between Drones and Space

Space systems are proving to be particularly effective in delivering solutions when combined with airborne systems (e.g. drones, balloons, etc.), especially where regional or local needs cannot be fulfilled by an existing ground infrastructure.

The first point to be addressed is how to best define a clean interface between these systems, in order to ensure full interoperability and integration of data streams in

[5]European Commission UAS. (2011). *Panel process—4th workshop, Societal impact on UAS – discussion paper,* 16 Nov 2011.

real time. The interface between drones and space-based infrastructure, particularly in disaster situations, is critical since in situ, drone, aviation, and space data has to be rapidly transformed into actionable information for those on the ground.

This is a prerequisite for allowing end users to receive the information in a format which can be directly received and used in the field also by mobile equipment. An example of how international suppliers of space data, are working together is given by the ICG of the United Nations in the field of GNSS, and by the GEOSS in the field of Earth Observation.

16.4.1 International Committee on Global Navigation Satellite Systems (ICG)

The ICG was established on a voluntary basis as an informal body for the purpose of promoting cooperation, as appropriate, on matters of mutual interest related to civil satellite-based positioning, navigation, timing, and value-added services, as well as compatibility and interoperability among the GNSS systems, while increasing their use to support sustainable development, particularly in the developing countries.

A Providers' Forum was established with the aim to promote greater compatibility and interoperability among current and future providers of the Global Navigation Satellite Systems (GNSS). The current members of the Providers Forum, including China, India, Japan, the European Community, the Russian Federation and the United States, addressed key issues such as ensuring protection of GNSS spectrum and matters related to orbital debris/orbit de-confliction. It also established principles of compatibility and interoperability.

16.4.2 GEOSS

In Earth Observation there are many systems of satellites, including the EU's Copernicus, NOAH in the US, and meteorological satellites to name a few. Although there is a large amount of data produced, these all have different standards, meaning that there is an effusion of different data formats. This limits interoperability.

GEOSS was conceived around the goal of ensuring that the end-user receives input in a standardized format, including both users from the public and the private sector. For GEOSS, however, the challenge is bigger, as the type of data gathered is not Position, Navigation, Time, but rather images. A further challenge is the high level of change in the instruments used, as each successive satellite aims to capture higher resolution images.

The challenge can be compared to creating interoperability between different operating systems on computers. It is possible to ensure that Windows, Apple and Linux will all be able to run software by the respective other, but every time the OS is updated to a newer version, the whole process must be undergone again. Instead of three OS, however, there are many more.

ESA alone has more than 10 Earth Observation satellites, of which only a fraction is part of the Copernicus system. In addition, there are military concerns over civilian access to high-resolution images. GEOSS provides a forum for these issues to be addressed. It promotes common technical standards so that data from the thousands of different instruments can be combined into coherent data sets. The 'GEOSS Portal' offers a single internet access point for users seeking data, imagery and analytical software packages relevant to all parts of the globe.

The question is whether these organizations might be extended, or whether they could serve as examples for new platforms. In the case of drones, the variety of instruments that can be employed to get multitudes of different data together with a rapidly growing private manufacturing sector make this challenging, but more and more necessary.

16.4.3 The International Charter on Space and Major Disasters

The Charter is a good example of a tool which may be employed on a political level for disaster situations, and it is a well-established process. It is based on a pre- defined list of appointed users, known as 'Authorized Users' (AUs).

Typically they are national civil protection, rescue, defence and security bodies from countries of Charter member agencies, able to request Charter support for emergencies in their own country, or in a country with which they cooperate for disaster relief.

An Authorized User can call a single number to request the mobilization of the space and associated ground resources associated with Charter members in order to obtain data and information on a disaster occurrence.

A 24-hour on-duty operator receives the call, checks the identity of the requestor and verifies that the User Request form, sent by the Authorized User, is correctly completed. The operator passes the information to an Emergency On-Call Officer who analyses the request and the scope of the disaster with the Authorized User, and prepares an archive and acquisition plan using available satellite resources.

Data acquisition and delivery takes place on an emergency basis, and a Project Manager, who is qualified in data ordering, handling and application, assists the user throughout the process. A Value Added Reseller further processes and interprets the data acquired over the area affected by the disaster and delivers the images to the End User.

The Charter has an agreement with UNOOSA (Vienna) and UNITAR/UNOSAT (Geneva) to provide support to UN agencies. UNOOSA and UNITAR/UNOSAT may submit requests on behalf of users from the United Nations. Some examples of use include: typhoons, earthquake in Haiti, volcanic eruption in Iceland, Ebola, hurricane over Fiji. One interesting step further might be to expand its mandate to sustainable development, and to ease access to the charter by NGOs, that currently have indirect access only through Authorized Users.

Whether the above coordination mechanisms could be expanded also to other (new) types of platforms worldwide requires a comprehensive assessment of the current and future evolution of systems, requirements and potential cooperation frameworks.

A key aspect is to ensure that the data originated by different platforms is harmonized, in a way that creates the most productive interface with the Charter on cooperation, to achieve the coordinated use of space facilities in the event of natural or technological disasters. Also, the interoperability requires some careful consideration, in particular with respect to systems that are provided with the same type of instruments, and others which have different instruments on board.

In fact, in a first instance, it is important to scrutinize the existing structures, identifying what functions can be addressed through them. Then it is necessary to determine the type of products and services that can meet the needs of end-users.

Finally, it is crucial that the integration process produces added value in the chain for both the users and the service providers.

It is then important to analyse how this valuable information can be placed at the service of the International Charter of Space and major disasters.

In such a case, the well-established process of disaster response coordinated with the Charter data, could be enriched with data provided by airborne platforms that have the added value of flexibility, modularity, accuracy, and targeted response in a localized area.

In all cases, while technical aspects could already be addressed in various fora, the political and regulatory frameworks still need to be well pondered, especially considering that a comprehensive information system, allowing better analysis and assessment of causes of hazard events or events affecting security, can prove to be particularly useful at this point in time.

The current regulatory framework is based on the fact that over the past century, aviation rules and regulations have developed based on the use of crewed aviation, and today military drones can fly within segregated areas.

The fast development of non-military drones has raised the need for their integration into non-segregated airspace in order to perform required missions. Since EU airspace is going to be reorganized following the EU's Single-Sky programme, drones' operations must be included in this new scenario, especially the liability related aspects.

16.5 Conclusion

Disaster prevention is one of the most demanding issues for Civil Protection agencies but with the available technologies many disasters cannot be forecasted (e.g. earthquakes, volcanic eruption). However, some might be monitored at the beginning of their occurrence (e.g. floods, fires), offering the opportunity to prevent or at least reduce the costs and loss of lives.

In post–disaster relief activities the rescue teams need quick information dissemination, coordination, communication, and connection. During subsequent days to a disaster, most of the traditional monitoring systems may not work because of bad weather or infrastructure damages, and sometimes this situation could persist for several days, endangering the rescue teams.

Drones in combination with satellites may help in providing these services and in updating the vulnerability maps of the impacted zones in a timely manner, which, due to difficulties in reaching the area, could be key in acquiring the evolving state, communicating to the central agencies and to rescue teams a clear and timely picture of the crisis situation.

The further integration of drones and satellite assets will have an increasing role in Civil Protection, Security and Environmental Control in Europe. Nevertheless, in order to achieve this objective there are still a number of limited but challenging issues to be overcome: drones cannot fly in controlled airspace; the specific missions and roles of drones in Civil Protection, Security and Environment protection are not yet addressed in a pragmatic and exhaustive way; European citizens are not fully aware of their capabilities and are concerned about the potential implications of their wider use.

Therefore it is important to further develop the sector, reflecting on the specific needs of the end-users, including civil protection authorities, security and environment protection entities, since there is still limited availability of civilian concepts of operations and requirements.

In this process, it is important to explore not only drone utilisation but also the needs of the stakeholders and any possible implications on current and future rules and laws.

Ultimately, the crucial point is that all these processes should be completely transparent to the end users, who are interested in receiving information which can be readily used in the field, irrespectively of its source.

Information must also be secured in a Long Term Data Preservation System. This data bank system shall be consultable at any time, allowing public authorities to take preventing measures based on historical data. Cybersecurity will then need to be considered as a key aspect, both during the data integration and during the data storage and dissemination processes.

The key recommendation is that all the stakeholders need to be involved in a dedicated platform to assess policy, technical and regulatory frameworks, and to build an end-to-end scenario, designed according to the needs and requirements of the end users, including proper coordination mechanisms among all the actors involved.[6]

This paper was presented at the 67th International Astronautical Congress, 26–30 September 2016, Guadalajara, Mexico, www.iafastro.org.

---

[6]United Nations General Assembly. (2015). *Seventieth session, Transforming our world: The 2030 Agenda for Sustainable Development*, UN A/RES/70/1.

Chapter 17
Australian Perspectives on Knowledge Transfer from Space Technologies to Global Health

Chandana Unnithan, Ajit Babu and Melanie Platz

Abstract Australia plays a unique role in the global space network with its exceptional location and the outstanding contributions to space science and international space programs. Its partnership with NASA dates back 50 years and continues to push boundaries of exploration in search of benefits to life on Earth. Afar from the global expeditions, the Australian Space Agency (ASA) was established by the government in July 2018, with a mandate on development and application of space technologies on Earth and assisting with the growth of the civic space and public health. Australia has significant experience in integrating space sourced data into communications, Earth Observations from Space, and Global Navigational Satellite Services (GNSS), with an advantage of its participation in the international space industry supply chain from the southern hemisphere. Strong capabilities are exhibited in photonics, quantum cryptography, optical design, adaptive optics, artificial intelligence, advanced analytics, and 3D-printing, which are space derived. Use cases presented in this chapter illustrate some of these areas. Digital Earth Australia and open Data Cube initiatives which enable space sourced data synthesis, The Mangrove Observing System and IoT integration applications in public health are presented in this chapter. Australian initiatives in capacity building and knowledge translation using space derived data, apps and capabilities were presented at the 56th UNCOPOUS meeting in Vienna, which have been detailed in this chapter.

17.1 Introduction

What was the rationale for Australia to invest in space and how does it connect with global health? And how do they relate to the Sustainable Development Goals (SDGs).

C. Unnithan (✉)
Torrens University Australia, Melbourne, Australia
e-mail: chandana.unnithan@gmail.com

A. Babu
St. Louis University, St. Louis, USA

M. Platz
Pedagogical University of Tyrol, Innsbruck, Austria

S. Ferretti (ed.), *Space Capacity Building in the XXI Century*, Studies in Space Policy 22,
https://doi.org/10.1007/978-3-030-21938-3_17

The information in this chapter attempt to address these questions. Australia's unique partnership with member states in the United Nations and the WHO, along with the Australia space agency's initiatives in civil space has positioned it in a distinctive role for knowledge translation and capacity building, particularly in digital global health. This chapter presents some latest insights and perpectives.

17.2 Australian Space Industry Capabilities

In the 2017 ACIL Allen Consulting report to the department of Industry, Innovation and Science in Australia,[1] it was noted that Australia has capabilities along most of the space industry supply chain. The strongest area was in space applications, where Australia has experience in integrating space sourced data into communications, Earth Observations from Space and Global Navigational Satellite Services (GNSS). The major advantage of Australia is its participation in the international space industry supply chain from the southern hemisphere. The well positioned ground stations across a 4000 km baseline are able to observe a number of satellites, space debris and weather. The report highlighted that while Australia has no capability in manufacturing of large satellites and has limited competence in design and specification of launch vehicles; it does have an emerging capability in the design and manufacture of Nano and Micro satellites, based in universities and by start-up companies. In space operations, Australian companies are strong on telemetry, tracking and command for satellite operations; and Australian universities also have satellite operation capabilities including ground stations to support Nano-satellite missions. Australian government agencies have well established ground station networks and infrastructure supporting Earth Observations from Space, meteorology, deep space exploration and astronomy.

For example, Optus Satellites have over 30 years of experience in operation of communication satellites from Australia.[2] They currently operate 5 Optus communication satellites and 2 NBN satellites and have supported more than 90 international missions. Lockheed Martin,[3] who have a facility at Uralla in New South Wales (NSW) provide telemetry, tracking and command (TT&C) services. Among Universities, the most notable is the Institute for Telecommunications Research (ITR)[4]

[1]ACIL ALLEN (2017). Australian Space Industry Capability, Department of Industry, Innovation and Science Canberra.
http://www.acilallen.com.au/cms_files/ACILAllen_SpaceIndustryCapability_2017.pdf. Accessed 22 March 2018.

[2]Optus satellite coverage. 2017. https://www.optus.com.au/about/network/satellite/coverage. Accessed 22 March 2018.

[3]Lockheed Martin STELaRLab (2017). Satellite Tracking. https://www.lockheedmartin.com.au/au/what-we-do/space-systems/satellite-tracking---control.html. Accessed 22 March 2018.

[4]Institute for Telecommunications Research (2017). University of South Australia. http://www.unisa.edu.au/Research/Institute-for-Telecommunications-Research/Research/Satellite-Communications/. Accessed 22 March 2018.

at the University of South Australia which provides commercial services as well as supporting research. The University of Tasmania[5] operates a 26 m satellite tracking antenna that is used to support commercial projects as well as radio astronomy. The Australian National University, the University of NSW and the University of Sydney have ground stations to support Nano satellite missions.

Conversely, the private sector also has ground stations to support operations from communications, earth observations from space and GNSS. ACIL Allen's review in 2017 identified 388 companies, 56 education and research institutions, and 24 government agencies that have space industry capabilities. The location of Australia supports low earth orbit satellite launches. Currently, an organisation is working on establishing a launch complex in the Northern Territory of Australia. In Western Australia, the Square Kilometre Array Project has advanced capabilities in astronomy.

Perhaps, the foremost strength of Australia is in *space applications*. Agriculture, mining, logistics, aviation and communications depend on space derived services for communication, imagery and positioning. Strong capabilities are also visible in photonics, quantum cryptography, optical design and adaptive optics amongst Australian companies. Integration of space derived imagery in weather forecasting, vegetation and land use monitoring, national security, emergency services and surveying/mapping is highly advanced. These are significant in times of environment related disasters. For example, Australia uses NovaSAR satellite images in South Australia, to manage bushfires and floods.[6]

ACIL Allen reported[7] that a highly developed activity in space as of 2016 was earth observations from Australia. Integrations of satellite imagery data into spatial applications are growing rapidly, driven by the demand from the Department of Defence, disaster management agencies, agriculture, vegetation mapping, ocean and atmosphere monitoring, design and development of built environments, finance and insurance. Data from Earth Observations from Space, both imagery and synthetic aperture radar (SAR), have been used in mapping, weather modelling and forecasting, ocean monitoring, vegetation mapping, agricultural production monitoring and emergency management and are now being incorporated into 3D models of the built environment. Earth Observations from Space are also gaining application in the finance and trade sectors to monitor crop production capacity.

Australia's geographic location incidentally positions it at the boundary of multiple international satellite positioning constellations. This aspect has driven the development of technologies for integration of multiple signals to improve accuracy and augmented by the integration of augmented GNSS systems, to enable delivery of quality solutions for industry applications in logistics, navigation, agriculture, mining and transportation.

---

[5]University of Tasmania (2017). Mt Pleasant Observatory. http://www.utas.edu.au/maths-physics/facilities/mt-pleasant-observatory. Accessed 22 March 2018.

[6]CSIRO (2017). AdelaideNow. http://www.adelaidenow.com.au/technology/science/csiro-leases-time-from-novasar-satellite-to-help-fight-sa-bushfires/news-story/6b2b6cc3a935e8e2837bb671161543c1. Accessed 22 March 2018.

[7]ACIL Allen (2016). SBAS and Australia. http://www.spaceindustry.com.au/Documents/SIAA%20SBAS%20and%20Australia%20Report.pdf. Accessed 22 March 2018.

The Bureau of Meteorology (BOM)[8] provides accurate and timely weather monitoring services on a non-commercial basis that industries such as aviation, maritime and agriculture depend on. After being processed and utilised for daily services and predictions, the data is archived to support climate monitoring and disaster responses. The service receives 95% of its data from satellites, that is accessed from 20 different instruments, despite BOM not owning any satellites. The organisation's capability is underpinned by a network of ground stations that receive the satellite data, a $70 million supercomputer to store the data and the daily collection of in situ data via weather balloons and drifting buoys.[9] The BOM assists Australians in dealing with the harsh realities of their natural environment, including drought, floods, fires, storms, tsunamis and tropical cyclones, thus contributing to the sustainable development goal 11 (SDG11). Through regular forecasts, warnings, monitoring and advice spanning the Australian region and Antarctic territory, it contributes to national social, economic, cultural and environmental goals by providing observational, meteorological, hydrological and oceanographic services; and by undertaking research into science and environment related issues in support of its operations and services.[10]

The development of new Space applications is supported by mature research programs within Australian universities, the Commonwealth Scientific and Industrial Research Organisation (CSIRO) and the Cooperative Research Centre for Spatial Information (CRCSI). There is a mature capability within GA and the BOM for the development of new processes and the delivery of public good services, and there is growing expertise within Federal and State Government Departments of Australia. Landgate, NSW Spatial Services, Queensland Government and Victoria Department Environment, Land Water and Planning, are essential participants in the CRCSI.[11]

The Northern Territory Department of Environment and Natural Resources provide bushfire mapping services using satellite data.[12] The satellite-based information they deliver is particularly important in central and northern Australia, where bushfire management information is as crucial for safeguarding productivity as rainfall data. Rehabilitation of legacy mines has become a key issue in the Northern Territory. The Legacy Mines Unit has installed equipment connected to satellites and broadband telemetry at hard-to-reach locations to remotely monitor conditions such as rainfall, humidity, and temperature. Australia leads the world in remote asset management, with autonomous trucks at iron ore mines in Western Australia being controlled remotely, 24 hours per day, 365 days a year, from thousands of kilometres away.

[8] Burearu of Meterology (2017). http://www.bom.gov.au. Accessed 22 March 2018.

[9] ACIL ALLEN (2017). Australian Space Industry Capability, Department of Industry, Innovation and Science Canberra. 2017.
http://www.acilallen.com.au/cms_files/ACILAllen_SpaceIndustryCapability_2017.pdf.

[10] Burearu of Meterology (2017). http://www.bom.gov.au/inside/index.shtml?ref=hdr. Accessed 22 March 2018.

[11] CRCSI (2017) Essential Participants, Support Partners, Stakeholders and International Partners. http://www.crcsi.com.au/partners/. Accessed 22 March 2018.

[12] https://denr.nt.gov.au/land-resource-management/water/water-information-systems/spatial-data-requests. Accessed 22 February 2019.

The same region is also expected to become host to the largest robot on Earth; a fully automated ore train, and automated drills. Woodside Australia, which has an existing partnership with NASA, is using artificial intelligence, advanced analytics, and 3D-printed parts for the remote maintenance of its offshore rigs.

Space capacity building is dependent on integrating capabilities from SMEs, universities, research centres and government entities in Australia. The Department of Defence Innovation Hub[13] is an important example of how government requirements can focus on capacity building. The Next Generation Technologies Fund also represents around $750 million over the next decade for strategic next generation technologies that have the potential to deliver new capabilities.[14]

The creation of the *Digital Earth Australia*[15] concept by Geoscience Australia creates important opportunities. The application provides access for a wider range of users of space sourced data, facilitating analysis and processing of the data into information. It is an analysis platform for satellite data and other forms of Earth observation data, operated by Geoscience Australia. It provides tools for industry and the government to improve policy and investment decisions and to drive the digital economy. It has broad applications relevant to monitoring and reporting, and helps in achieving SDGs relating to agriculture, environmental issues, climate change, and disaster management. The *Open Data Cube* (ODC) is a free, open source initiative to provide a data architecture solution that has value to its global users and increases the impact of satellite images. Using the ODC, products developed in Australia under Digital Earth Australia can be deployed internationally. ODC deployments are underway in South Africa, Vietnam, Colombia, and Canada, to name a few.[16] It is gaining recognition internationally as an opportunity for improving the application of space-based data in many instances. With the emergence of the Internet of Things (IoT), the use of space-derived data and services has become ubiquitous and any restriction on access will be a disadvantage. Increasingly, the development of the smart healthy cities, reflecting on SDG11, depends on this assured access to Space derived data. Some pertinent forays include: Australia's Mangrove Observing System[17] over the Terrestrial Ecosystem Research Network (TERN) has provided open access to decades of newly acquired field and EO data. TERNs 2017 LiDAR and aerial imagery dataset over the *Gulf of Carpenteria* represents the largest single dedicated acquisition of such data over a mangrove ecosystem, and helps researchers understand how this dynamic ecosystem is responding to environmental change in Australia. It allows for informing environmental policy in Australia and allows the

[13]Defence Innovation Hub (2017) https://www.business.gov.au/Centre-for-Defence-Industry-Capability/Defence-Innovation/Defence-Innovation-Hub. Accessed 22 March 2018.

[14]NTF (2017) Next generation technologies fund. https://www.business.gov.au/Centre-for-Defence-Industry-Capability/Defence-Innovation/Next-Generation-Technologies-Fund. Accessed 22 March 2018.

[15]Digital Earth Australia (2017) http://www.ga.gov.au/about/projects/geographic/digital-earth-australia. Accessed 22 March 2018.

[16]Open Data Cube Initiative (2018) http://www.ga.gov.au/dea/odc. Accessed 22 February 2019.

[17]https://sdgs.org.au/project/australias-mangrove-observing-system/ Accessed 22 February 2019.

nation to contribute to international environmental conventions including the United Nations Sustainable Development Goals.

17.3 Linking SDGs and Public Health

The Australian Government has made significant contributions toward achievement of SDG11 through key domestic initiatives such as the Smart Cities Plan, the Smart Cities and Suburbs Program and the National Cities Performance Framework.[18] This has been achievable through civic engagement and active participation. For example, from the 17 SDGs, the main areas of focus in relation to public health and well-being in Australia are to prevent domestic violence (focused on gender inequality), mental illnesses (caused mainly from social isolation) and non-communicable diseases (diabetes, cancer, heart diseases, asthma that may be lifestyle diseases or of genetic origin), leading to sustainable healthy cities development.[19]

Mental health caused by social isolation may emerge from two different streams of immigration i.e. the large number of international (linguistically diverse) students/skilled migrants and (2) similarly linguistically diverse refugees from many countries. *How is Australia utilising Space technology applications*? Australia has highly developed satellite communications as well as a saturated marketplace with smartphones. Leveraging on these aspects, to address the SDG goals, many apps have been developed that synthesise the capabilities to promote public health through civic participation. We present some examples.

17.3.1 Angelhands—Apps

MyWitness app turns the smartphone into a digital witness that enables the recording of sounds from the immediate surroundings and notifies loved ones when help is needed. It combines video and GPS technologies and stores this data safely in the cloud. The *iMatter app*, launched by Doncare Community Services help young women understand the warning signs of abusive and controlling behaviour, to promote self-esteem. *Daisy* is a smartphone app by the national organisation 1800RESPECT to provide a range of resources to assist local victims. It brings together information on services in each State. The *Circle of 6* app for iphone and Android makes it quick and easy to reach 6 friends in your circle. 2 touches let the circle know the location and how they can help, with touch icons representing actions for privacy.

[18]SDG11 (2017) https://cities.infrastructure.gov.au/sustainable-development-goal-11. Accessed 22 March 2018.

[19]Conversation (2017) SDGs a win-win for Australia. http://theconversation.com/sustainable-development-goals-a-win-win-for-australia-47263. Accessed 22 March 2018.

17.3.2 Wearable Devices (MIoT/AI)

Wearable sensory devices (mobile IoT) to remotely manage patients' chronic conditions were trialled in Australia since 2007. In 2017, through a collaboration between Griffith University, Huwaei Australia and Tonwo Health Clinic Technology,[20] Narrowband Internet devices (NB-IoT chips) are able to remote-monitor patients wearing the devices. Patients can self-monitor their blood pressure, glucose levels, BMI, heart rate, pulse oximeter and heart conditions, and send the data through on an application platform.

mCareWatch[21] is a device that connects through the ConnectiveCare platform that helps people with dementia, disabilities, their family and carers, those with chronic conditions or general health concerns, and people who wish to improve and maintain their health fitness. When a person connects, a wide range of useful functions that support health, safety and well-being are presented, with speed and accuracy, helping citizens live healthy and safe lives. ConnectiveCare functions range from measuring and reporting hear rate, blood pressure, blood sugar through to activity and GPS tracking, geo-fencing, medical alerts, medication and appointment reminders.

In 2018, after the National Electronic Health Care record systems,[22] tracking and tracing of patient data to provide end-to-end medical care online is now possible, enabled by satellite communications. There are connected Medicare services, (GP systems, hospitals, emergencies), immunisation, dental, imaging, lab reports, aged care services, etc. Conversely, Public Health Information Development Unit (PHIDU)[23] is an initiative that has access to real-time data on a national scale (24.6 million population). It is committed to research and dissemination of information on a broad range of health and other determinants across the lifespan. The PHIDU data synthesis leverages building infrastructure where it is required (including building public hospitals and care facilities). Both these initiatives would derive greater benefit from space derived data, to build future medical facilities and train new medical personnel.

---

[20]DiabetesQLD (2018) Wearable devices to remotely manage patient chronic conditions—a media report. https://www.diabetesqld.org.au/media-centre/2017/december/wearable-devices-to-remotely-manage-patient-chronic-conditions.aspx. Accessed 22 March 2018.

[21]Mcarewatch (2018) About mcarewatch. https://mcarewatch.com.au/about-us/. Accessed 22 March 2018.

[22]https://www.healthcareit.com.au/article/australia-leads-world-personal-control-electronic-health-records. Access 22 February 2019.

[23]http://phidu.torrens.edu.au Accessed 22 February 2019.

17.4 Space and Global Health—Future Directions

The UNISPACE III resolution[24] entitled, "The Space Millennium: Vienna Declaration on Space and Human Development states that "... *activities of the United Nations Programme on Space Applications should improve public health services by expanding and coordinating space-based services for telemedicine.*" The program on space applications promotes space-based solutions to healthcare by providing capacity building opportunities in the areas of telehealth and assists Member States in using satellite remote sensing, global positioning, GIS and satellite communications to integrate ecological, environmental and habitation data. This data is then used for predictive modelling in disease surveillance.

A need emerged for an integrated global health alert system, to combat global epidemics. To best prepare emerging health threats, the Australian Government Health Security Initiative for the Indo-Pacific region was launched in October 2017.[25] It is contributing to the prevention and repression of infectious disease threats with the prospective to cause social and economic harm on a national, regional or global scale. It is being implemented to promote global and regional cooperation, catalyse international responses to Australian identified needs, applying Australian unique strengths in health security and space.

In 2019, at the meeting of the working group on Space and Global Health held during the Scientific and Technical Sub Committee meeting of the Committee on Peaceful Uses of Outer Space (COPUOS) in Vienna, it was proposed to build a pioneering master's program on space and global health from an Australian University inviting collaborating universities globally via the working group on Space and Global Health. Collaborating universities will provide modular content into the master's course, which will be partially based on open data sources and Space derived data to assist emerging public health and medical practitioners.

17.5 Conclusion

Australia is well positioned for research and exploration, with its unique geographic location, knowledge capabilities in the sector, and global partnerships. Conversely, the advanced knowledge capabilities within the technology sector, particularly in the IoT field that seamlessly connect to public health initiatives in the nation exhibit the potential for knowledge sharing amongst global communities leveraging the potential of Space and Global Health linkages.

[24]UNOOSA (1999) http://www.unoosa.org/oosa/en/ourwork/psa/schedule/1999/unispace-iii.html. Accessed 22 March 2018.

[25]DFAT (2018). Indopacific Security. http://indopacifichealthsecurity.dfat.gov.au/Pages/default.aspx. Accessed 22 March 2018.

Australian initiatives for capacity building and knowledge translation from space technologies to global health were presented at the 56th session of the Scientific and Technical Sub Committee of the United Nations Office of Outer Space Affairs in Vienna[26] Currently, Australia is poised to leverage on its strengths and capabilities, from Space derived data to enhance life on earth, moving forward.

[26]Australian initiatives on Capacity Building and knowledge transfer from space technologies to global health (2019). http://www.unoosa.org/documents/pdf/copuos/stsc/2019/tech-14E.pdf Accessed 22 February 2019.

Chapter 18
Shaping the Future of Global Health: A Review of Canadian Space Technology Applications in Healthcare

Aranka Anema, Nicholas D. Preston, Melanie Platz and Chandana Unnithan

Abstract Technology innovation is exponential globally and the global health sector is imbibing these rapidly. In Canada, the growth in space technology innovations have been driven by the UN 2030 sustainable development goals (SDGs) and applied to public health and foreign aid priorities and policies. In this chapter, we explore the shared vision of space and global health in Canada, through an appraisal of contributions.

18.1 Introduction: Era of Exponential Technologies

We are experiencing a global revolution in technology innovation that is growing at an exponential rate. Leading futurist Ray Kurzweil has suggested that while human development was historically characterized by local and linear shifts in knowledge and technology, the past 100 years have been defined by technological innovations that are both global and exponential.[1] Proponents of 'exponential technologies' suggest that these trends lead to rapid and exponential increases in innovation. This jump from linear to exponential technologies over recent years is illustrated by the comparison of the NASA Apollo Guidance Computer to today's iPhone. The Apollo guidance computer was revolutionary at its time in terms of computing power; it

---

[1] Kurzweil R. 2005. The Singularity Is Near: When Humans Transcend Biology. Penguin Publishing, USA.

[2] O'Brien F. The Apollo Guidance Computer, Architecture and Operation". Springer, 2010.

A. Anema (✉)
Global Health PX, Vancouver, Canada
e-mail: aranka.anema@gmail.com

N. D. Preston
University of Victoria, Victoria, Canada

M. Platz
Pedagogical University of Tyrol, Innsbruck, Austria

C. Unnithan
Torrens University Australia, Melbourne, Australia

S. Ferretti (ed.), *Space Capacity Building in the XXI Century*, Studies in Space Policy 22,
https://doi.org/10.1007/978-3-030-21938-3_18

cost millions to build and was the size of a car.[2] In 2018, the iPhone is 120 times computationally faster, costs a couple hundred to manufacture and fits into a back pocket.

Technology entrepreneurs have postulated a theoretical framework to further understand the concept of exponential technologies, known as the 'Six Ds'.[3] Once a technology has been '*digitized*', or information-enabled, it has the potential to scale exponentially by virtue of being easily accessed, shared and distributed. The development cycle of exponential technologies can be '*deceptive*' during its initial periods of growth, and its fullest potential difficult to predict owing to unprecedented leaps in innovation. Exponential technologies are often '*disruptive*' by virtue of their ability to outperform traditional solutions in effectiveness and cost, opening new market opportunities. Technologies are being '*demonetized*', as software and hardware development becomes cheaper to manufacture, to the point of being free of cost. Finally, exponential technologies are '*dematerialized*', with digitization of functionality removing the need for physical products; and they are increasingly being '*democratized*', by way of becoming widely available to potential users. Examples of exponential technologies include: 3D printing, artificial intelligence (AI), industrial robots, robotics, digital medicine, drones, internet connectivity, genetic sequencing, cell phones, synthetic biology and nanotechnology.

Over the recent years, we have witnessed a proliferation of exponential technologies in medicine and healthcare that are poised to transform health globally. Some salient examples include wearable technologies, which have seen annual revenue market growth from $0.75 billion in 2012 to $5.8 billion in 2018[4]; human genome sequencing, which has dropped from $1 billion in 2001 to $500 in 2017[5]; This trend is largely fuelled by private sector start-ups, with latest market insights estimating there are now over 100 new entrepreneurial ventures transforming healthcare with AI.[6] Looking to the 2020s, the Internet of Things (IoT) will generate exponential amounts of data from devices and sensors. 5G networks will be 10–100 times faster than current mobile broadband networks, enabling transfer of vast amounts of high-fidelity, real-time data. And, once powered by the analytic abilities of AI and machine

[3]Ramirez VA. The 6 Ds of Tech Disruption: A Guide to the Digital Economy. November 22 2016. Singularity Hub. Available at: https://singularityhub.com/2016/11/22/the-6-ds-of-tech-disruption-a-guide-to-the-digital-economy/.

[4]Wearable Technology—The Next Mobility Market is Booming—Wearable Tech—goo.gl/Mx27nG.

[5]Shendure, Jay, Shankar Balasubramanian, George M. Church, Walter Gilbert, Jane Rogers, Jeffery A. Schloss, and Robert H. Waterston. "DNA sequencing at 40: past, present and future." Nature 550, no. 7676 (2017): 345.

[6]CB Insights. From Virtual Nurses To Drug Discovery: 106 Artificial Intelligence Startups In Healthcare Feb 03, 2017. Available at: https://www.cbinsights.com/research/artificial-intelligence-startups-healthcare/.

learning, data will inevitably become a new form of capital with blockchain technology as the mechanism for securely storing, distributing, managing and deploying that capital.[7]

18.2 Leadership in Space Technology for Global Health

Canada has played a leading role internationally in the translation of space science to terrestrial applications in health. In 2001, the United Nations Committee on the Peaceful Uses of Outer Space (UN-COPUOS) established the Action Team 6 (AT6) as a mechanism for implementing recommendations of the third United Nations Space Conference UNISPACE III (1999). Co-chaired by Canada, an Action Team 6 Follow-Up Initiative (AT6FUI) was established with the World Health Organization (WHO) in 2007 to refine domestic and international priorities for public health applications of space technologies. In 2011, AT6FUI held a meeting on 'Space technology for Public Health Actions in the Context of Climate Adaptation" together with UNOOSA and ESA, to foster knowledge exchange, partnerships and discussion around satellite technology applications to climate change and public health.[8] In 2015, The United Nations inaugurated an expert group on space and global health under the leadership of the Canadian government, the WHO and the Canadian Space Agency (CSA). This initiative is part of a broader goal to apply space technology to socioeconomic development, articulated in the United Nations Committee on the Peaceful Uses of Outer Space 2015 (UN-COPUOS) Thematic Priority #5: Strengthened space cooperation for global health. It outlines priority objectives around the use of space technologies and information systems to promote cooperation in emergencies, epidemics and natural disasters.[9,10] Most recently, in 2017, the CSA and National Research Council of Canada's Industrial Research Assistance Program (IRAP) co-hosted the Space, Health and Innovation conference which aimed to highlight latest research and

[7]Lyons P and Hartani R. The Internet Is Fast Approaching 50. How Might It Look At 60? January 23, 2018. Forbes Magazine. Available at: https://www.forbes.com/sites/worldeconomicforum/2018/01/23/the-internet-is-fast-approaching-50-how-might-it-look-at-60/#6e01768a2f99.

[8]Committee on the Peaceful Uses of Outer Space Scientific and Technical Subcommittee Forty-eighth session, Vienna, 7–18 February 2011. Item 5 of the provisional agenda. Implementation of the recommendations of the Third United Nations Conference on the Exploration and Peaceful Uses of Outer Space (UNISPACE III). Available at: http://www.unoosa.org/pdf/limited/c1/AC105_C1_L305E.pdf.

[9]Committee on the Peaceful Uses of Outer Space Scientific and Technical Subcommittee Fifty-first session. Vienna, 10-21 February 2014. Item 6 of the provisional agenda. Space technology for socioeconomic development in the context of the United Nations Conference on Sustainable Development and the post-2015 development agenda. Available at: http://www.unoosa.org/pdf/limited/c1/AC105_C1_2014_CRP24E.pdf.

[10]DiPippo S. United Nations Office for Outer Space Affairs (UNOOSA). UNISPACE +50. UN/WHO/Switzerland Conference: Strengthening Space Cooperation for Global Health. UN-SPACE Open Session. 24 August 2017. Available at: http://www.unoosa.org/documents/pdf/unspace/iam/2017/ois-01E.pdf.

development (R&D) and applications of space technologies into healthcare across Canada.[11]

In this Chapter, we propose a shared vision for Canadian space science contributions to global health and highlight future directions for Canada's contribution to global health technologies towards achieving the United Nations 2030 Sustainable Development Goals (UN SDGs). We review the Canadian Space Agency's (CSA) space technology innovations as applied to domestic healthcare and public health and Canada's foreign aid priorities.

18.3 Space Technology Applied to Health in Canada

CSA's mission is to advance the knowledge of space through science and use its discoveries for the good of Canadians and all of humanity. One of its priorities is to use space science and technology for the development of innovative applications on Earth. CSA is actively positioning the private sector to lead domestic space industry innovations and to bring cutting-edge technologies and skills to international markets. Several of CSA's space technologies have been adapted for use in terrestrial healthcare and public health, paving inroads into 'exponential medicine'. Notable examples include spin-off robotics technologies, communication and observation satellites, global positioning systems and remote sensing devices.

18.3.1 Robotics

Canadarm2 was designed by MacDonald, Dettwiler and Associates Ltd (MDA) for the CSA to manoever payloads onto the International Space Station (ISS), and is a prime example of how Canadian space technology has been adapted for use in terrestrial healthcare.[12] Launched into space in 2001, Canadarm2 has catalyzed several spinoff technologies now applied in specialized surgery and diagnostics across Canada. One example of a spinoff technology is NeuroArm, the world's first robot capable of performing surgery inside magnetic resonance imaging (MRI) machines. NeuroArm uses miniaturized tools such as laser scalpels with pinpoint accuracy and is capable of performing soft tissue manipulation, needle insertion, suturing and cauterization.[13] Another adaptation of Canadarm2 for healthcare is the Image-Guided

[11]Canadian Space Agency (CSA). Space, Health and Innovation: Emerging challenges, new opportunities and benefits to society. November 29–30, 2017, John H. Chapman Space Centre, Saint-Hubert, Quebec.

[12]Canadian Space Agency (CSA). Canadarm2 Spinoff Technology Transforming Surgery on Earth. Available at: http://www.asc-csa.gc.ca/eng/iss/canadarm2/canadarm2-spinoff-technology-transforming-surgery-on-earth.asp.

[13]Canadian Space Agency (CSA). 2015. Robotic arms lend a healing touch: neuroArm and its legacy. Available at: http://www.asc-csa.gc.ca/eng/canadarm/neuroarm.asp.

Autonomous Robot (IGAR), a digital surgical tool aimed at improving access and precision of minimally invasive surgeries for breast cancer. Working with an MRI scanner, IGAR allows radiologists to identify surgical access points and precision pathways for automated robotic biopsies and ablation. The diagnostic tool has shown to be highly cost-effective and is enabling a new era of minimally invasive excision and treatment of small tumours.[14] Canadarm2 has inspired the development of KidsArm, a next-gen surgical tool built at SickKids Hospital's Centre for Image-Guided Innovation & Therapeutic Intervention (CIGITI). KidsArm is the first image-guided robotic surgical arm in the world specifically designed for pediatric surgery, and aims to assist doctors in fetal, cardiac, neurological and urological surgeries.[15] Most recently, Canadarm2 technology is now being adapted for use in digital microscopy to improve hospital diagnoses and clinical workflow.

18.3.2 Satellites

Canada currently has five commercial satellites in operation (Anik F1, Anik E1, Anik E2, Nimik and MSAT). Most recently, CSA successfully launched The Canadian Maritime Monitoring and Messaging Microsatellite (M3MSat) in 2016.[16] CSA is incubating CubeSat technologies in partnership with universities across Canada[17] and together with the University of Alberta successfully deployed Ex-Alta 1 CubeSat into space which offers potential to catalyze faster, smaller, smarter and cheaper commercial technologies for terrestrial applications. CSA's satellites are being used for both communication and observation purposes, contributing to a global knowledge economy and enabling advanced support in search and rescue teams; ships and aircraft geopositioning information; and cross-geographic communication purposes.

In public health, satellite communications have enabled to e-health, telehealth and telemedicine by enabling digital or telephone communication in rural and remote areas. These capabilities have increased access to healthcare education and delivery in settings with limited telephone infrastructure, allowing for novel applications domestically in Canada's northern geographies as well as internationally. CSA satellites have been used for Earth observation and used to track environmental determinants of health, climatic and weather variables essential to understanding trends in vector-borne and water-borne infectious diseases; zoonotic disease and human-wildlife interaction; agricultural productivity and food security; land-use land-cover

---

[14]Canadian Space Agency (CSA). 2015. Robots from space lead to one-stop breast cancer diagnosis and treatment. Available at: http://www.asc-csa.gc.ca/eng/canadarm/igar.asp.

[15]Canadian Space Agency (CSA). 2015. Canadian Space Technology to Help Sick Children. Available at: http://www.asc-csa.gc.ca/eng/canadarm/kidsarm.asp.

[16]Canadian Space Agency (CSA). 2016. Maritime Monitoring and Messaging Microsatellite (M3MSat). Available at: http://www.asc-csa.gc.ca/eng/satellites/m3msat/default.asp.

[17]CubeSats. Last modified 2017-12-12. Available at: http://www.asc-csa.gc.ca/eng/satellites/cubesat/default.asp.

change and ecosystem health services; and poverty, urbanism and the social determinants of health.[18] Global Positioning System (GPS) technology has now become ubiquitous in hand-held mobile phone technology, allowing public health officials and scientists to further pinpoint the precise location of reported risks and gather insights into human behaviour and movement. When combined with population, transportation and infrastructure data, satellite observations can provide real-time tracking, visualization and spatial analysis of public health risks. The Public Health Agency of Canada (PHAC) has used satellite technology to assess domestic water contamination risks in recreational lakes, to evaluate the risk of vector-borne diseases, such as Lyme disease and West Nile virus, and to monitor the spread of pandemic disease outbreaks.[19]

18.3.3 Remote Sensing

MacDonald, Dettwiler and Associates Ltd. (MDA) and CSA collaborated to develop RADARSAT-2 one of world's most advanced commercially available Earth remote sensing technologies. Generating data over expansive territories of land over time, RADARSAT-2's remote sensing produces a high-quality data product that has many applications in marine surveillance, disaster management and defense. RADARSAT-2 has strengthened capabilities for mapping, allowing for the creation of Digital Elevation Models (DEM)s, the detection and mapping of centimetre-scale movements at the Earth's surface (InSAR), and the extraction and identification of features to support environment management and security. The RADARSAT Constellation is CSA's latest evolution of the RADARSAT Program, and aims to ensure data continuity, improve operational use of Synthetic Aperture Radar (SAR) and improve system reliability. The three-satellite configuration is planned for launch in 2018, and will provide daily revisits of Canada's land and marine territory, as well as 90% of the world's surface.[20]

Canada has used RADARSAT-2 technologies to monitor changes in vegetation, humidity in soil, crop productivity and disease transmission patterns. Currently, Canadian researchers are using RADARSAT to improve detection and prevention of urban health risks posed by high heat, air pollution and emergence of vector-borne diseases.[21] RADARSAT-2 images are used to characterize the pattern of urban landscapes to predict areas susceptible to higher heat and health-related risks. It is being

[18]Olson SH, Benedum CM, Mekaru SR, Preston ND, Mazet JAK, Joly DO, and JS Brownstein. 2015. Drivers of Emerging Infectious Disease Events as a Framework for Digital Detection. Emerging Infectious Diseases. 21 (8), 1285.

[19]Space technologies for Health. *Bull World Health Organ* 2015; 93:519–520.

[20]Canadian Space Agency (CSA). RADARSAT Constellation. Available at: http://www.asc-csa.gc.ca/eng/satellites/radarsat/default.asp.

[21]Canadian Space Agency (CSA). Using satellites to identify potential health risks in cities. Updated: 2017-05-19. Available at: http://www.asc-csa.gc.ca/eng/blog/2017/05/19/using-satellites-to-identify-potential-health-risks-in-cities.asp.

used to identify areas where air is stagnating, posing potentials risks for cardiovascular and respiratory diseases, and to locate sites prone to mosquito breeding and transmission of vector-borne diseases. Satellite imagery has also proven useful for observing changes in ice cover and its impact on food security and health in the north, resulting from both transportation of supplies, drowning risk, and access to hunting grounds.

18.4 Synergies with Foreign Aid

Global Affairs Canada manages Canada's diplomatic and consular relations, promotes the country's international trade and leads Canada's international development and humanitarian assistance. One of the hallmarks of Canada's foreign aid approach is the promotion of knowledge transfer and capacity building. CSA-incubated science and technology has been applied in both domestic and international contexts to sustainably improve the lives of vulnerable groups, especially for children, women, Indigenous populations and the elderly. We reviewed space technology applications relevant to Canada's foreign aid priorities and highlight how these may strengthen Canada's global health role in areas of pandemic preparedness, food and nutrition security, disease prevention, disaster relief and human security. CSA innovations in space science and technology have significantly contributed to Canada's foreign aid capabilities and have the potential to shape Canada's future contributions to global health.

The following case studies highlight pandemic preparedness, food and nutrition security, disease prevention and disaster relief.

18.4.1 Pandemic Preparedness

Global Affairs Canada has an overt focus on pandemic preparedness for influenza, including minimizing serious illness and overall deaths by reducing the spread of infection through promotion of individual and community actions; protecting the population through strategic provisioning and positioning of pandemic vaccines and implementation of other public health measures; and providing treatment and support for large numbers of persons while maintaining other essential health care. A key aspect of this is ensuring evidence-informed decision making by supporting partnering countries in the rigorous collection, analysis and dissemination of surveillance and scientific information for use by health professionals and public health officials. Collectively these activities contribute to Canada's efforts towards achieving UN SDG 3: ensuring healthy lives and promote well-being at all ages; and SDG11: make cities and human settlements inclusive, safe, resilient and sustainable.

Canada has been at the forefront of global pandemic preparedness under the WHO International Health Regulations (IHRs). The IHRs (2005) govern international surveillance and response to public health emergencies of international concern[22] and require States Parties abilities to monitor, detect, assess and report public health hazards in a way that does not adversely impact cross-border travel and trade.[23] A comprehensive evaluation of the functioning of the WHO's IHR(2005) detection process emphasized the need for increased capacity building of States Parties in surveillance,[24] and notably in low-income countries and fragile states.[25] Canada pioneered digital biosurveillance for enhancing early detection and response to public health emergencies through the inception of the Global Public Health Intelligence Network in the late 1990s.[26] Many public health events notifiable under the IHR(2005), such as SARS, H1N1, MERS-CoV and Ebola, have been identified through informal sources.[27,28,29] Web-based queries and participatory systems also produce cost-effective data for syndromic surveillance.[30,31]

New data products derived from CSA Earth observation satellites enable further risk characterization based on population, climatic and environmental determinants

---

[22]World Health Organization (WHO). Fifty-eighth World Health Assembly Resolution WHA58.3: Revision of the International Health Regulations. Available at: http://www.who.int/ipcs/publications/wha/ihr_resolution.pdf.2005.

[23]World Health Organization (WHO). WHO's Interim Guidance for the Use of Annex 2 of the IHR (2005): Decision instrument for the assessment and notification of events that may constitute a public health emergency of international concern. Available at: http://www.who.int/ihr/Annex_2_Guidance_en.pdf.2008.

[24]Anema A, Druyts E, Hollmeyer HG, Hardiman MC, Wilson K. Evaluation of the Functioning of the International Health Regulations (2005) Annex 2. Global Health. 2012; 8(1).

[25]Sturtevant JL, Anema A, Brownstein JS. The new International Health Regulations: considerations for global public health surveillance. Disaster Med Public Health Prep. 2007 Nov; 1(2):117–21.

[26]Dion M, AbdelMalik P and Mawudeku A. Big Data and the Global Public Health Intelligence Network (GPHIN). Canada Communicable Disease Report (CCDR). Volume 41–9, September 3, 2015. Available at: https://www.canada.ca/en/public-health/services/reports-publications/canada-communicable-disease-report-ccdr/monthly-issue/2015-41/ccdr-volume-41-9-september-3-2015-data/ccdr-volume-41-9-september-3-2015-data-1.html.

[27]Olowokure B, Pooransingh S, Tempowski J, Palmer S, Meredith T. Global surveillance for chemical incidents of international public health concern. Bull World Health Organ. 2005; 83(12): 928–34.

[28]Barboza P, Vaillant L, Mawudeku A, Nelson NP, Hartley DM, Madoff LC, Linge JP, Collier N, Brownstein JS, Yangarber R, Astagneau P. Early Alerting Reporting Project of the Global Health Security Initiative. Evaluation of Epidemic Intelligence Systems Integrated in the Early Alerting and Reporting Project for the Detection of A/H5N1 Influenza Events. PLoS One. 2013; 8: 1–9.

[29]Anema A, Kluberg S, Wilson K, Hogg RS, Khan K, Hay SI, Tatem AJ, Brownstein JS. Digital surveillance for enhanced detection and response to outbreaks. *Lancet Infect Dis*. 2014 Nov; 14(11):1035–7.

[30]Hulth A, Gustaf R, Annika L. Web queries as a source for syndromic surveillance. PloS One. 2009; 4(2): e4378.

[31]Freifeld CC CR, Mekaru SR, Chan EH, Kass-Hout T, Lacucci AA, Brownstein JS. Participatory Epidemiology: Use of Mobile Phones for Community-Based Health Reporting. PloS Med. 2010; 7(12): e1000376.

of health, and predictive models for pandemic preparedness planning.[32] The field of tele-epidemiology refers to epidemiology and space technologies applied to human and animal health and is driving new understandings of environmental determinants of infectious disease outbreaks. CSA's RADARSAT-2 imagery has been used in Canada to effectively monitor vectors associated with West Nile Virus and Lyme disease, and other similar applications.[33] Beyond our borders, satellite imagery and tracking is providing insights into the seasonality of avian migration in relation to the spread of influenza and other pathogens including malaria and cholera. These processes have the potential to connect Canada to disease outbreaks in neighbouring countries via international flyways.[34] Partnerships between CSA, WHO, and the Public Health Agency of Canada are advancing applications of Earth observation for public health decision making and creating a basis for increased tele-epidemiology applications in global health, including for monitoring of public health emergencies under the IHR, such as Ebola and Zika. These partnerships are further essential for informing public health decision-support systems to facilitate timely response and support to States Parties requiring emergency assistance.

Canada is now poised to enter the era of low earth orbit (LEO) satellites, which will contribute to disaster-resilient connectivity, spanning crises from public health emergencies to natural disasters.[35] These LEO communication solutions by groups such as OneWeb, followed by a larger SpaceX fleet, will allow direct support from satellite to commodity cell phones. Canadian companies such as Telesat Canada and Kepler Communications are also in the milieu. This will provide high-bandwidth voice and data connectivity that is largely impervious to terrestrial disruptions to grid or cable infrastructure. There is funding for LEOs in the 2018 budget for $100 million over the following 5 years.[36]

---

[32]Brazeau, S., Kotchi, S.O., Ludwig, A., Turgeon, P., Pelcat, Y., Aube, G., Ogden, N.H. Tele-epidemiology and public health in the Canadian context. European Space Agency (Special Publication) 2016.

[33]Brazeau S, Aubé G, Turgeon P, Kotchi S, Michel P. Tele-Epidemiology: Advancing the Application of Earth Observation to Public Health Issues in Canada. 02 May 2014. Available at: https://earthzine.org/2014/05/02/tele-epidemiology-advancing-the-application-of-earth-observation-to-public-health-issues-in-canada/.

[34]Loiseau C, Harrigan RJ, Cornel AJ, Guers SL, Dodge M, Marzec T, et al. (2012) First Evidence and Predictions of Plasmodium Transmission in Alaskan Bird Populations. PLoS ONE 7(9): e44729. https://doi.org/10.1371/journal.pone.0044729.

[35]Z. Qu, G. Zhang, H. Cao, and J. Xie, "LEO satellite constellation for internet of things," IEEE Access, vol. 5, pp. 18 391–18 401, 2017.

[36]Budget for LEOs 2018. http://business.financialpost.com/telecom/ottawas-bet-on-low-earth-orbit-satellites-a-positive-step-for-rural-internet-industry.

18.4.2 Food and Nutrition Security

Canada has played a longstanding leadership role in the promotion of global food and nutrition security. It is the founding donor of Nutrition International (formerly the Micronutrient Initiative) and is the largest donor to vitamin A programs internationally. Global Affairs works in close partnership with domestic and international agencies such as UNICEF, Helen Keller International, World Health Organization, FHI360, Care Canada, Save the Children, World Vision, HealthBridge, Effect-Hope, and Action Against Hunger, to deliver essential nutrition services. It promotes private investment in agriculture through its active role in the New Alliance for Food Security and Nutrition, a global commitment launched in 2012 by the G-8. Further, Canada is a long-term supporter of the HarvestPlus and efforts to biofortify staple foods with essential micronutrients. Collectively, these efforts contribute to the UN SDG 2: To end hunger, achieve food security and improved nutrition and promote sustainable agriculture.

CSA's remote sensing capability RADARSAT-2 has several powerful features that respond directly to the needs of the agricultural sector and can be leveraged to enhance Canada's impacts in foreign aid. Dual-polarization and quad-polarization modes of RADARSAT-2 allow mapping of crop characteristics over large spatial areas and monitoring of changes in soil and crop conditions over time. Canadian support to foreign aid country partner countries should center around capacity building for the citizen use and uptake of RADARSAT-2 data to mitigate agricultural risks and promote increased production towards improved food security. Capacity building within government, non-profits and the entrepreneurial community could catalyze new innovations in agritech that would have the dual benefit of growing local economies and promoting social impact.

CSA's work in food science has led to development of numerous compact, lightweight and nutritious foods through processes of irradiation, rehydration and thermostabilization. Future deep-space missions to Mars will additionally require biological studies into the growing of plants in outer space. CSA's space tech innovations around food products have the potential to translate to hard-to-reach or austere environments on Earth, such as complex emergencies characterized by armed conflict and food insufficiency. CSA's development of food rations for space travel has generated an understanding of human physiology and diet that may be transferable to Earth in the form of innovative nutritious, low-cost food technologies. There is both a need and an opportunity for Canada to lead the way in the R&D of novel food rations, such as locally produced ready-to-eat-therapeutic-foods (RUTF), that can save lives in humanitarian aid settings with limited or no reliance on water and energy inputs. Innovations in cellular agriculture, spearheaded by the entrepreneurial community, are creating novel plant-based proteins as well as 'clean meat' in a trend that is poised to revolutionize food-tech globally. Future advances in cellular agriculture, inspired by the need for sustainable space exploration and colonization, could revolutionize food security and democratize access to a healthy, environmentally sustainable, and low-cost diet.

18.4.3 Disease Prevention

Global Affairs Canada is committed to reducing the burden of key diseases causing widespread morbidity and mortality, specifically diarrhea, HIV/AIDS, malaria and tuberculosis (TB), and works in close partnership with its government and development partners to strengthen health systems. Capacity building of front-line community health works and key personnel constitutes a key aspect of Global Affairs work to support improved prevention and treatment of diseases. CSA space technologies such as robotics, satellite communications and biological tools all have relevant applications. Canada's international public health services and are ideally positioned to improve health of vulnerable groups living in remote and isolated low-resource settings.

Successful applications of satellite communication technologies in Canada's rural and remote Northern Indigenous communities for diseases such as HIV/AIDS, TB, and community vaccination programs provide an evidence-base for scaling out these applications to international settings. CSA's communications satellites have enabled eHealth, or information technologies and patient monitoring systems that improve healthcare accessibility and efficiency. They have enhanced telehealth capabilities, such as video conferencing for patient-provider specialist consultations, and improved access to vaccination and emergency services.[37] Telemedicine applications, such as tele-operated surgical systems, are now possible using both satellite communications and robotic technology developed for Canadarm2. UNOOSA has highlighted the importance of scaling capacity building in telecommunications infrastructure to enable e-learning, e-training and telemedicine and for collection of health in sub-Saharan Africa and Asia Pacific.[38] Voice-based telemedicine, the emergence of immersive applications, combined with robotic advances, democratize access to care in rural and northern communities, and can extend the reach of health experts to the last mile. CSA has developed several space health technologies that may have applications in global health.[39] Examples of this include MicroPREP, an automated "lab-on-a-chip technology" that isolates macro-molecules such as DNA, proteins or rare cells as a first step towards personalized or precision diagnostics[40]; and Bio-Analyzer, a portable diagnostic tool to test blood, saliva, and urine, which

[37]Khan I, Ndubuka N, Stewart K, McKinney V and Mendez I. The use of technology to improve health care to Saskatchewan's First Nations communities. Can Commun Dis Rep. 2017;43(6):120–4.

[38]Committee on the Peaceful Uses of Outer Space Scientific and Technical Subcommittee Forty-eighth session, Vienna, 7–18 February 2011. Item 5 of the provisional agenda. Implementation of the recommendations of the Third United Nations Conference on the Exploration and Peaceful Uses of Outer Space (UNISPACE III). Available at: https://www.unoosa.org/pdf/limited/c1/AC105_C1_L305E.pdf.

[39]Rai, A., Robinson, J.A., Tate-Brown, J., Buckley, N., Zell, M., Tasaki, K., Karabadzhak, G., Sorokin, I.V., Pignataro, S. 2016. Expanded benefits for humanity from the International Space Station. Canadian Space Agency. Available at https://ntrs.nasa.gov/archive/nasa/casi.ntrs.nasa.gov/20170006172.pdf:

[40]Canadian Space Agency (CSA). MicroPREP: New advances in sample purification. Updated: 2017-11-29. Available at: https://www.asc-csa.gc.ca/eng/sciences/microprep.asp.

may have future applications in disaster relief situations and in medical care in remote settings.[41]

18.4.4 Disaster Relief

Global Affairs deploy humans on short notice anywhere in the world in response to situations ranging from natural disasters to complex humanitarian emergencies. The Disaster Assistance Response Team (DART) is a multidisciplinary organization composed of military members and civilians, and works in cooperation with global agencies to stabilize the primary effects of the disaster; prevent the onset of secondary effects of a disaster; support the deployment of humanitarian and sustainable recovery programs. CSA's and Global Affairs' work is particularly aligned in the area of disaster relief through operational partnerships. CSA is a founding member of the International Charter "Space and Major Disasters",[42] which pools space technologies from countries around the world to facilitate emergency response to natural disasters and support relief efforts. Spillover technologies from these applications include water making and purification advances from the space sector, and hence the potential to provide safe drinking water to reduce the burden of disease during disaster operations. *Water making* has been a specific impact of the DART program.

By providing accurate and timely information and connecting emergency response teams, satellites play a critical role supporting first responders and search and rescue teams in emergency situations. CSA's RADARSAT-2 imagery has regularly provided support to rescue teams on the ground since the Charter was founded in 2000. Recent examples of this include earthquake response in Nepal in 2015; response to Typhoon Haiyan in the Philippines in 2013; the earthquake in Haiti in 2010; the Gulf of Mexico oil spill in 2010; the tsunami in Indonesia and Thailand in 2004; and over 500 other disasters spanning the globe. To date, CSA has contributed over 1500 RADARSAT-1 and RADARSAT-2 data acquisitions in response to these disasters.

18.5 Future Directions

Canada's leadership in space technology is uniquely positioned to make a transformative impact in global health. Innovations in robotics, satellite and radar imagery have catalyzed novel applications in healthcare within Canada and have the potential to inform the country's future directions in foreign aid. Applications of space

[41]Canadian Space Agency (CSA). Bio-Analyzer: Instant biomedical results from space to Earth. Updated: 2017-12-07. Available at: https://www.asc-csa.gc.ca/eng/iss/bio-analyzer.asp.

[42]Available at: https://disasterscharter.org/web/guest/home;jsessionid=E835145BC8FC857472ECBD4F03E1C911.jvm1.

technology are particularly well suited to Global Affairs priorities in pandemic preparedness, food and nutrition security, disease prevention and disaster response, and should form an integral part of Canada's foreign aid strategy. As these 'exponential technologies' continue to be tested, validated and scaled worldwide, Canada will need to consider how it can best position the use of digital data generated by these tools to foster accountability and transparency of investments towards the UN SDGs. Partnerships between government, the private and the non-profit sectors will continue to be essential to achieve these goals. With political support and catalytic investments in healthtech, such as Grand Challenges Canada, the SD Tech Fund and the Innovation Superclusters Initiative, Canada aims to commercialize new technology products, processes and services that will improve global health for all.

Acknowledgements The authors wish to thank Audrey Barbier from the Canada Space Agency and Isabelle P Roy from Global Affairs Canada, for their contributions and review of this chapter.

Chapter 19
Open Community Approaches (OCA) for Interfacing Space and Global Health

Melanie Platz, Chandana Unnithan and Engelbert Niehaus

Abstract In this contribution, the Open Community Approach for capacity building at the interface between Space and Global Health is analysed. Therefore, the term "Open Community Approach" is defined and regarded in several contexts. Wikiversity as collaborative development environment for Open Educational Resources is thematised. Recommendations for the structure of Open Educational Resources for Space and Global Health are derived from capacity building programs published by the World Health Organisation and finally, an Australian perspective on the Open Community Approach is offered. Australia is chosen as an example of a country with diverse demographic multi-ethnic, multi-lingual, and multicultural communities, distanced from each other due to the unique geo-location.

19.1 Introduction

The United Nations Office for Disaster Risk Reductions define capacity building and capacity development as follows:

> Capacity development is the process by which people, organizations and society systematically stimulate and develop their capacities over time to achieve social and economic goals. It is a concept that extends the term of capacity building to encompass all aspects of creating and sustaining capacity growth over time. It involves learning and various types of training, but also continuous efforts to develop institutions, political awareness, financial resources, technology systems and the wider enabling environment.[1]

[1]The United Nations Office for Disaster Risk Reduction. *Terminology*, https://www.unisdr.org/we/inform/terminology, 2017. Accessed 20 March 2018.

M. Platz (✉)
Pedagogical University of Tyrol, Innsbruck, Austria
e-mail: melanie.platz@ph-tirol.ac.at

C. Unnithan
Torrens University Australia, Melbourne, Australia

E. Niehaus
University of Koblenz-Landau, Mainz, Germany

S. Ferretti (ed.), *Space Capacity Building in the XXI Century*, Studies in Space Policy 22,
https://doi.org/10.1007/978-3-030-21938-3_19

In this chapter, the Open Community Approach for capacity building at the interface between Space and Global Health is analysed. Therefore, the term "Open Community Approach" (OCA) is defined and regarded in several contexts. *Wikiversity* as collaborative development environment for Open Educational Resources is thematised. Recommendation for the structure of Open Educational Resources for Space and Global Health are derived from capacity building programs published by the World Health Organisation and finally, an Australian perspective on the Open Community Approach is offered. Australia is chosen as an example of a country with diverse demographic multi-ethnic, multi-lingual, and multicultural communities, distanced from each other due to the unique geo-location.

19.2 The Open Community Approach

The Open Community Approach (OCA) was proposed in the *Action Team 6 Follow-Up Initiative*[2] in 2012, which was transformed in 2015 to the *Expert Focus Group on Space and Global Health (EFG-SGH)* and in 2019, to the *Working Group on Space and Global Health (WG-SGH)*.[3] The analysis of OCA for capacity building at the interface between Space and Global Health leads to some logical implications, that are elaborated as connected sequences of arguments in this chapter.

To set the premise, we define the *Open Community Approach*:

> Open Community is a generalisation of the concept of Open Source to other collaborative efforts. The term "open" for an open community refers to the opportunity for anyone to join and contribute to the collaborative effort. The direction and goals are determined collaboratively by all members of the community. The resulting work ("product") is made available under a free license, so that other communities can adapt and build on them.[4]

In the context of EFG-SGH, the "product" of the Open Community is an "improved global health infrastructure by application of space technologies".

The OCA applies to:

- (OD) Open Data,
- (OSS) Open Source Software and
- (OER) Open Educational Resources/Open Content

whenever it is applicable. Nevertheless, it is obvious that not all data can be open particularly in the health domain. Data privacy protection or governmental data security regulations have to be applied as part of national policies or regulations specified by the World Health Organisation.

---

[2]Expert Focus roup—Space & Global Health. *Overview*, http://at6fui.weebly.com/, 2012. Accessed 20 March 2018.

[3]United Nations Office of Outer Space Affairs. Space for Global Health, http://www.unoosa.org/oosa/en/ourwork/psa/globalhealth/index.html, 2019. Accessed 03 March 2019.

[4]Niehaus, E. "Definition: Open Community (OC)", *Expert Focus Group—Space & Global Health*, http://at6fui.weebly.com/open-community-approach.html, 2013. Accessed 20 March 2018.

Communities that can benefit from capacity building have different requirements and constraints defined by the availability of resources and socio-cultural environments for which the capacity building is designed. Therefore, a “One Size Fits All” approach for capacity building and learning has to be replaced by tailored educational resources. To enable adaption of the OER, the geo-location of an OER-user can be helpful to ensure the attribution of regionally and culturally suitable and relevant content. The analysis of the target group and geographical region for which the OER will be adapted within a capacity building course made available. For example via Wikiversity, has to consider the following aspects:

- Does the community have access to the resources that are mentioned in the capacity building course? (e.g. access to the crop health map (NDVI) for applications in precision agriculture)
- Are the used images referring to the geographical region in which the OER will be applied?
- Is it possible to translate the OER into the local or regional language?
- Is the proposed workflow, in e.g. a video of a capacity building program, compatible with the local working environment?

19.2.1 OCA as Safety Net for Capacity Building

The OCA is a kind of “safety net” for capacity building, if the financial resources are insufficient (e.g. in developing countries). The application of the OCA in EFG-SGH has the objective to support capacity building, even if there is no or limited resources available, especially in risky situations. In this context, Open Source and Open Content plays a major role to provide access to resources. The OER are used in the context of risk management. Simplified, risk can be explained as the product of probability of a harmful event and impact a harmful event in the *One Health* context.

$$Risk = Probability \times Impact$$

Risk mitigation strategies can work on reducing the probability of an event (e.g. reduction of the infection rate) or on the impact (e.g. improved treatment after a patient was infected). The multiplicative composition of risk shows also the relevance for the access to risk mitigation resources.[5]

[5]Example: A risk mitigation strategy “A” might reduce the probability of an event by 70% and a risk mitigation strategy “B” might reduce the probability by 15% only. But if the risk mitigation strategy “A” can only be applied by a small number of people and the majority of the population can access risk mitigation strategy “B”, then risk mitigation might be more effective with “B” due to the higher impact.

19.2.2 OCA for Capacity Building to Support Small and Medium Business Enterprises

Small and Medium Business Enterprises (SMEs) cannot afford development units in many countries. Conversely, innovation is triggered by small creative teams. Shared knowledge allows the access to the most recent scientific results and the OCA for capacity building supports SMEs in the use Space Technology for their services in general. The flexibility of SMEs to react to new scientific and technical results or customer demands in a shared economy, allows to build on the most recent technology and scientific results with OER for these results, even if a research and development unit consists of a single person or a few passionate developers. In the context of space technology, SMEs can benefit from OER by the application of Space Technology for their business models to create new services.

19.2.3 OCA in a Humanitarian Context

In the context of Global Health, humanitarian benefits are of particular interest.[6] The Open Community Approach (OCA) maximises access to the data and/or creates it by collaborative mapping. For example, the *Humanitarian Open Streetmap Team (HOT-OSM)* has an Open Data Policy. In comparison to Commercial Data Harvesting, the data is owned by the community (e.g. Wikiversity), instead of the organisation. The objective of HOT-OSM is to provide mapping products that are helpful for humanitarian purposes. In the global health context, a Malaria map[7] is an example and a public transport map (e.g. for Managua[8]) can be mentioned. Collaborative Mapping is used to create that map and the mapping result is for the benefit of civil society.

[6] UNOOSA. *Thematic priority 5: Strengthened space cooperation for global health.* Document A/AC.105/1172: http://www.unoosa.org/oosa/oosadoc/data/documents/2018/aac.105/aac.1051172_0.html, 2018. Accessed 20 March 2018.

[7] HOT-OSM. *Malaria Elimination Campaign.* https://www.hotosm.org/projects/malaria_elimination_campaign, 2018. Accessed 20 March 2018.

[8] HOT-OSM. *A crowd-sourced public transportation map for Managua.* https://www.hotosm.org/updates/2016-01-07_a_crowd_sourced_public_transportation_map_for_managua, 2016. Accessed 20 March 2018.

19.2.4 OCA in the Context of Virtual Conferences by the WG-SGH

In the WG-SGH, a number of ideas are being piloted that allow participation of globally distributed networks at local or regional meeting points.[9] This approach reduces the carbon footprint of the semi virtual meeting by tele-conferencing facilities and enables participation of developing countries, even if the financial resources do not permit a physical participation, at a certain location. In accordance with the United Nations document A/AC.105/1175, an Open Universe Initiative[10] was proposed which assures the support of citizen science approaches and Open Data access in the context of Space Technology. Other open data policies (e.g. the European Union Copernicus Programme[11]) support the OCA also in the health domain.

19.3 OCA and Wikiversity

Creative Commons[12] encourages licensing models that allow reuse and adaptation. Contrary to the encyclopaedic objective of Wikipedia, Wikiversity[13] is a collaborative development environment for Open Educational Resources (OER) where most of the resources are licensed under a Creative Commons license, thus allowing for distribution, remix, and reuse of materials for the target group.

19.3.1 Access and Maintenance of a Wiki Infrastructure

Wikiversity is maintained by the Wikimedia Foundation.[14] Wikipedia is the well-known "digital encyclopedia" which is listed often amongst the first entries of search engine queries. Wikiversity is the product for "capacity building and learning" and the Wiki community is considerable. Maintenance of a web infrastructure (patch

---

[9]Platz, M., Rapp, J., Größler, M. & Niehaus, E. "Open Community Approach for Capacity Building to improve Public Health through Collaborative Mapping for Risk Management and tailored allocation of available resources". *Post AT6FUI-Meeting 2016*, http://at6fui.weebly.com/uploads/1/5/2/6/15264308/easychair_at6fui_platz.pdf, 2016. Accessed 20 March 2018.

[10]UNOOSA. *Report on the United Nations/Italy Workshop on the Open Universe initiative.* Document A/AC.105/1175: http://www.unoosa.org/oosa/oosadoc/data/documents/2018/aac.105/aac.1051175_0.html, 2018. Accessed 20 March 2018.

[11]ESA. „About Sentinel Online". *Sentinel Online*, https://sentinel.esa.int/web/sentinel/about-sentinel-online. Accessed 20 March 2018.

[12]Creative Commons. *What we do*, https://creativecommons.org/about/. Accessed 20 March 2018.

[13]Wikiversity. https://www.wikiversity.org. Accessed 20 March 2018.

[14]Wikimedia Foundation. https://wikimediafoundation.org/wiki/Home. Accessed 20 March 2018.

management of servers, backup management, etc.) is performed by the Wiki foundation. To build an own wiki infrastructure is technically possible, but it duplicates the efforts to maintain the server infrastructure and splits wiki communities' efforts, for reviewing, authoring and community based quality assurance, to different resources. Furthermore, "reinventing the wheel" weakens the collaborations. That is, similarly to the existing Wikipedia infrastructure, development of concurrent wikis to Wikiversity resources are not useful in terms of maintenance costs. Furthermore, the community of practice of the WG-SGH in Wikiversity will not sacrifice the strength of a linked community for the ownership of a small wiki web portal, with a non-sustainable small community. Embedded in a larger community, it becomes easier for the community to contribute with cross-disciplinary improvement of joint resources and support user-driven innovation with Space Technology in the context of Global Health.[15] Risk Management is one underlying Module.[16] The Sustainable Development Goals[17] define the key objectives for the United Nations activities. In Wikiversity, learning resources can be assigned to SDGs by categories. The categories automatically aggregate all contents that address a specific SDG. As an example, Niehaus created a small sample for "Water", because water is a crucial resource for life and Global Health and the content is very simple, so that it is easy to understand[18] (in comparison to spectral analysis of Satellite Images, for example). The Water learning resource was created in Wikiversity and linked to the SDG6 Clean Water and Sanitation.

19.3.2 Wikiversity's Non-formal Education Portal

Non-formal education is the official umbrella term for all education which takes place outside the formal sector, where formal education comprises the provision of education at pre-school, primary, secondary and tertiary levels. Non-formal education includes learning alongside and outside schools and universities, such as sports and hobbies. It also includes much of professional and vocational education, adult literacy training and lifelong learning. The production of OER goes beyond even the organizations that create the materials. There are thousands of producers involved in the open education movement, though many of them participate without realising it. Every time material is created and licensed under a Creative Commons or other

[15]User:Bert Niehaus (Creator). "Expert Focus Group for Space and Global Health." *Wikiversity*, https://en.wikiversity.org/wiki/Expert_Focus_Group_for_Space_and_Global_Health. Accessed 20 March 2018.

[16]User:Bert Niehaus (Creator). "Risk Management". *Wikiversity*, https://en.wikiversity.org/wiki/Risk_Management. Accessed 20 March 2018.

[17]Griggs, D., Stafford-Smith, M., Gaffney, O., Rockström, J., Öhman, M. C., Shyamsundar, P., … & Noble, I. "Policy: Sustainable development goals for people and planet". *Nature*, *495*(7441), 2013, 305.

[18]User: Bert Niehaus (Creator). "Water". *Wikiversity*, https://en.wikiversity.org/wiki/Water. Accessed 20 March 2018.

"open" license, there is potential for that material to be used in an educational setting. For example, many of the images found in Utah State University's OpenCourseWare courses[19] were found on the *Flickr*[20] image website using the Creative Commons search. These images were originally taken for some other purpose, but because the creator chose a Creative Commons license, they were easily and legally re-used on an OpenCourseWare site. Project Gutenberg,[21] a collection of over 20,000 public domain books, is another example of how open educational resources can be re-purposed for education. Two public domain plays from the readings section of the Utah State University OCW Theatre Arts course have been made available in their entirety in this manner. Creative Commons and other open licenses allow material to have the potential to be readily re-used in an OpenCourseWare course or other educational product. "Creative Commons defines the spectrum of possibilities between full copyright—all rights reserved—and the public domain—no rights reserved"[22] In most cases, the only restrictions are that the original producer be given attribution, that the work may not be used for commercial purposes, and that adaptations of the work be shared with the community.

19.4 Application of WHO Structure for OER

The recommendation for the structure of OER for Space and Global Health can be derived from capacity building programs published by the World Health Organisation. The capacity building program "Clean Care Safer Care" initiative[23] offers a background for learning resources. Clean Care Safer Care targets the risk mitigation for health care-associated infections (HCAIs). The reduction of HCAI in health-care delivery is a global challenge that is relevant for developed and developing countries. Furthermore, scientific evidence and rationale are provided for the capacity building programme from WHO that serves as a foundation and as the rationale for the application of the methodology in health care facilities and home care services. An information centre provides access to resources, that support health care facilities to implement the methodology.

Similar to this WHO approach for HCAIs, the development of OER for Space and Global Health requires a health problem that affects humankind on a global scale. For

[19]Utah State University. *Utah State OpenCourseWare*, http://ocw.usu.edu/. Accessed 20 March 2018.

[20]Flickr. *About Flickr*, https://www.flickr.com/about. Accessed 20 March 2018.

[21]Project Gutenberg. *Free ebooks - Project Gutenberg*, http://www.gutenberg.org/. Accessed 20 March 2018.

[22]Caswell, T., et al. „Open content and open educational resources: Enabling universal education." *The International Review of Research in Open and Distributed Learning*" 9.1, 2008.

[23]Pittet, D., & Donaldson, L. "Clean care is safer care: the first global challenge of the WHO World Alliance for Patient Safety". *American journal of infection control, 33*(8), 2005, 476–479. See also: World Health Organisation, Clean *Care Safer Care*, http://www.who.int/gpsc/background/en/, 2005. Accessed 20 March 2018.

Space and Global Health, we chose Chronic Kidney Disease of non-traditional causes (CKDnt). Traditional causes for Chronic Kidney Disease are mainly hypertension and obesity, while the causes cannot be responsible for most of the cases in Sri Lanka, Nicaragua, El Salvador, Guatemala, mainly found in male farm workers. The key questions are:

- Are nephrotoxic agrochemicals used?
- Are the farm workers exposed to these agrochemicals (hard manual work)?
- What are applicable risk mitigation strategies incorporating all other known risk factors (e.g. heat stress)?
- What is the official number of cases that are reported via the national health system and what fraction of population cannot access the national health system for medical support due to geographical distance or financial constraints? Especially with a low income the probability increases that CKD cases will not appear in the official statistics.

Currently, the CKDnt can be regarded as a multi-factorial disease with a high prevalence in the male population working in agriculture. The application of a precautionary principle could lead to the application of a precision agriculture approach that aims for the reduction of applied agrochemicals by keeping the same harvest yield. Precision agriculture uses remote sensing data, for example the Normalized Difference Vegetation Index (NDVI), to detect crop health. The detected spatial patterns of crop health determine the application rate of agrochemicals. The methodology is widely used in agricultural production with GPS and automated application according to the NDVI. Low-Cost precision farming transforms the application to smartphones and manual work. The transformation follows the same principles as the application with tractor high-tech agricultural production: apply less agrochemicals for the same harvest yield. NDVI maps can be encoded in a JSON database with geocoded application rates. Access to the GPS sensor can be accessed by a WebApp which in turn evaluates the NDVI map at the geolocation.[24]

19.4.1 Open Source WebApp and Linux Distributions

The WebApp itself can be shared as HTML5 bundle zip via blue tooth without internet connectivity if the produced data and maps are generated from open data (satellite data), or from collected ground truth data via the smartphone itself, prior to the application of agrochemicals. The open data policy (e.g. EU Copernicus Programme) allows a participatory approach for generation of pesticide application maps with Open Source operating systems (e.g. the OSGEO Linux distribution[25]). The Linux distribution downloads available remote sensing data and calculates the

[24]w3schools.com. *HTML5 Geolocation*, https://www.w3schools.com/html/html5_geolocation.asp. Accessed 20 March 2018.

[25]OSGEO Live Linux Distribution, https://live.osgeo.org/. Accessed 20 March 2018.

NDVI. Applied on open research at a university methodology, algorithms are Open Source, and recently announced or published optimization of crop health detection could encourage users to create new releases of the algorithm or to remix existing and available libraries in a new version of the OSGEO Linux distribution and publish the Linux distribution on their own site or on Open Source Linux.[26] Open and Sustainable Learning accompanies the Open Source content with OER that support the learners in applying the provided technology. Commercial settings of the publications of educational and/or scientific resources create either costs for publication on the author site or costs for readers to access the knowledge. These commercial limitations lead logically to publication in open access journals that neither charge the author, nor the reader, nor learner. The remix requirement of the OCA leads furthermore to an open publication strategy for the developed technologies that leverage open licenses and encourage users to build and share custom collections of open materials. Wikiversity can be regarded as an open repository of learning materials produced for self-learners, students and non-formal educational settings, for example farm workers. At the same time, it is designed for faculty members to build on existing technologies and existing scientific results.

19.4.2 Scientific Publications as OER

The WikiJournal for Medicine[27] was one the first peer reviewed journals in Wikiversity. Other WikiJournals followed that development to offer publication options for other scientific disciplines as well. Applications for Space and Global Health can be published, if the authors provide free access for authors and readers, without financial constraints as a scientific asset.

Scientific publications are educational resources for tertiary education. Researchers learn from the recent scientific results and build new experiments on the recent scientific and technical evolution. The government pays for commercially published scientific results multiple times:

- Government pays for the researcher that generates new scientific results,
- Government pays for the researchers that perform the peer-reviewing of submitted scientific results,
- Government pays for the access of the publication in the libraries of universities so that other researchers can build their new results on it.

The Wiki Journal concept in Wikiversity fulfills these OCA requirements and provides access to publication facilities at no cost for authors and readers.

---

[26]PINGUY. *Pinguy Builder—An App to Backup/Remix *buntu*, https://pinguyos.com/2015/09/pinguy-builder-an-app-to-backupremix-buntu/, 2015. (Tool to create a tailored Linux Distribution). Accessed 20 March 2018.

[27]User:Mikael Häggström (Creator). "WikiJournal of Medicine". *Wikiversity*, https://en.wikiversity.org/wiki/WikiJournal_of_Medicine. Accessed 20 March 2018.

A major concern of scientists in a Wiki environment is that a specific article will be changed over time and does not show the referenced content after a period of time. These concerns are addressed with the Wikipedia/Wikiversity citation module. The underlying MediaWiki, that is used for Wikipedia and Wikiversity has a versioning system by default to rollback disruptive changes or any other unwanted alteration of the Wikipedia or Wikiversity content. The versioning system of the MediaWiki assigns a unique ID to all versions of MediaWiki documents. This technique allows the citation of a single version of a page in Wikiversity or Wikipedia. The link will always show the same reference content at a specific time index.

19.5 An Australian Perspective on OCA

In Australia, holistic health approaches to managing population health are a key priority, as the population is living in remote locations. Therefore, there is a need to understand where and how people live to inform health planning. Epidemiological data that records location of people with different health conditions helps in healthcare infrastructure planning. Spatial data analysis helps the country in understanding such questions as "is the cluster of disease just due to chance?" and optimally, implementing strategies for prevention. As a geographic region that is relatively isolated in the world, there are specific needs in Australia that can be supported with ICTs for capacity development.

19.5.1 The SERVAL Project

The SERVAL project is such an approach that fundamentally believes that communications must be available to the financially disadvantaged, who may need it the most. SERVAL is a telecommunications system comprising minimum two mobile phones, which are able to work outside of regular mobile phone tower ranges, using the SERVAL App and SERVAL Mesh.[28] Typically, mobile phones will not work without cellular infrastructures. The SERVAL Mesh allows mobile phones to form spontaneous networks, consisting only of two mobile phones. It allows people who are close by to keep communicating in times of emergencies, without the cellular infrastructure. In Australia, 75% of the geographic space still lacks mobile tower coverage. Therefore, allowing mobile phones to form stand-alone networks is a cost-effective method that can be offered to communities in remote areas, in times of need.

In this century, private conversations with medical service providers, friends and families for discussing medical issues are vital. The SERVAL Mesh foundation was engineered to support security, with end-to-end encryption using strong 256-bit ECC

[28]Serval Project. *What is the Serval Project?* http://servalproject.org, 2017. Accessed 20 March 2018.

cryptography, and these calls works on low cost Android phones. This would render telemedicine services possible in regions that are not only remote, but also prone to environmental challenges. In October 2017, the SERVAL team set up the mesh extender capability within the New Zealand Red Cross, for their IT and telecoms emergency response unit (IT&TERU).[29]

In late 2018, a project funded by the Humanitarian Innovation Fund (HIF) through the UK foreign aid program was completed by the SERVAL project team which reduced the costs of early warning systems significantly. The aim was to enable communications for early warning on tsunamis, cyclones, bush fires and other hazards into very remote locations (in particular the Indo-Pacific region). The team addressed the problem from the perspective of working in resource-constrained environments.

An Off-the-Shelf satellite TV receiver part was combined with signal processing by Othernet.is[30] so that no dish was required. It was pointed to approx 10° of the geostationary satellite which provided a low-bit-rate digital broadcast signal that can be received over a wide area. Added to this was the direct-leasing of satellite capacity, so that the operating cost of the system was kept low, and there was no monthly charge per user which meant that the solution could be scaled up, region wide. A cheap automotive air-horn was then used to hear the alarm within a village. The result was that the receiver hardware was 100 times cheaper than older approaches and could be installed by communities themselves which otherwise had to rely on big programs. Furtheremore, a low-power FM radio transmitter could be installed so that news, weater, climate change mitigation and other information could be received in each village. When the unit stops working (when the project funding ends), the village community could organise themselves to replace and reinstall the transmitter, without having to rely on any aid program. In summary, the solution was more appropriate to the needs of the Asia Pacific region. Conversely, the solution offered great potential for bush-fire early-warning systems in Australia, where cellular networks are often displaced or removed by large bush fires. Households in these areas could install the cheap solution themselves without financial stress.

19.5.2 Spatial Analysis and Mapping for Public Health in Australia

Spatial analysis and mapping includes community profiling and demographic analysis to understand the varied population groups and resources in specific catchment areas and identify locations with selected attributes.[31] Location analysis is

---

[29]Gardner-Stephen, P. „Setting up Mesh Extender capability within NZ Red Cross“. *Enabling Communications, Anywhere, Anytime*, https://servalpaul.blogspot.com.au/2017/10/setting-up-mesh-extender-capability.html, 2017. Accessed 20 March 2018.

[30]https://othernet.is.

[31]Data Analysis Australia. *Spatial Analysis and Mapping*, https://www.daa.com.au/expertise/mapping/. Accessed 20 March 2018.

then used to optimise the location of services and infrastructure by identifying sites with surrounding populations or resources that best match a set of criteria. Spatial demand modelling renders it possible to determine where services and infrastructure is required the most, by locating high demand areas or gaps in health services. For example, governments in the health sector have long term plans in capital investment (such as community health centres or hospitals). The infrastructure has to be suitable in design and location to meet the needs of the community, at present and in future. In Western Australia, such a mapping exercise was done to support the health infrastructure.[32]

In 2017, Australia adopted the International Open Data Charter.[33] This fortifies the Australian commitment to the open data agenda to enable sharing and learning best practices on international data initiatives, in collaboration with research, private and non-government sectors, as well as the Australian citizens. As the immigrant population rises and the public health infrastructure needs to be rebuilt according to the needs of new citizens, the usage of open data enabled by space technologies, enables healthier, smart cities development. Since the concentration of immigrant population aims at living in the cities, this approach helps further in mapping the location and health requirements of such population groups.

The newly formed Australia Space Agency,[34] along with the government, has many initiatives that assist in disaster management. *The Digital Earth Australia*[35] (DEA) is an analysis platform for satellite data and other forms of Earth observation data, operated by Geoscience Australia. It has broad applications relevant to monitoring, reporting, and disaster management. DEA is Australian government's implementation of the open source analysis platform developed as part of the *Open Data Cube* (ODC)[36] initiative—a free and open source initiative to provide a solution that has value to its global users and increases the impact of satellite images. ODC deployments are underway in South Africa, Vietnam, and Colombia, to name a few.

To enable open data sharing, many of the state governments in Australia have begun offering free WiFi in cities.[37] In alignment with the Living Labs approach, (promoted by WG-SGH) in public health risk mitigation,[38] this approach to offer free WiFi enables co-creation of data by citizens, to build health infrastructures, as required by them, enabled and supported via space technologies and mapping.

---

[32]Department of Health (Western Australia). *Health Status data for Western Australia*. https://data.gov.au/dataset/health-status-data-for-western-australia, 2016. Accessed 20 March 2018.

[33]samira.hassan. "Australia adopts the International Open Data Charter". *Australian Government*, https://blog.data.gov.au/news-media/blog/australia-adopts-international-open-data-charter, 2017. Accessed 20 March 2018.

[34]Australia Space Agency: https://www.industry.gov.au/strategies-for-the-future/australian-space-agency Accessed 22 February 2019.

[35]Digital Earth Australia: http://eos.ga.gov.au. Accessed 22 February 2019.

[36]Open Data Cube, http://www.ga.gov.au/dea/odc. Accessed 22 February 2019.

[37]Invest Victoria. "Victoria's free public WiFi and future plans". *The Victorian Connection*, http://connection.vic.gov.au/victorias-free-public-wifi-and-future-plans/, 2017. Accessed 20 March 2018.

[38]Expert Focus Group—Space & Global Health. *Living Labs*, http://at6fui.weebly.com/living-labs1.html. Accessed 20 March 2018.

19.6 Conclusion

The logical analysis has shown, that the application of the OCA leads to the development of OER in a versioned repository like Wikiversity. It is a repository for OER and at the same time a WikiJournal environment with the application of peer-reviewed journals. Scientific publications can be regarded as educational resources for tertiary education, in which the evolution of science is represented by the evolution of a scientific publication in a versioning system. Static links for specific versions allow a citation concept that assures the content display of the cited version at the time index of citation. Nevertheless, the OER analysis of this chapter itself would lead to the recommendation, that this manuscript should be published in a WikiJournal.

According to this logical dilemma, all images that could improve the comprehension of this text are published on Wikiversity,[39] so that the scientific community will not lose the right to build and remix the content of the OER. Furthermore, the logical conclusion of the mathematical definition of risk is, that the OER and the scientific results should be published under an open access license, so that the scientific community can access the resources without financial constraints and the access to the content is maximized.

OCA and utilisation of spatial data assists Australia in managing the health infrastructure for their diverse communities, living in remote, distant locations from each other. OCA based technologies such as SERVAL are extremely useful in connecting people in times of emergencies, particularly in environmentally challenging situations. Similarly, initiatives such as DEA and ODC offer satellite-based GIS solutions that are invaluable in disaster management. Conversely, as research has shown, mapping the varied communities through free Wi-Fi renders co-creation and evaluation of health infrastructure needs, enabled by space technologies, for the future.

[39]Wikiversity. Expert Focus Group for Space and Global Health, https://en.wikiversity.org/wiki/Expert_Focus_Group_for_Space_and_Global_Health. Accessed 03 March 2019.

Part III
Climate Change and Resilient Societies

Chapter 20
The World Meteorological Organization and Space-Based Observations for Weather, Climate, Water and Related Environmental Services

Werner Balogh and Toshiyuki Kurino

Abstract This chapter provides an overview of the space-related activities of the World Meteorological Organization (WMO) and its Space Programme. The WMO Space Programme was established by the fourteenth World Meteorological Congress in 2003. Its overall objective is to increase the effectiveness and contributions of space-based observing systems to WMO Programmes and to coordinate the related meteorological and environmental satellite matters and activities. As such, the WMO Space Programme acts as a bridge between satellite operators and users and is supported by a dedicated WMO Space Programme Office (SAT) consisting of the Space-Based Observing System Division (SBOS) and the Satellite Data Utilization Division (SDU). A major aim of the Programme is to help achieve the WMO Vision for the WMO Integrated Global Observing System (WIGOS) in 2040 by implementing the space-based observing system component of WIGOS, and to enhance the capacity of WMO Members to translate space-based data and products into societal benefits. The objective of the Programme is achieved through strong partnerships with the Coordination Group for Meteorological Satellites (CGMS) and the Committee on Earth Observation Satellites (CEOS) and their respective working groups and subsidiary bodies.

Disclaimer: The views expressed herein are those of the author(s) and do not necessarily reflect the views of the United Nations or of any of its specialized agencies. The designations employed and the presentation of material in this publication do not imply the expression of any opinion whatsoever on the part of the Secretariat of the United Nations or of any of its specialized agencies concerning the legal status of any country, territory, city or area or of its authorities, or concerning the delimitation of its frontiers or boundaries.

W. Balogh (✉) · T. Kurino
World Meteorological Organization (WMO), Geneva, Switzerland
e-mail: wbalogh@wmo.int

S. Ferretti (ed.), *Space Capacity Building in the XXI Century*, Studies in Space Policy 22,
https://doi.org/10.1007/978-3-030-21938-3_20

20.1 Introduction

The World Meteorological Organization (WMO)[1] originated from the International Meteorological Organization (IMO), which was founded as one of the first international organizations, following the Vienna International Meteorological Congress held in 1873. The purpose of IMO was to facilitate the exchange of weather information between its Member countries.

In 1950, the IMO was formally transformed into the World Meteorological Organization and became part of the United Nations system as a specialized agency, where it acts as the authoritative voice of the United Nations on weather, climate and water and related environmental services. It also provides the framework for international cooperation on related matters.

As of 2019, the membership of WMO consists of 193 Member States and Territories, grouped into six Regional Associations.[2] The main partners for WMO in its Member countries are the National Meteorological and Hydrological Services (NMHSs).

Prior to the beginning of the space age, weather information relied on data collected from observations made at land, at the sea and in the air. For many locations, in particular at sea and in the air, the available data is rather scarce and consequently limits the accuracy of weather forecasts. However, this situation began to change with the arrival of Earth-orbiting satellites and their ability to gather environmental data globally.

20.2 The World Weather Watch and Space-Based Observations

Soon after the launch of the first artificial satellite, Sputnik, on 4 October 1957, discussions began at the United Nations on how to ensure that outer space would be used for peaceful purposes only. It was decided to establish a dedicated committee of the United Nations General Assembly, which was tasked with discussing issues related to international space cooperation, reaching agreement on a legal framework for activities of countries in outer space and considering how the work of the United Nations itself could benefit from space technology and its applications.

That committee, to be known as the United Nations Committee on the Peaceful Uses of Outer Space (UNCOPUOS), held its first ad hoc meeting in 1958 and WMO participated in that meeting. Based on the report of the meeting, United States President John F. Kennedy proposed in 1961 to launch "cooperative efforts between all the nations in weather prediction and eventually in weather control" making use of

---

[1] See https://www.wmo.int.

[2] The six Regional Associations consists of: Region I: Africa, Region II: Asia, Region III: South America, Region IV: North America, Central America, Caribbean, Region V: South-West Pacific, Region VI: Europe.

space-based observations from satellites. This translated into a request to WMO by UNCOPUOS and later endorsed by the General Assembly, to prepare in greater detail a plan for the use of meteorological satellites.[3,4] WMO responded to the Committee with a proposal for a global programme to advance atmospheric science research and to develop improved weather forecasting capabilities using space technology.[5]

In 1963, the General Assembly endorsed the efforts towards the establishment of such a global programme under the name World Weather Watch (WWW) and under the auspices of WMO to include the use of satellite as well as conventional data, with data centres to facilitate the effectiveness of the system.[6] In the same year, the fourth World Meteorological Congress (Cg-IV) established the WWW with three focus areas:

(1) Global observational data coverage, to be provided by a Global Observing System (GOS);
(2) data processing systems, to be provided by a Data- processing and Forecasting System (DPFS); and
(3) a world-wide coordinated telecommunication system, to be provided by the Global Telecommunications System (GTS).

The first low Earth orbiting weather satellites were launched in the early 1960s and demonstrated the capability to provide useful measurements from space. They were followed in 1966 by the first Geostationary satellite with experiments to collect weather data and to provide data-relay and re-transmission to end-users using a Weather Facsimile (WEFAX) service.[7] By the end of the 1960s, weather satellites were mainly providing cloud images to support weather forecasts, but the potential benefits of satellite data for numerical weather prediction were already noted, including the use of satellite data for climatology and for the refinement of atmospheric models.

This potential began to be fully realized in the 1970s with the deployment of more advanced operational meteorological satellites. Today, the GOS consist of numerous satellites positioned in various orbits to provide continuous, global coverage of our planet (see Fig. 20.1).

---

[3]General Assembly resolution 1721 (XVI) C, http://www.unoosa.org/pdf/gares/ARES_16_1721E.pdf.

[4]General Assembly resolution 1802 (XVII), http://www.unoosa.org/pdf/gares/ARES_17_1802E.pdf.

[5]See Tillmann Mohr, "The Global Satellite Observing System: a Success Story", WMO Bulletin n°: Vol 59 (1) - 2010, https://public.wmo.int/en/bulletin/global-satellite-observing-system-success-story and WMO, "First report on the advancement of atmospheric sciences and their application in the light of developments in outer space", WMO, 1962, https://library.wmo.int/index.php?lvl=notice_display&id=10240.

[6]General Assembly resolution 1963 (XVIII) III, http://www.unoosa.org/pdf/gares/ARES_18_1963E.pdf.

[7]Low Earth Orbit (LEO): TIROS-1 (US, 1960), Meteor (Soviet Union, 1964), Geosynchronous Equatorial Orbit/Geostationary Earth Orbit (GEO): ATS-1 (US, 1966).

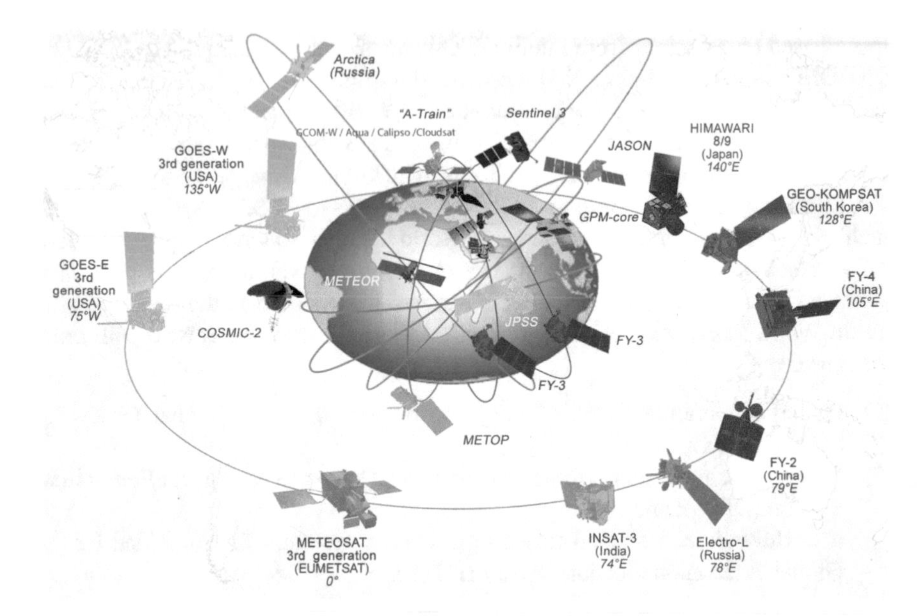

Fig. 20.1 The space-based component of the WMO Integrated Global Observing System in 2021 (*Note* The copyright for this figure is with WMO)

The Coordination Group for Meteorological Satellites (CGMS)[8] was established in 1972 to coordinate the space-based observing system component of the GOS. Under its guidance the system is evolving into a well-planned system of meteorological and environmental satellites, integrated with in situ based observation networks and supporting, in addition to the WWW, a growing range of WMO application programmes, coordinated and managed under the framework of the WMO Integrated Global Observing System (WIGOS).

New meteorological and environmental satellites are launched on a continuous base and join the global observing system. The majority of the data derived from these space-based observing systems is shared among WMO Members under open data policies and standards, disseminated through affordable and easily accessible communication networks, coordinated under the framework of the WMO Information System (WIS), which has evolved from the GTS. The DPFS is evolving into the seamless Global DPFS (GDPFS) and continues to provide the backbone for ingesting the data into processing systems and models for operational applications and services.

In addition to improving daily weather forecasts—the original aim of the WWW—satellites and their data today also provide information for policy- and decision-making related to weather, climate, water and related environmental applications, which is essential for successfully implementing the global development agendas,

[8] See https://www.cgms-info.org.

including the 2030 Agenda for Sustainable Development, the Sendai Framework for Disaster Risk Reduction and the Paris Climate Agreement.

Without question, the World Weather Watch and the other activities that have evolved from it are major success stories demonstrating the sharing of space benefits among all countries.

20.3 The WMO Space Programme

With the growing importance of the space-based observing system component, the fourteenth World Meteorological Congress in 2003 decided to initiate a cross-cutting, dedicated WMO Space Programme to increase the effectiveness and contributions from satellite systems to WMO Programmes and to coordinate the related environmental satellite matters and activities.[9,10]

The Space Programme acts as a bridge between satellite operators and users and is supported by a dedicated WMO Space Programme Office (SAT) and its Space-Based Observing System Division (SBOS) and Satellite Data Utilization Division (SDU).

Its objectives are achieved through strong partnership with the Coordination Group for Meteorological Satellites (CGMS) and the Committee on Earth Observation Satellites (CEOS)[11] and their respective working groups and subsidiary bodies.

It is built around four main components together covering the full value chain of satellite data (see Fig. 20.2):

(1) Integrated space-based observing system;
(2) Access to satellite data and products;
(3) Awareness and training; and
(4) Space weather coordination.

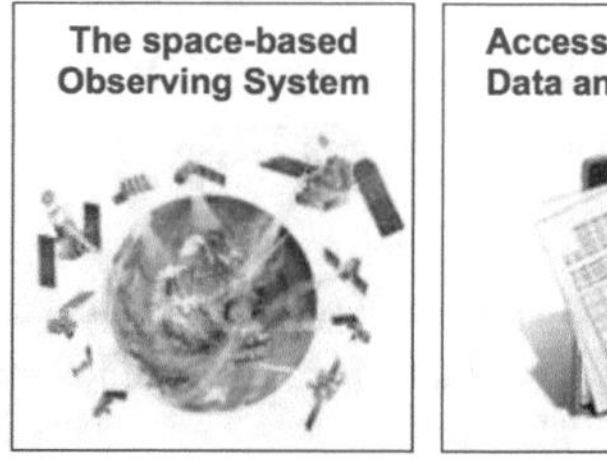

Fig. 20.2 WMO space programme components (*Note* The copyright for this figure is with WMO)

[9]WMO Resolution 5 (Cg-XIV)—WMO Space Programme.

[10]See http://www.wmo.int/sat .

[11]See http://ceos.org.

20.3.1 Integrated Space-Based Observing System

The WMO Space Programme contributes to the coordination and planning of the space-based component of WIGOS.

The design of the space-based observing system is informed by a systematic process called the "Rolling Review of Requirements" (RRR) through which WMO gathers observing system requirements from users of the WMO application areas and of additional cross-cutting activities, such as for urban applications and polar and high mountain observations.[12]

The technology-free observing system requirements are then stored in a database called "Observing System Capability Analysis and Review Tool" (OSCAR) and through gap analysis matched against the existing surface and space-based observing system capabilities to help identify missing observational gaps and to provide guidance on how to close them.[13]

The overall process is guided by a description of the surface- and space-based component of the desired, future integrated global observing system, which is contained in a document "Vision for WIGOS in 2040".[14] It replaces the "Vision for the Global Observing System in 2025", which was adopted in 2009. In many ways the 2025 Vision foreshadowed the development of WIGOS, whereas the current document anticipates a fully developed and implemented WIGOS framework that supports all activities of WMO and its Members within the general areas of weather, climate and water and related environmental observations.

As of 2019, the space-based observing system component of WIGOS is undergoing dynamic development with the launch of several next-generation meteorological satellites in recent years that are now operationally deployed in LEO and GEO orbit positions. Further satellites are under development and will be added to WIGOS over the coming years.

To monitor, improve and harmonize data quality from operational environmental satellites for climate monitoring and weather forecasting, the Global Space-based Inter-Calibration System (GSICS) was initiated in 2005 by the WMO Space Programme and CGMS.[15]

Through the Satellite User Readiness Navigator (SATURN) database, WMO and satellite operators provide helpful transition information to assist end-users with the

[12]Presently, WMO defines 14 application areas: (1) Global numerical weather prediction, (2) High-resolution numerical weather prediction, (3) Nowcasting and very short range forecasting, (4) Sub-seasonal to longer predictions, (5) Aeronautical meteorology, (6) Forecasting atmospheric composition, (7) Monitoring atmospheric composition, (8) Atmospheric composition for urban applications, (9) Ocean applications, (10) Agricultural meteorology, (11) Hydrology, (12) Climate monitoring, (13) Space weather, (14) Climate science. See https://community.wmo.int/rolling-review-requirements-process.

[13]See http://oscar.wmo.int.

[14]The vision document is updated every few years. Its current version is available from https://community.wmo.int/vision2040.

[15]See https://gsics.wmo.int.

change to the next generation of meteorological satellites.[16] A Product Access Guide (PAG) facilitates access to satellite-based geophysical datasets.[17]

20.3.2 Access to Satellite Data and Products

The benefits and capabilities of these new, powerful satellite systems that form part of the space-based component of WIGOS are made available to all WMO members. The open access and the sharing of data from the observations is mandated by several WMO resolutions.[18]

The data is provided through a wide range of dissemination channels, including those offered by the WMO Information System (WIS) with the Global Telecommunications System (GTS) at its core.

Many developing countries in the WMO regions of Africa, Asia and the Pacific and of the Americas and the Caribbean receive technical and financial support from operational satellite operators and have been provided with local direct-broadcast receiving stations.

The WMO Space Programme contributes through its Satellite Data Dissemination Strategy (SDDS) and coordinates the Direct Broadcast Network (DBNet) which promotes a set of operational arrangements for the real-time acquisition of LEO satellite data through a worldwide network of local, Direct Broadcast receiving stations.

Initiatives, such as the Sustained, Coordinated Processing of Environmental Satellite Data for Climate Monitoring (SCOPE-CM),[19] Sustained, Coordinated Processing of Environmental Satellite Data for Nowcasting (SCOPE-Nowcasting),[20] and the Space-based Weather and Climate Extremes Monitoring Demonstration Project (SEMDP) aim to facilitate the use of satellite data and products and to develop operational services.

20.3.3 Awareness and Training

The full benefits of the space-based data and products can only be realized if users are aware of its existence and are trained to access and utilize it.

For this purpose, WMO and CGMS established the WMO-CGMS Virtual Laboratory for Education and Training in Satellite Meteorology[21] to raise awareness on

---

[16]See https://www.wmo-sat.info/satellite-user-readiness/.

[17]See https://www.wmo-sat.info/product-access-guide/.

[18]WMO Resolutions 25 (Cg-XIII), 40 (Cg-XII), and 60 (Cg-XVII).

[19]See http://www.scope-cm.org.

[20]See https://community.wmo.int/activity-areas/wmo-space-programme-wsp/SCOPE-Nowcasting.

[21]See https://www.wmo-sat.info/vlab/.

satellite capabilities and promote satellite-related education, with a focus on developing countries, and to promote availability and utilization of satellite data and products for weather, climate, water and related applications. VLab activities are integrated into the WMOLearn initiative of the WMO Global Campus, a collaborative network of WMO Member institutions and NMHSs involved in the development and delivery of education and training.[22]

So called Satellite Data Requirements (SDR) Groups have been established to provide a direct link between satellite operators and users, for users to convey their satellite data and product needs. SDR Groups meet regularly and conduct regional SDR survey, which complement a global SDR survey conducted every four years by the WMO Space Programme.

20.3.4 Space Weather Coordination

Space weather phenomena are triggered by events occurring on the Sun and in interplanetary space and eventually produce impacts in the natural Earth environment ranging in size from the global to the regional scale. Space weather disturbances can affect critical technologies with potential harmful consequences for the global economy.

Noting the considerable impact of space weather on meteorological infrastructure and important human activities and acknowledging the potential synergy between meteorological and space weather services for operational users, WMO Members agreed in 2008 that WMO should support international coordination of space weather activities.

As a result, space weather observations are integrated into WIGOS and the related requirements are developed within the WMO Rolling Review of Requirements (RRR). Space weather data sharing and management is taking place within the context of the WMO Information system (WIS).

A specific objective of the coordination is to improve space weather warnings in major application areas, including in aviation in cooperation with the International Civil Aviation Organization (ICAO), which has led to the selection of providers for space weather information services for aviation.

20.4 Earth System Approach and Global Development Agendas

In 2019, WMO is undergoing its first governance reform since its establishment in 1950. At the Eighteenth World Meteorological Congress, WMO Members decided

[22] See http://learningevents.wmo.int.

on a wide range of transformational measures to prepare WMO for anticipated future challenges, including a reform of its constitutional bodies.[23]

Under the governance reform, the future work of WMO will be based on an Earth system approach, covering meteorology, climatology, hydrology, oceanography, seismology, volcanology, air quality, greenhouse gases, space weather and related multi-hazard and impact-based seamless services, whether over land, at sea, in the air or in space.

This new approach shall also ensure the streamlining of WMO contributions to global development agendas. Sustainable Development has been called the greatest challenge for humankind in our present times. Fortunately for our Planet and for future generations, Member States of the United Nations in 2015 committed to achieving the 2030 Agenda for Sustainable Developments and its 17 Sustainable Development Goals (SDGs).[24] WMO contributes to 12 out of the 17 SDGs and together with UN Environment (UNEP) acts as co-custodian for SDG 13 on Climate Action.[25]

A major focus related to SDG 13 and to the Paris Agreement is the implementation of the architecture for climate monitoring from space in close cooperation with the joint CEOS-CGMS Working Group on Climate (WGClimate)[26] and with the Global Climate Observing System (GCOS).[27]

GCOS is an integrated, long-term approach to systematically observe the Earth's changing climate and to identify the requirements of measurements in support of adaptation measures and how they support observations at local and regional levels. Many of these observations do not yet exist and specific actions are required to develop guidance for the provision of high-resolution global or regional datasets from satellite products or by downscaling of model results.

More specifically, the GCOS Implementation Plan is based on a set of identified Essential Climate Variable (ECVs). Soon, measurements of anthropogenic greenhouse-gas fluxes, made from space-borne platforms will augment the bottom-up approaches of the Intergovernmental Panel on Climate Change (IPCC) guidelines and allow improved integrated estimates of emissions, in line with the requirements of the Paris Agreement for a global stocktake with a five-year repeat. The first global stocktaking in 2023 will benefit from prototype systems that are expected to develop into a more operational system thereafter.

GCOS activities are closely coordinated with the space agencies that are developing and operating relevant space-borne platforms, including the Committee on Earth Observation Satellites (CEOS) and the Coordination Group for Meteorological Satellites (CGMS).

[23]See https://public.wmo.int/en/governance-reform.

[24]See http://sustainabledevelopment.un.org.

[25]See https://public.wmo.int/en/our-mandate/what-we-do/wmo-contributing-sustainable-development-goals-sdgs.

[26]See http://climatemonitoring.info.

[27]See https://gcos.wmo.int.

Since 1993, WMO has been issuing an annual Statement on the State of the Climate.[28] The latest, provisional Statement on the State of the Climate in 2018 shows that the past four years have been the warmest on record, with many high impact weather events which bear the hallmarks of climate change. The global average temperature is nearly 1 °C above the pre-industrial era.

Increasingly the preparation of the annual Statement is informed by data from the space-based observing system. Some of its findings would not be possible without the data gathered by Earth Observation satellites. The collection of space-based climate data over the past four to five decades is now starting to pay off as we start seeing certain climate trends in that data.

While the accuracies of nowcasting and 10 to 14-day weather forecasts continue to improve due to the integration of data from a wide range of satellites, recent research is also coming to the conclusion that near-term, decadal, climate model predications are becoming more accurate with the improved inclusion of observational data, in particular also, from space-based real time observations.[29]

Earth observations also contribute to the Sendai Framework for Disaster Risk Reduction. The WMO Space Programme supports the WMO Disaster Risk Reduction Programme and other related programmes that consider the impact of weather, climate and water on infrastructures and human populations.

20.5 Conclusion

The arrival of space-based observations fundamentally changed the field of meteorology by considerably enhancing the scientific data basis of the discipline. It is expected that the accuracy of weather forecasts will continue to improve with additional observational data becoming available.

WMO and its Members are now working to apply the experience gained in the field of Numerical Weather Prediction to other WMO applications areas for weather, climate, water and other environmental services.

Achieving the global development agendas will require sound policy and decision making, informed by science-based data and information.

Initial assessments of our progress towards meeting the SDGs of the 2030 Agenda for Sustainable Development show that most data are available for economic SDG indicators, fewer data for social and least data for environmental SDG indicators.

To help change this situation will be a major challenge for the future work of WMO and its WMO Space Programme, in which the evolving space-based observing system will continue to play an increasingly important role and will remain a central essential element of WMO activities.

---

[28] See https://public.wmo.int/en/our-mandate/climate/wmo-statement-state-of-global-climate.

[29] See https://public.wmo.int/en/media/news/near-term-climate-prediction-'coming-of-age'.

Chapter 21
Earth Observation Capacity Building at ESA

Francesco Sarti, Amalia Castro Gómez and Christopher Stewart

Abstract Among the wide range of space programmes that ESA has developed in its efforts to contribute to sustainable development, a relevant example is Earth Observation (EO). Defined as the process of acquiring observations of Earth's surface, interior and atmosphere via remote sensing instruments, EO has a major potential to inform and facilitate international development in a globally consistent manner. Thanks to EO satellites, data at various scales (global, national or local) can be available in a continuous and consistent way.

21.1 Introduction to Earth Observation and to ESA's Education, Training and Capacity Building Activities

Among the wide range of space programmes that ESA has developed in its efforts to contribute to sustainable development, a relevant example is Earth Observation (EO).

Defined as the process of acquiring observations of Earth's surface, interior and atmosphere via remote sensing instruments, EO has a major potential to inform and facilitate international development in a globally consistent manner. Thanks to EO satellites, data at various scales (global, national or local) can be available in a continuous and consistent way. This is crucial to understand the Earth as a system and to predict the evolution and impact of climate change. Furthermore, knowledge based on EO satellites supports political efforts in the implementation of conventions and protocols related to sustainable development, and plays a crucial role in mitigating the negative effects of natural disasters. EO brings the capacity to expand monitoring capabilities and to reduce the associated costs. The data acquired can be reused for other purposes, and the analysis is objective and repeatable. It

F. Sarti (✉) · A. Castro Gómez · C. Stewart
European Space Agency, Frascati, Italy
e-mail: Francesco.sarti@esa.int

S. Ferretti (ed.), *Space Capacity Building in the XXI Century*, Studies in Space Policy 22,
https://doi.org/10.1007/978-3-030-21938-3_21

is a unique resource for planning, implementing and evaluating projects related to sustainable development.[1,2]

ESA carries out a wide variety of activities in EO education, training and capacity building, in response to the increasing requirements from a very diverse assortment of users worldwide.[3] Users may range from students of primary school to professional researchers and to operational service providers at both a technical and managerial level. Requests may require addressing a large audience for general EO outreach or may be intended to train a small group of specialists on a particular application.

The different types of activities aim to satisfy in the most efficient manner the wide range of requests. A crucial activity is online education and e-learning, where the most relevant tools are Massive Online Open Courses (MOOCs). The objective of MOOCs is to reduce the barriers many non-technical users perceive, and provide basic information regarding the data, their uses and challenges. Other e-learning tools include web tutorials, software, ibooks and tablet apps. The organisation of training courses is also an essential activity, where participants benefit from direct interaction with experts (Fig. 21.1). Participants can be school teachers who receive basic-level training or young researchers and academics attending advanced training. Participants can also come from industry for trainings targeting a specific sector or region of the world. Besides trainings, ESA also produces and distributes printed

Fig. 21.1 Example of a training course in ESA/ESRIN

[1]European Space Agency (2019) Earth Observation Science for Society. https://eo4society.esa.int/. Accessed 10 January 2020.

[2]Aschbacher J., Santi C.B., Rathgeber W. (2018) Space Agencies' Perspective on Space for Sustainable Development. In: Al-Ekabi C., Ferretti S. (eds) Yearbook on Space Policy 2016. Yearbook on Space Policy. Springer, Cham. 10.1007/978-3-319-72465-2_9.

[3]Current and recent ESA training activities in EO (face-to-face courses, MOOCs, webinars, educational tools) can be viewed on https://eo4society.esa.int/training-education/ (Fig. 21.2).

material (atlases, posters, flyers, training manuals), supports outreach events and contributes to international collaboration. Cooperation is essential to make the best use of resources and avoid duplication of efforts, therefore ESA collaborates with many other organisations that carry out EO education through international societies, committees and associations.

Capacity Building (CB) in Earth Observation is carried out for a variety of sensors, techniques and/or themes. In this chapter we detail several examples of "thematic" initiatives, projects and activities, with a CB component, related to the topics of water, food security, sustainable development and agriculture. On top of thematic activities, we will also describe "geographic" activities, namely for Africa, which cover almost all themes mentioned earlier, again with a CB component.

21.2 Water

The unprecedented growth of population in our planet and the widespread use of unsustainable water management practices are rapidly depleting water resources. Human populations exploit water resources at a very intensive pace, therefore water resources are over-exploited and polluted at scales never witnessed before. This dynamic makes water scarcity a looming reality, with a magnitude and spatial reach that pose a real concern for development and peace between nations.[4] More than 40% of the global population is affected by water scarcity, and by 2050 it is estimated that at least one in four people may live in a country suffering from shortages of fresh water that could be recurring or even chronic.[5] Furthermore, around 70% of the freshwater extracted is used for agricultural irrigation, and this number is expected to rise by 20% by 2050.[6]

In 2015, the widespread and growing concerns related to water scarcity translated into several actions at the level of the international community. This year was a landmark year for international policy shaping, where the World Economic Forum declared the water crisis as a top global risk and only a few months later, the United Nations released the 2030 Agenda for Sustainable Development. The 2030 Agenda was adopted by all United Nations Member States and is centred around the 17 Sustainable Development Goals (SDGs), an urgent call for action by all countries. In the topic of water (SDG number 6), it addresses not only sanitation and drinking water quality, but also how water is managed and governed.[7]

---

[4]European Space Agency (2020) Earth Observation for Sustainable Development – Water Resources Management Initiative. http://eo4sd-water.net/. Accessed 02 January 2020.

[5]United Nations (n.d.) Sustainable Development Goals. https://www.un.org/sustainabledevelopment/water-and-sanitation/. Accessed 02 January 2020.

[6]See Footnote 5.

[7]United Nations (n.d.) Sustainable Development Goals Knowledge Platform. https://sustainabledevelopment.un.org/?menu=1300. Accessed 02 January 2020.

WE WORK WITH

ACT NOW

Fig. 21.2 Centralized web pages addressing current and recent ESA training activities in EO (face-to-face courses, MOOCs, webinars, educational tools) developed within the "EO Science for Society" programmatic element

Fig. 21.3 Centralized web pages addressing current and recent ESA training activities in EO (face-to-face courses, MOOCs, webinars, educational tools) developed within the "EO Science for Society" programmatic element

This is an acknowledgement of the fact that a water crisis can be seen as a management crisis, where a key mitigating factor is the application of an integrated management of water resources (IWRM) at multiple levels.[8]

Implementing and monitoring IWRM practices in a successful way needs access to reliable data and information on the status of water and the challenges linked to it. It can be difficult to obtain such information, because problems in the collection, analysis and use of data related to water can be problematic, thus affecting the policies and management decisions. There is a growing awareness that EO technology can serve these needs by filling the existing information gaps.[9]

---

[8]See Footnote 5.

[9]See Footnote 5.

The European Space Agency (ESA) has a long history working closely with different institutions and building capacity to harness the benefits of Earth Observation for water management. Some example initiatives that will be detailed in this chapter include the TIGER Initiative, the Earth Observation for Africa (EOAFRICA) initiative, the Hydrology Thematic Exploitation Platform (Hydrology TEP) and the Water Management and Resources domain within the Earth Observation for Sustainable Development (EO4SD) initiative.

21.2.1 The TIGER Initiative

ESA has been running the TIGER initiative for 10 years, promoting Earth Observation as a tool for the improvement of the integrated management of water resources. This initiative had a strong capacity building component. Being focused on Africa, this initiative is detailed later in Sect. 21.4.1 (The TIGER initiative and the EOAFRICA initiative), among ESA Capacity Building initiatives for Africa.

21.2.2 EO for Sustainable Development (EO4SD)—Water Management and Resources Domain

The Earth Observation for Sustainable Development initiative aims to support the uptake in sustainable development of information derived from EO. One of the domains is Water Management and Resources.

The initiative is based around large scale demonstrations in different regions of the world (Africa, Asia, Latin America) within the operations related to water from major International Financing Institutions (World Bank, Asian Development Bank, Inter-American Development Bank, and Global Environmental Facility).

This domain addresses key water related priority areas (e.g. river basin characterization and monitoring, hydrological management, water supply and sanitation, etc.).[10]

It aims to support the development of human, technical and institutional capacity to enable an independent and sustainable use of EO data from the stakeholders. Efforts will focus on the formation of operators, technicians, scientists and decision-makers. The content of the training courses will be developed in close cooperation with the involved stakeholders and will include elements from the technical perspective or from the water management perspective, depending on the specific audience.

The capacity building materials will be accessible via the Canvas Distance Education portal. In there, participants will obtain information before a course starts and

[10]See Footnote 5.

also after it has finished. The portal includes a forum where users can communicate and share questions, comments, or assignment results.[11]

21.2.3 Hydrology Thematic Exploitation Platform (Hydrology TEP)

The Thematic Exploitation Platform for Hydrology is a project by ESA, developed to create an heterogeneous community of users including scientists, service providers and river basin organisations, and to enable them to access, use and exchange relevant data, as well as exchange services and knowledge.

The platform is composed by several thematic services: the Water Observation and Information System, the Flood Monitoring Service, the Hydrological Modelling service, the Water Quality Service, the Water Level Service and the Small Water Body Mapping Service.

The capacity building role of this platform stems from its collaborative nature, which is enhanced by the fact that users can have access to tools and datasets from other participants or other platform services, and by the fact that they can create their products without having a local infrastructure. Therefore the platform enables them to remain updated about the latest developments and to build their products in an efficient way. Besides this aspect, a set of introductory online training tutorials guide new users around the platform, and at the start of the project there were two dedicated face-to-face training events.[12]

21.3 Agriculture and Food Security

Food security plays a critical role in addressing the future needs of the planet and of humanity, not only related to the production of food but also to health and to the preservation of the environment.

The Sustainable Development Goals tackle the complex challenges humanity faces in our interconnected planet. Those goals cover a whole range of policy areas.[13] Nine of the SDGs are directly or indirectly connected with agriculture, highlighting

---

[11]European Space Agency (2020) Earth Observation for Sustainable Development—Water Resources Management Initiative. Capacity Building. http://eo4sd-water.net/content/capacity-building. Accessed 03 January 2020.

[12]European Space Agency (n.d.) Hydrology Thematic Exploitation Platform. https://hydrology-tep.eu/. Accessed 02 January 2020.

[13]Farming First (n.d.) The story of agriculture and the Sustainable Development Goals. https://farmingfirst.org/sdg-toolkit#section_1. Accessed 02 January 2020.

the multi-dimensional importance of agriculture and the importance of a systems approach.[14] Almost all of the SDGs are relevant to the development of aquaculture.[15]

Global population has grown from an estimated 2.6 billion people in 1950 to more than 7 billion people in 2015. Projections from the UN indicate the trend will continue, with global population reaching 9.7 billion in 2050 and 11.2 billion by the end of this century.[16] In the context of this dramatic growth, the big achievement of modern agriculture has been to keep the global levels of hunger generally constant over that period.[17] However, chronic hunger still affects an estimated 805 million people.[18] This constitutes a major challenge to the global agricultural sector, which will need to raise the production of food around 70% by 2050, compared to the production levels in 2005.[19] However, agriculture relies on practices which often have negative consequences on the quality of water and on biodiversity, accentuated by these dramatic projections.[20]

It is estimated that the majority of growth in crop production will come from an increase in cropping intensity and from higher yields. Despite this, arable land is also expected to expand by around 70 million hectares, mostly in developing countries in the sub-Saharan Africa and Latin America.[21]

The demand for water is also set to increase, given that agriculture is responsible for the large majority (70%) of withdrawals of freshwater in the globe.[22] Greenhouse gas emissions from agriculture constitute a quarter of the net emissions at a global level,[23] and in 2050 this number is expected to increase by an additional 30%.[24]

---

[14]Michalopoulos S (2016) Agriculture holds key to UN Sustainable Development Goals. Available via EURACTIV. https://www.euractiv.com/section/agriculture-food/news/agriculture-holds-key-to-un-sustainable-development-goals/. Accessed 02 January 2020.

[15]Food and Agriculture Organization of the United Nations (2017) The 2030 Agenda and the Sustainable Development Goals: The challenge for aquaculture development and management. http://www.fao.org/cofi/38663-0a3e5c407f3fb23a0e1a3a4fa62d7420c.pdf. Accessed 02 January 2020.

[16]United Nations (n.d.) Global Issues: Population. https://www.un.org/en/sections/issues-depth/population/. Accessed 02 January 2020.

[17]See Footnote 12.

[18]Food and Agriculture Organization of the United Nations (2014) World hunger falls, but 805 million still chronically undernourished. http://www.fao.org/news/story/en/item/243839/icode/. Accessed 02 January 2020.

[19]Food and Agriculture Organization of the United Nations (2017) Water for Sustainable Food and Agriculture. http://www.fao.org/3/a-i7959e.pdf. Accessed 02 January 2020.

[20]European Space Agency (2020) Earth Observation for Sustainable Development—Agriculture and Rural Development Initiative. https://www.eo4idi.eu/. Accessed 02 January 2020.

[21]Food and Agriculture Organization of the United Nations (2009) Global agriculture towards 2050. http://www.fao.org/fileadmin/templates/wsfs/docs/Issues_papers/HLEF2050_Global_Agriculture.pdf. Accessed 02 January 2020.

[22]See Footnote 16.

[23]See Footnote 18.

[24]Food and Agriculture Organization of the United Nations (2014) Agriculture, Forestry and Other Land Use Emissions by Sources and Removals by Sinks. http://www.fao.org/3/a-i3671e.pdf. Accessed 02 January 2020.

The challenge remains in finding a sustainable path to increase the supply of food, agriculture commodities[25] and aquaculture.[26] Modern agriculture relies more and more on the availability of accurate, diverse and continuous information.[27] There is a growing awareness in the international community regarding the positive role satellites can play in this regard, not only in improving the efficiency of crops and reducing their impact, but also in support to regulatory, policy, planning and protection functions.[28,29] However, there is a large gap between the EO specialists who can generate very sophisticated information from satellites (mainly related to precision farming and enhanced water management, but also crop monitoring) and the tools end users rely on. Access to processed geospatial data is often difficult for end users.[30]

Earth Observation can also provide useful information supporting aquaculture practices by providing observation of parameters such as water temperature, concentrations of chlorophyll and the presence of harmful algae blooms.[31]

In this context, we present several initiatives launched by ESA to promote the use of E-based geo-information for agricultural management and food security.

21.3.1 The Sentinel-2 for Agriculture Project (Sen2Agri)

The Copernicus Sentinel-2 satellites are providing unique and new perspective on our land and vegetation. Thanks to their large and systematic imaging capacity and their spectral richness, it is designed to offer continuity to previous missions (US Landsat and French Spot missions), as well as to expand on them. This translates into in increased precision for products and into services with a higher reliability.[32]

With the objective to demonstrate, in close cooperation with the users, the benefits of the Sentinel-2 mission for a wide range of crops and agricultural practices, ESA launched the Sentinel-2 for Agriculture (Sen2Agri) project. Sen2Agri is an open

---

[25] See Footnote 18.

[26] See Footnote 14.

[27] See Footnote 18.

[28] European Space Agency (n.d.) Business Applications. Food and Agriculture. https://business.esa.int/projects/theme/food-agriculture. Accessed 02 January 2020.

[29] European Space Agency (2017) Taking farming into the space age. https://www.esa.int/Applications/Observing_the_Earth/Taking_farming_into_the_space_age. Accessed 03 January 2020.

[30] See Footnote 24.

[31] European Commission (2017) Making Earth Observation data accessible to aquaculture. In AQUA-USERS: AQUAculture USEr driven operational Remote Sensing information services. https://cordis.europa.eu/article/id/196591-making-earth-observation-data-accessible-to-aquaculture Accessed 10 January 2020.

[32] European Space Agency (2018) Copernicus Sentinel-2 leads precision farming into new era. In: Sentinel Online. https://sentinel.esa.int/web/sentinel/missions/sentinel-2/news/-/article/copernicus-sentinel-2-leads-precision-farming-into-new-era. Accessed 03 January 2020.

source system that enables the exploitation of Sentinel-2 for operational monitoring of agriculture at national scale. Although the primary objective was to provide the user community with validated algorithms that could be used to derive products for crop monitoring,[33] Sen2Agri is in itself a capacity building effort. It was launched as an important contribution to the Research and Development and national capacity building components of the Group on Earth Observations Global Agricultural Monitoring Initiative (GEOGLAM), as well as a contribution to the Joint Experiment for Crop Assessment and Monitoring (JECAM) network activities.[34]

The project is organised around demonstration sites (national or local), where the specific requirements of users are a valuable input in the assessment of the suitability of the Sen2Agri system and its products.[35] In the national demonstration countries (e.g. Mali, Ukraine), a series of training events were organised together with the installation of servers and software. The relevant stakeholders were involved and they were trained to use the Sen2Agri system for their particular environment.[36]

The development of algorithms was done in close cooperation with the scientific community (GEOGLAM and its JECAM development group), and a number of papers have been published in literature related to this work.[37]

Besides these efforts in the national demonstration countries, a set of webinars are available to the public explaining how to install the software, and users can collaborate and exchange experiences in a forum. On top of the on-line webinars, of the on-line training material (both on the Sen2-Agri web pages)[38] and of the dedicated training events already mentioned, the use of the Sen2-Agri system has also been included as part of the latest ESA "9th Advanced Training course on Land Remote Sensing/Agriculture" run in Belgium in September 2019, which included a number of participants from Sen2-Agri user communities.

---

[33] Sen2-Agri Consortium (2018) Sentinel-2 for Agriculture: The Sen2-Agri System. http://www.esa-sen2agri.org/. Accessed 03 January 2020.

[34] Joint Experiment for Crop Assessment and Monitoring (2015) Sen2-Agri. http://jecam.org/experiment/sen2-agri/. Accessed 03 January 2020.

[35] Defourny P,Bontemps S, Bellemans N et al. (2019) Near real-time agriculture monitoring at national scale at parcel resolution: Performance assessment of the Sen2-Agri automated system in various cropping systems around the world. Remote Sensing of Environment, Volume 221, 551–568. https://doi.org/10.1016/j.rse.2018.11.007.

[36] See Footnote 29.

[37] See Footnote 29.

[38] See Footnote 29.

21.3.2 Earth Observation for Sustainable Development (EO4SD)—Agricultural and Rural Development Domain

Over the past decade, ESA has been working in close cooperation with Multilateral Development Banks and their Client States to exploit the benefits Earth Observation brings to a global sustainable development. The recent EO4SD initiative has the goal to increase the uptake of information based on Earth Observation for development operations. Within its Agriculture and Rural Development Cluster project, the objective is to demonstrate that information derived from Earth Observation can significantly enhance the effectiveness of the interventions and investments related to agriculture that the Multilateral Development Banks carry out.[39]

The Agricultural and Rural Development domain of EO4SD has a capacity building plan that targets the specific needs of the main stakeholders. It is a way to ensure that all personnel involved (from the Multilateral Development Banks, from their client countries or from the different organisations involved in the projects financed) is engaged and aware. Each type of stakeholder (e.g. Project Managers, National stakeholders in client countries) has different specific needs, and activities are tailored to them. Decision makers (e.g. staff in Multilateral Development Banks) need to understand the capabilities of Earth Observation as well as the availability and use of different types of data. On the other hand, the users and beneficiaries of Earth Observation information (e.g. public institutions in client countries) need to increase their human and institutional capacity by training professionals that could assist them operationally. These activities include online learning courses and workshops of different durations and levels.[40]

21.3.3 Sentinels for Common Agriculture Policy (Sen4CAP)

Agriculture is a sector that depends heavily on the weather and climate, and where producing goods (i.e. cereals) takes time, hence creating a time gap between demand and supply. This puts farmers in a vulnerable position despite their importance. Given the uncertainties surrounding the agricultural sector, and given the impact agriculture has on the environment, the public sector plays a very important role. Launched in 1962, the Common Agricultural Policy (CAP) of the European Union is key in ensuring farmers maintain a decent standard of living while being able

[39] See Footnote 18.

[40] European Space Agency (2020) Earth Observation for Sustainable Development—Agriculture and Rural Development Initiative. Capacity Building. https://www.eo4idi.eu/capacity. Accessed 02 January 2020.

to improve their productivity in a sustainable way.[41,42] The Common Agricultural Policy includes measures such as income support (direct payments to ensure stability and remuneration for adopting sustainable practices), market measures (dealing with sudden drops in demand or falls in prices) and rural development measures. CAP is financed as part of the EU budget and payments are managed at the national level by each EU country.[43] When a farmer submits a declaration, the national Paying Agency of the country has to decide whether the farmer is compliant and requirements are met, in order to release funds.[44]

The current legal framework will be modernised and simplified by 2020.[45] This constitutes an important reform, where one of the objectives is to improve the Integrated Administration and Control System (IACS), which manages the majority of the CAP budget. The IACS also supports farmers in submitting their declarations. The 2020 reform aims to make IACS more cost-effective, and EO in general and the Sentinels in particular are expected to play a significant role in this modernization and simplification effort.[46]

In this context, the Sen4CAP project has the objective to provide an open-source Earth Observation processing system with validated algorithms, products, workflows and best practices to the European and national stakeholders. A number of seven pilot countries have been selected, to ensure the heterogeneity of agricultural practices, landscape and climate that is present in the EU is well represented. In 2019 the Sen4CAP products and use cases have been systematically demonstrated over more than 15 million parcels in the project's pilot countries. ESA works closely with the Paying Agencies of these countries, to ensure their requirements shape the way the space capacity is integrated in their system.[47] The first version of the Sen4CAP Earth Observation processing system is freely available from the project website to the public since the end of 2019.

The Sen4CAP project constitutes a capacity building effort at an institutional level, where the main capacities derive more from the cooperative aspect as well as from the transfer of technical expertise. The project includes training sessions in the pilot Paying Agencies as well as overall training for the rest of the community via the website forum and regular webinars.[48]

---

[41] European Commission (n.d.) The common agricultural policy at a glance. https://ec.europa.eu/info/food-farming-fisheries/key-policies/common-agricultural-policy/cap-glance_en. Accessed 03 January 2020.

[42] Sen4CAP consortium (2017) SEN4CAP—Sentinels for Common Agriculture Policy. http://esa-sen4cap.org/. Accessed 03 January 2020.

[43] See Footnote 34.

[44] See Footnote 35.

[45] European Commission (n.d.) Legislative proposals. https://ec.europa.eu/info/food-farming-fisheries/key-policies/common-agricultural-policy/future-cap_en Accessed 09 January 2020.

[46] European Commission (2018) Modernising the CAP: satellite data authorised to replace on-farm checks. https://ec.europa.eu/info/news/modernising-cap-satellite-data-authorised-replace-farm-checks-2018-may-25_en Accessed 09 January 2020.

[47] See Footnote 35.

[48] See Footnote 35.

21.3.4 The GeoRice Project

There are a number of ESA EO projects related to Food Security, though not always with a specific capacity building component. As an example, we can mention the ongoing ESA project for Regional Rice Monitoring in South-East Asia with Sentinel-1, called "GeoRice".[49] This project is driven by the need of countries to continuously monitor their rice production, in the context of human and climate-induced production variability. EO products included by the project are (i) dynamic rice crop maps and annual anomalies; (ii) dynamic maps of phenological stages.

Initially, the project was conducted in the Mekong Delta, one of the major rice production regions in the world. Later, the project has been extended to 5 full countries in South-East Asia (Vietnam, Laos, Cambodia, Thailand, and Myanmar). The project includes user workshops (since 2016) in order to provide larger awareness and training. The plan is to release the GeoRice algorithms in 2020 as open source, so that they can be used by the user community, also for training and capacity building purposes.

21.3.5 The Food Security Thematic Exploitation Platform (Food Security TEP)

The second UN Sustainable Development Goal is related to ending hunger and malnutrition. However, in a context where global population is growing, increasing food supply presents a challenge. This situation highlights the need to support future sustainable and efficient farming and aquaculture, with the objective to ensure food security. Different factors are involved and the stakeholders involved is very heterogeneous.

Food production systems need to optimise their use of resources and maximise yield while reducing their impact on the environment, in a context of increasingly unstable climatic conditions. Small scale farmers remain vulnerable to hazards and need better access to financing and insurance. And international/governmental stakeholders, who define early mitigation measures, need information on upcoming risks to be reliable, spatially detailed and delivered in a timely manner.

In this context, ESA launched the Food Security Thematic Exploitation Initiative, which provides a platform for efficient use of EO data to a heterogeneous user community related to agriculture and aquaculture. The Food Security TEP will grant timely access to satellite datasets (mainly from Copernicus Sentinel-1 and -2 satellites) and to processing tools related to various topics. It will allow to process large datasets in the cloud and to analyse the results directly. Three interfaces will be available: a free and open one for processing by expert users, a mobile version for on-site visualization of products via smartphone, and one for customer-tailored services,

---

[49] GeoRice Consortium (n.d.) GeoRice. Regional Rice Monitoring in South-East Asia with Sentinel-1. http://georice.net/ Accessed 09 January 2020.

managing confidential data in a secure environment. The project has implemented several service pilots, including crop monitoring in Europe and Africa, as well as aquaculture focused on Africa.

The Food Security TEP will eventually allow interaction with the user community, thus improving the use of satellite data for food security.[50]

In order to demonstrate the ability of the Food Security TEP to support agriculture and aquaculture with tailored EO based information services, three different Service Pilots were implemented.

The First agricultural Service Pilot included crop monitoring and the derivation of advanced biophysical parameters and yield predictions to increase efficiency of agricultural production, including irrigation, on farm level in Europe and Zambia, as a first application in Africa.[51]

The Second agricultural Service Pilot[52] focused on the improvement of financial services for farmers in Africa and deals with High Resolution Crop parameters to optimize Area Yield Index insurance, as well as improved access to credits for smallholder farms. In collaboration with WFP, capacity building of two local companies[53] dealing with farmers in Kenya has been performed, to make them aware of the potential of such an EO-based service.

The Third Service Pilot for aquaculture developed satellite EO aquaculture applications for the coastal region of Tanzania, Africa. During this Service Pilot,[54] local university staff were trained about the potential and the use of this TEP.

More extensive training activities based on this TEP are expected to take place in 2020.

21.4 Africa

ESA EO capacity building initiatives related to Africa cover basically all themes mentioned in the previous paragraphs: water management and hydrology, food security and agriculture, sustainable development.

---

[50]European Space Agency (n.d.) Food Security Thematic Exploitation Platform. https://foodsecurity-tep.net/ Accessed 09 January 2020.

[51]European Space Agency (2018) Tackling drought issues for food security https://www.esa.int/Applications/Observing_the_Earth/Tackling_drought_issues_for_food_security Accessed 10 January 2020.

[52]European Space Agency (n.d.) Food Security Thematic Exploitation Platform. Service Pilot 2: Micro-finance. https://foodsecurity-tep.net/service-pilot-2 Accessed 09 January 2020.

[53]a local insurance provider (PULA) providing affordable insurance to farmers in Kenya and a local company (FarmDrive) facilitating access to micro-financing for farmers in Kenya by providing assessments of small-scale agricultural lending risk to financial institutions.

[54]European Space Agency (n.d.) Food Security Thematic Exploitation Platform. Service Pilot 3: Aquaculture in Tanzania. https://foodsecurity-tep.net/service-pilot-3 Accessed 09 January 2020.

21.4.1 The TIGER Initiative and the EOAFRICA Initiative

The Johannesburg World Summit on Sustainable Development (WSSD) of 2002 stressed the critical situation of Africa regarding widespread water scarcity and the need for support to mitigate it. The effectiveness and the sustainability of water resources management, at both national and regional scales, is hampered by important information gaps. Recognizing that Earth Observation technology has advantages that can help overcome problems in the collection, analysis and use of geo-information related to water, ESA launched in 2002 an initiative known as TIGER, to promote, with the use of Earth Observation, an improved an integrated management of water resources.

This initiative was founded in the context of the Committee for Earth Observation Systems (CEOS), and is an international collaboration between an large number of partners. TIGER is a user driven initiative, where African experts and decision makers guided the strategy and provided local expertise. Over more than 150 African institutions (water authorities, universities and technical centres) collaborated with TIGER in different research and development projects or in training activities.

Through TIGER it was possible to facilitate access to EO data and to carry cross-cutting activities of coordination and outreach. The major action lines of TIGER were on one hand Capacity Building and Training, and on the other hand Information and Knowledge Networking.

The goal of the Capacity Building and Training actions was to support the African partners in the consolidation of a critical mass of technical centres, to enable them to independently exploit EO technology and to disseminate it at various scales, for various purposes (scientific research and management). The needs varied for each research project and for each end user or water authority, therefore dedicated training sessions were set up. African scientists could also get scientific support for the projects they conducted, and one-year fellowships in European research institutes could be funded for project leads. The TIGER research projects generated a range of scientific publications which accumulated for example in the open access special issue[55] on "Earth Observation for Water Resource Management in Africa".

In the Information and Knowledge Networking line, African stakeholders could come together in TIGER workshops that were organised on a bi-annual basis. Furthermore, by participating in international symposia (such as the 6th Water Forum in 2012 or the ESA Living Planet Symposium in 2016) they could exchange experiences and get further involvement.

The developments created from TIGER resulted in a portfolio of services, which are open to the public. These include catchment characterisation and base mapping, or results related to water quality, for example. Alongside this, an open-source Water

[55] Koetz B, Vekerdy Z, Menenti M et al. (2016) Special Issue "Earth Observation for Water Resource Management in Africa", Remote Sensing Journal. https://doi.org/10.3390/books978-3-03842-154-2.

Observation and Information System (WOIS) was developed.[56,57] The fact that it is open to the public is an important characteristic in the efforts to create a community of developers. For example it enables the production and application of a range of EO information products, it is built around the open source QGIS software and allows users to develop their own algorithms and integrate them as a plug into share it with the public.

Building on TIGER, ESA is starting in 2020 the EO African Framework for Research Innovation, Communities and Applications initiative (EO AFRICA), where the first thematic priorities will be water and food security, always in accordance with the Sustainable Development Goals. But ESA's contribution will be significantly different in this initiative, because of various aspects such as innovative tools like cloud computing, digital learning, open access Sentinel data and following an open science approach. There will be a partnership with GMES and Africa of the African Union Commission and capacity building in the form of an African MOOC, webinars and training courses. Besides that, users will have support to access Copernicus data via a Copernicus Data Access Cooperation Agreement. Finally the exploitation of EO data will be facilitated with a cloud computing approach, by bringing the algorithms to the data (e.g. with the Data and Information Access Services (DIAS) platform, procured by ESA on behalf of the European Commission's Copernicus Programme) or with cloud cooperation via the ESA Thematic Exploitation Platforms (TEPs).[58,59]

21.4.2 ESA Land MOOC: Africa Run

Since a few years, ESA has started delivering EO training and CB also online and via dedicated interactive MOOC's (massive open online courses). They are not supposed to replace face-to-face training courses, but rather to complement them, allowing to reach thousands of participants at once. After several sensor-oriented MOOC's,[60] explaining the principles of optical remote sensing or optical remote sensing, we have just started the procurement of a new, larger and thematic MOOC on Land

[56]European Space Agency (n.d.) The TIGER Water Observation Information System (WOIS). http://www.tiger.esa.int/page_eoservices_wois.php Accessed 09 January 2020.

[57]Guzinski, R, Kass, S, Huber, S et al. (2014) Enabling the Use of Earth Observation Data for Integrated Water Resource Management in Africa with the Water Observation and Information System. Remote Sens. 2014, 6, 7819–7839. https://doi.org/10.3390/rs6087819.

[58]European Commission (2018) The DIAS: User-friendly Access to Copernicus Data and Information. https://www.copernicus.eu/sites/default/files/Copernicus_DIAS_Factsheet_June2018.pdf Accessed 09 January 2020.

[59]GMES4Africa (2019) The African Union Commission conducts hands-on training on applications of cloud computing in Morocco. https://gmes4africa.blogspot.com/2019/11/the-african-union-commission-conducts.html Accessed 09 January 2020.

[60]European Space Agency (n.d.) Online EO for everyone: MOOCs, webinars and video tutorials. https://eo4society.esa.int/training-education/online-eo-for-everyone-moocs-webinars-and-video-tutorials/. Accessed 10 January 2020.

Applications of EO from Space, promoting the potential and fostering the use of all possible relevant EO data, in particular data freely available from the Sentinels and from the ESA EO archive, in support of a wide variety of scientific, operational and societal Land Applications and Services. Because of the large number of Land Applications, techniques and EO data relevant for this course, a total of five runs are foreseen. A first, general run, should recall the principles of Remote Sensing from space (optical and radar) explaining how it can be used for a number of Land Applications. Other runs will target more in detail one or more EO Land Applications selected among Land Cover, Forests and Natural Resources, Agriculture (including crop classification and crop monitoring, in support to agricultural management and to Food Security) and many others. One of the runs will be devoted to regional Land Applications for Africa, possibly related to Sustainable Development (like in the programme "GMES and Africa" of the African Union Commission and the European Commission) such as Agricultural applications.

Overall (with its different re-runs), the MOOC should be useful not only in an academic environment but also in the context of Capacity Building, giving insight on the selection of relevant EO data, on the available techniques used and on the available tools and toolboxes (with focus on those made available by ESA). It should also address the use of relevant cloud computing environments (like DIAS, RUS, etc.) given the challenge in terms of download/transmission and processing with new-generation satellite data.

21.4.3 The African Space University' Space Sciences Institute

The Pan African University is an initiative from the African Union Commission consisting of an academic network of five African Institutes which develop education and research programmes around different thematic areas. The final objective is to promote higher education and research in the continent via various Master and Doctoral programmes. Students are offered full scholarships and are offered the possibility to study remotely via the Virtual and E-University project. One of the thematic areas is Space Sciences, which will be hosted in South Africa by an institute to be defined, and will count with the support from both ESA and the European Commission.[61]

[61] Pan African University (n.d.) History. https://www.pau-au.africa/about/background. Accessed 03 January 2020.

21.5 Conclusion

ESA involvement in education and training is part of its mandate. Its Earth Observation programme includes a number of education, training and capacity building activities addressing a large variety of users. Providing an exhaustive list of all ESA projects on EO education, training and capacity building performed so far would be beyond the scope of this chapter. Therefore only a limited number of representative recent activities that ESA has developed in its effort to contribute to sustainable development by means of EO have been described here.

Acknowledgements We thank Benjamin Koetz, Espen Volden, Marcus Engdhal and Marc Paganini for their insight and expertise, and Antonios Mouratidis, whose work was very valuable in the preparation of this chapter. We are grateful to Philippe Bally and Theodora Papadopoulou for the support received. Finally, we thank Yves-Louis Desnos and Diego Fernández for their leadership and making this chapter possible.

Chapter 22
Space for Cities: Satellite Applications Enhancing Quality of Life in Urban Areas

Grazia Maria Fiore

Abstract Globally, 50% of the world's population lives in urban areas. In Europe, 72% of people live in cities, and this percentage is expected to rise to 80% by 2050. Cities are the powerhouses of economic growth, innovation and employment. Nevertheless, growing urban agglomerations also generate challenges in terms of physical and administrative infrastructures, environmental sustainability, social inclusion, and health. As growing urbanisation is challenging the way we live and interact with the natural environment, Eurisy launched an initiative to promote the use of satellite applications to make our cities healthier, cleaner, safer, and more efficient. The initiative aims to foster the exchange of expertise and know-how among city managers, SMEs and stakeholders, to identify challenges to the access and use satellite data and signals, and to make recommendations to service providers, space agencies and stakeholders on how to facilitate the use of satellite-based services at the city level. Indeed, satellite imagery is already employed by city managers, for example to identify urban heat islands, to make predictions about the impact of different traffic scenarios on air quality and to intervene on areas where construction materials retain too much heat. Satellite navigation is crucial in providing real-time information on public transport and numerous apps rely today on satellite navigation signals, e.g. to help persons with disabilities in their daily movements. Satellite communication is also used in cities, e.g. to connect rescue teams when other connections are down, or to perform health checks in public spaces. Satellites alone cannot fight global warming or inequalities, but they offer data and signals that can improve life in cities and that should hence be fully exploited. Through the "Space for Cities" initiative, Eurisy aims at promoting the use of satellite-based services to make our cities healthier, safer, more inclusive and resilient, as envisaged by the United Nations Sustainable Development Goal 11.

Eurisy is a non-for-profit association based in France. Founded by space agencies, Eurisy promotes peaceful uses of space technology to enhance social and environmental well-being.

G. M. Fiore (✉)
Eurisy, Paris, France
e-mail: grazia.fiore@eurisy.org

S. Ferretti (ed.), *Space Capacity Building in the XXI Century*, Studies in Space Policy 22,
https://doi.org/10.1007/978-3-030-21938-3_22

22.1 Why Do Cities Matter?

In Europe 72% of people live in cities, and this percentage is expected to rise to 80% by 2050. Globally, 50% of the world's population lives in urban areas.

Cities are the powerhouses of economic growth, innovation and employment. 85% of the EU Gross Domestic Product originates from cities, and local governments alone are responsible for 44% of public investments in the EU-28.[1] The concentration of people is functional to economic growth and increasing human capital. Cities offer more jobs, better wages and intense cultural exchanges.

Nevertheless, growing urban agglomerations also generate challenges in terms of physical and administrative infrastructures, environmental sustainability, social inclusion, and health. Indeed, in Europe cities use 80% of the produced energy.[2] In 2015, European capital cities generated between 270 kg (Dublin) up to 666 kg of waste per capita (Luxembourg), with the average at 445 kg.[3]

Globally, cities are responsible for 70% of greenhouse gas emissions.[4] Furthermore, cities risk becoming "inequality traps".[5] The Gini coefficient—measuring inequality on a scale of zero to one—shows a higher level of inequalities in cities than on the national average. Indeed, also major European cities have experienced an increase in spatial segregation during the last decade, with growing dissimilarities in levels of income, employment status or educational attainment.[6]

22.2 Why Satellite Applications for Cities?

In 2015, Eurisy implemented a Survey for public administrations using satellite-based services. Out of the replies analysed, 18% were submitted by local authorities.

[1]OECD (2016), Subnational governments in OECD countries: Key data (brochure), OECD, Paris, www.oecd.org/regional/regional-policy Database: http://dx.doi.org/10.1787/05fb4b56-en © OECD 2016. Accessed 12 Mar 2018.

[2]European Commission, Regional Policy, InfoRegio (2017) Urban Development. http://ec.europa.eu/regional_policy/en/policy/themes/urban-development/. Accessed 12 Mar 2018.

[3]European Commission—DG ENV, Assessment of separate collection schemes in the 28 capitals of the EU. Final Report. Reference: 070201/ENV/2014/691401/SFRA/A2. 13 November 2015 Brussels. Accessed 12 Mar 2018.

[4]United Nations Human Settlements Programme, Global Report on Human Settlement 2011. Hot Cities: battle-ground for climate change. http://mirror.unhabitat.org/downloads/docs/E_Hot_Cities.pdf. Accessed 12 Mar 2018.

[5]Bill Below, OECD Directorate for Public Governance and Territorial Development, Habitat III and the challenge of urbanisation in five charts, 20 October 2016, OECD Insights, Debate the issue. http://oecdinsights.org/2016/10/20/habitat-iii-and-the-challenge-of-urbanisation-in-five-charts/. Accessed 12 Mar 2018.

[6]OECD (2016), Making Cities Work for All: Data and Actions for Inclusive Growth, OECD Publishing, Paris. http://dx.doi.org/10.1787/9789264263260-en. Accessed 12 Mar 2018.

The results of the survey[7] suggest that local authorities might need more support than regional and national administrations to use satellite services.

Indeed, responses show that national administrations have been using satellite-based services before regional and local authorities, although this trend has decreased over time.

Moreover, considering previous knowledge of satellite applications, it seems that public managers at the local level have been less prepared to use these tools than their colleagues at the national and regional levels.

Surveyed local authorities seem to have been also less involved in demonstration projects than their peers at the national and regional levels. Almost half of the respondent national authorities and 38% of the regional administrations have been able to profit from demonstration projects. In contrast, only 26% of responding local authorities reported to have accessed such schemes. Finally, respondent local authorities declared to have had less access to data or expertise free of charge to implement and operate satellite-based services than regional and national authorities.

To support the use of satellite applications in cities, Eurisy launched the "Space for Cities" initiative, to understand what are the needs of cities to which satellite applications can contribute, raise awareness on available satellite-based services and promote the development of new services adapted to the operations of public administrations, SMEs and NGOs operating in cities.

22.3 Satellite Applications for Safe and Resilient Cities

22.3.1 Disasters and Security

The concentration of human activities in cities caused an important reduction of green areas within and around human settlements, which makes cities much more vulnerable to climate change and natural disasters.

Satellite-based services have already proven their added-value in improving the resilience of urban areas to natural disasters. Indeed, satellites provide information that can be useful to better prevent and monitor the impacts of natural hazards, such as floods, on urban settlements. Weather forecast is today mainly based on satellite information, which allows to foresee hazards and to adapt the urban infrastructure accordingly. Satellite imagery allows city managers to monitor the vegetation status along watercourses which might overflow in case of heavy rains; they also allow to evaluate slope risks and to monitor soil sinking with centimetre accuracy.

[7]Eurisy (2016), Satellites for Society: Reporting on operational uses of satellite-based services in the public sector, Paris (France), March 2016. ISBN 978-2-9551847-1-4. https://www.eurisy.org/data_files/publications-documents/28/publications_document-28.pdf?t=1467808834. Accessed 12 Mar 2018.

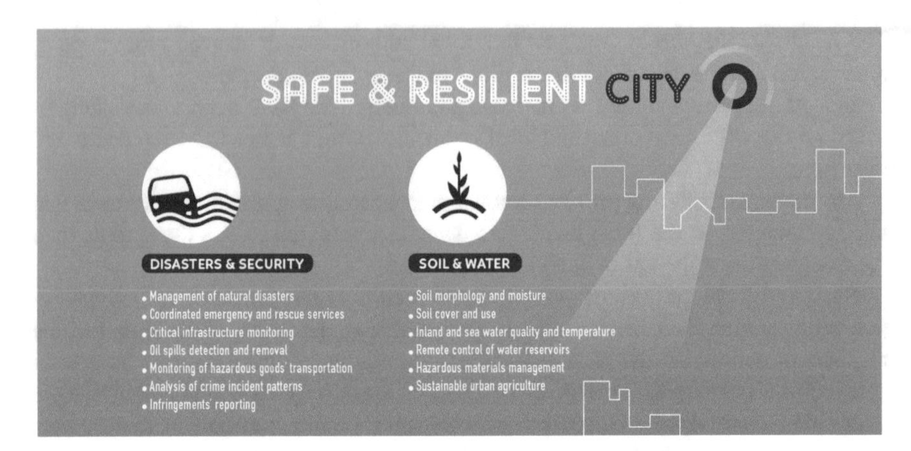

Fig. 22.1 Satellite Applications for Resilient Cities, *Source* L'Atélier de l'Estuaire

Furthermore, satellites can improve the capacities of **rescue teams**, allowing for precise coordination thanks to real-time geo-positioning, and they can ensure connectivity in case other connections are down. Finally, satellite-based maps are today largely used to generate post-disaster maps for a better planning of interventions. International mechanisms, such as the International Charter "Space and Major Disasters" and the Copernicus Emergency Management Service are already available to local and regional managers to better face such situations.

Satellites also provide innovative means to enhance security in cities. Many local administrations are already using data based on satellite navigation to study crime patterns and to implement preventive security measures according to residents' movements. At the same time, satellite navigation is allowing local police to process infringements more efficiently, while saving red tape.

22.3.2 Soil and Water

According to the International Ecocity Framework and Standards (IEFS), in a sustainable urban settlement "soils within the city and soils associated with the city's economy, function, and operations meet their ranges of healthy ecosystem functions as appropriate to their types and environments; fertility is maintained or improved".[8]

The soil allows for life on Earth: it provides a habitat for plants, allows us to produce food, and it is essential for the infiltration and cleansing of water, for micro climate regulation and for providing protection against flooding. However, particularly in **urban areas,** soil is being sealed off with increasing housing and infrastructure.

Satellite imagery offers very precise information on the degree of soil sealing. Earth Observation has allowed scientists to map soil sealing at neighbourhood and

[8]International Ecocity Standards. http://www.ecocitystandards.org/. Accessed 12 Mar 2018.

building levels in several European cities. Using satellite imagery, scientists could prove that temperatures in cities' green areas are much lower than in highly sealed built up areas. They were also able to find correlations between soil sealing and floods, since sealing prevents water from infiltrating the soil. This means that soil sealing is likely to exacerbate the effects of heat waves in cities. Satellites can provide city managers with the sort of reliable and comparable information needed to better plan and manage the urban space by boosting urban green.

22.4 Satellite Applications for Healthy and Inclusive Cities

22.4.1 Health

Sustainable cities are concerned with the health and happiness of their inhabitants. Policies based on the integrated urban development approach aim at enhancing the economic growth of cities, while respecting the environment and increasing residents' satisfaction with their lives. Indeed, if the effects of pollution, exclusion and inequalities are more severely felt in cities, it is also in cities that the most innovative solutions can be deployed.

Satellite imagery offers data about air quality (i.e. temperature, pollution, presence of pollens and other allergenic substances). This information can help prevent and manage respiratory illnesses. Indeed, in some cities, information on air quality is sent directly on residents' smartphones or is showcased on local television channels.

A sustainable city is also one in which people with discapacities and the elderly are provided with opportunities to move freely and have a fulfilling life. Satellite navigation is already embedded in online portals and smartphone apps which provide

Fig.22.2 Satellite Applications for Healthy and Inclusive Cities, *Source* L'Atélier de l'Estuaire

guidance to people with impaired mobility and it is integrated in systems allowing autonomous healthcare. Satnav also supports hospitals and emergency services, by enabling them for increased coordination and response.

22.4.2 Engagement

An integrated urban development requires policies which not only shape the urban infrastructure and services, but also residents' culture and behaviour. In a sustainable city, local authorities promote transparent flows of information and make sure that residents are involved in the design and implementation of local policies.

Satellite services can support engagement in cities. For instance, satellite navigation is already embedded in apps allowing residents to give feedback to their city administrations on different issues, such as damage to the city infrastructure. Satellite imagery is also integrated into geographic information systems (GIS) offering information on the urban environment, assets and policies.

22.4.3 Culture

Cities are ecosystems with their very specific culture. Local historical, natural and intangible heritage must be protected and promoted, not only for ethical reasons, but also to stimulate the local economy.

Satellite imagery offers precious information on the position and status of archaeological remains, helps monitoring historical buildings and landscapes and provides a layer for augmented reality apps and games. Satellite navigation is instead widely embedded in tourist apps offering information on cities' attractions and events or in geolocated outdoor serious games.

22.5 Satellite Applications for Clean Cities

In sustainable cities, patterns of production and consumption shall be conceived to have a minimum impact on the environment. This includes ensuring that emissions do not affect the quality of the air within the city or in the atmosphere, that the soil status and fertility are not endangered, that water sources are healthy, and that the energy consumed and produced does not exacerbate the effects of climate change. Resources should be sourced, distributed and consumed without affecting human health or ecosystems, and—where possible—recycled and reallocated according to the principles of circular economy.[9]

---

[9]International Ecocity Standards. http://www.ecocitystandards.org/. Accessed 12 Mar 2018.

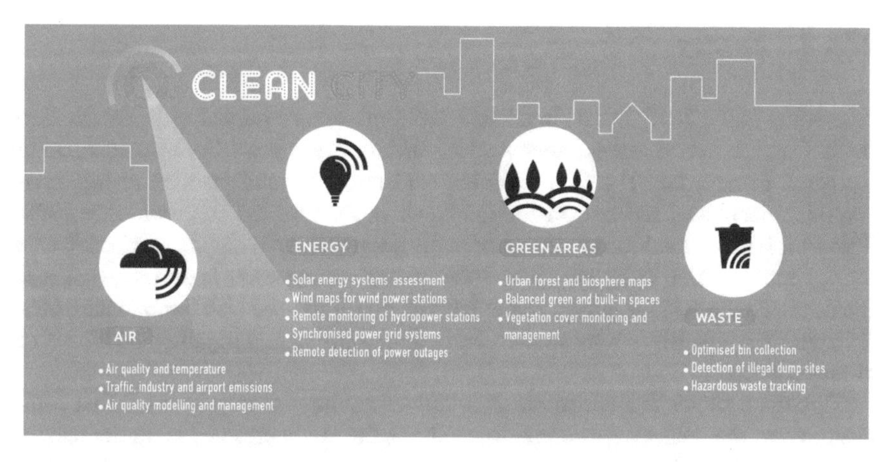

Fig. 22.3 Satellite Applications for Clean Cities *Source* L'Atélier de l'Estuaire

22.5.1 Air

According to the World Health Organisation,[10] pollution represents today a greater threat than Ebola and HIV, and is worldwide responsible for one in four deaths among children aged under five. The EU has developed standards and instruments to ensure good air quality by tackling a wide range of pollution sources such as urban traffic, domestic heating, power plants and industrial activities. A Partnership on Air Quality has also been created under the Urban Agenda for Europe.

Satellite imagery is today widely employed to provide meteorological information. But that is not all. Earth observation data allows to measure and monitor the temperature and composition of the air and to make predictions on the movements of pollutants in the air. It is also employed to identify urban heat islands (spots in which temperatures are more elevated than in the rest of the city) and to create models to test the effects of different traffic scenarios on air quality.

The Copernicus Atmosphere Monitoring Service uses a comprehensive global monitoring and forecasting system that estimates the state of the atmosphere and that can be used to make air quality predictions in Europe and in the main European cities.

[10]World Health Organization 2017, Don't pollute my future! The impact of the environment on children's health. Geneva. Licence: CC BY-NC-SA 3.0 IGO. http://apps.who.int/iris/bitstream/10665/254678/1/WHO-FWC-IHE-17.01-eng.pdf. Accessed 12 Mar 2018.

22.5.2 Energy

According to the 2015 Trends in Global CO2 Emissions[11] published by the Netherlands Environmental Assessment Agency and the European Commission's Joint Research Centre, in 2014 CO_2 emissions did not grow and primary energy consumption decreased, as compared to the previous year, for the first time since 1998. This shift is being made possible thanks to the progressive development of renewable energy sources such as hydropower, solar energy, wind power and biofuels. These important changes in the energy sector have been emphasised also through the establishment of a new EU energy strategy and policy, aiming at mitigating the effects of climate change.

This trend shows that economic growth does not have to rely on fossil fuel combustion and that energy consumption can be optimised instead of being increased. Many European cities have demonstrated their commitment to reducing their carbon emissions by at least 40% by 2030 by joining the Covenant of Mayors for climate and energy.

Satellite applications can support city administrations re-thinking the management of natural resources by providing additional tools to optimise energy consumption, and to enable the use of renewable and green energies. This is demonstrated by the several operational cases in which satellite services are used to foresee the potential of photovoltaic plants, to support smart grids and to monitor wind and hydropower systems remotely.

The research on green energies is still far from being concluded. Satellites are expected to play an increasing role in both the implementation and functioning of green energy systems.

22.5.3 Green Areas

Cities are not only made of buildings, people and infrastructure. Indeed, cities with the highest quality of life are well known for their open spaces. Every city is an ecosystem, and maintaining its good status is crucial for the health and happiness of city residents. This implies granting a good balance between green and built areas, sustaining biodiversity, restoring natural habitats and providing ecological corridors for wild species.

The European Union has a Strategy on Green Infrastructure. This can be defined as a strategically planned network of high quality natural and semi-natural areas with other environmental features, which is designed and managed to deliver a wide range

[11]PBL Netherlands Environmental Assessment Agency, Trends in global CO2 emissions: 2015 Report, The Hague. http://edgar.jrc.ec.europa.eu/news_docs/jrc-2015-trends-in-global-co2-emissions-2015-report-98184.pdf?utm_source=Dec+2016+Newsletter&utm_campaign=Ecocities+Emerging+Dec+2016&utm_medium=email. Accessed 12 Mar 2018.

of ecosystem services and protect biodiversity in both rural and urban settings.[12] Indeed, green areas are not only spaces for recreational activities; they also play a role in preserving natural environments, absorbing CO_2 emissions, improving air quality, and even preventing natural disasters, for example reducing rainfall runoff.

Moreover, green areas and infrastructure play a major role in contrasting urban heat islands, since they cool the air temperature, with positive effects on vulnerable people, particularly during heat waves. Furthermore, green areas represent assets that can contribute to other policy areas, creating jobs and opportunities for community development for example.

Satellite imagery carries information to both design and manage green areas. It helps city planners to decide on where to place new parks, provides information on vegetation types and status, allows for the mapping and monitoring of habitats, and it supports policies aimed at reducing air temperature and pollution by restoring and protecting natural lands, ecological reserves, wetlands, and other green areas within and around cities.

22.5.4 Waste

Municipal waste accounts for only about 10% of the total waste generated in Europe. However, it has a very high political profile because of its complex character, due to its composition, its distribution among many sources of waste, and its link to consumption patterns. In 2015, in the EU-28 cities generated 477 kg of waste per capita, 44 kg less than in the year 2000.[13]

Cities (and local authorities) are generally responsible for collecting the household part of municipal solid waste. Satellites can help city managers in their efforts to reduce the impact of human consumption on the environment, people's health, and the city's ecosystem.

Satellite imagery helps spotting illegal dump sites within and around cities, which could harm soils, water, and eventually the food we consume. Satellite navigation is instead already used to optimise bin collection services, track hazardous waste and build connected bins.

As the initiatives to better manage waste are flourishing among European cities, we expect in the next years the creation and further development of new and existing services relying on satellite data.

---

[12]European Union, 2013, Building a Green Infrastructure for Europe. ISBN 978-92-79-33428-3. http://ec.europa.eu/environment/nature/ecosystems/docs/green_infrastructure_broc.pdf. Accessed 12 Mar 2018.

[13]Official Journal of the European Union, 22.11.2008, Directive 2008/98/EC of the European Parliament and of the Council of 19 November 2008 on waste and repealing certain Directives. http://eur-lex.europa.eu/LexUriServ/LexUriServ.do?uri=OJ:L:2008:312:0003:0030:EN:PDF. Accessed 12 Mar 2018.

22.6 Satellite Applications for Efficient Cities

Sustainable cities include social, economic and environmental considerations in designing and managing the urban infrastructure, its roads and transport networks, buildings, green areas, and public services. Integrated development strategies take into account diverse territorial needs within and outside the city, including its peripheral areas and surroundings. Satellites can support city administrations to better manage urban assets and services, and also to better understand the interconnections among the different assets and services of the city and between the city and its hinterland.

22.6.1 Urban Planning

Growing urbanisation makes it necessary for city managers to have a precise overview of soil status and of land uses. A smart sustainable city is indeed one in which residential areas, green spaces and services are equally distributed, and a good proportion exists between the green volume and the built "grey" volume, which is a necessary condition not only to protect cities' ecosystems and biodiversity, but also to grant a healthy life to their inhabitants.

As compared to other surveying techniques, satellite imagery offers a unique overview of land uses which is objective and comparable over time: satellite-based maps are already used in several cities worldwide to map land features and uses, create and update cadastral maps, plan and monitor the access to services and green areas, monitor soil status and subsidence and even to evaluate property taxes, among many other uses. Satellite imagery and navigation are used to create urban 3D maps,

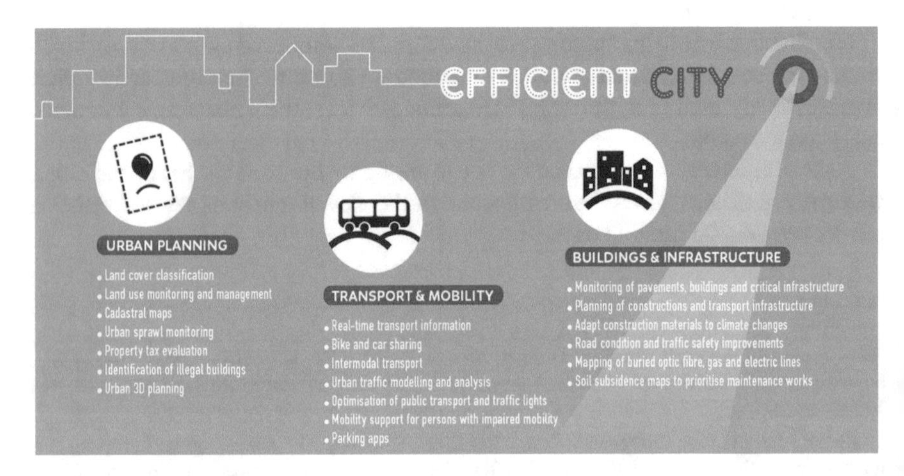

Fig. 22.4 Satellite Applications for Efficient Cities, *Source* L'Atélier de l'Estuaire

allowing public managers to visualise and virtually test different scenarios for the future development of the city.

Satellite imagery is, or it will be, an integral part of cities' Geographical Information Systems (GIS). Where different GIS exist for different uses, they can potentially be combined by integrating layers of satellite imagery for the use of different municipal services.

22.6.2 *Transport and Mobility*

Efficient and reliable transport systems are of paramount importance for cities. They are fundamental for cities' liveability and also for their economic competitiveness, fostering both business and tourism. The sector includes both private and public forms of transport (including trains, buses, trams, ferries, tube lines, cars, cycling and walking) and the possibility to shift among them. Indeed, being able to move around easily and having access to reliable information on itineraries, timings and traffic status facilitate the everyday life of cities' inhabitants. Optimising transport and mobility while encouraging "ecofriendly" ways of transport also helps decrease pollution rates, with positive effects on residents' physical and mental health. Indeed, urban mobility accounts for 40% of all CO_2 emissions of road transport and up to 70% of other pollutants from transport.[14]

Satellites can facilitate transport in cities in several ways and many of their applications, notably those based on satellite navigation, have already proven their efficacy. Indeed, satellite navigation is used to monitor the position of public buses, trams and shared car and bike systems in many European cities, providing information on timings and itineraries in real time. Satnav is also used to optimise traffic lights and to monitor traffic fluxes, collecting data that traffic managers will use to improve and regulate the circulation. It is also employed to monitor the transport of dangerous goods, minimising the risks of their transit within cities.

Satellite navigation, in combination with Earth observation, is used to monitor the effects of traffic on the street pavement. In the near future, Satnav will be an essential component of unmanned vehicles, as it is already the case in some pioneering cities. The versatility and wide availability of satellite navigation signals offer concrete opportunities to support urban transport and more innovative solutions are expected to emerge in the coming years.

[14]European Commission (2018), Mobility and Transport, Clean Transport Urban Transport Urban Mobility. https://ec.europa.eu/transport/themes/urban/urban_mobility_en. Accessed 12 Mar 2018.

22.6.3 Buildings and Infrastructure

The first concern of city administrations when dealing with built areas is the quality of the soil. To make sure that new and existing buildings and infrastructure are safe and sustainable, it is necessary to have a precise understanding of the hydrogeological features of the land and its changes. As an example, in cities built on soft soil (with a high concentration of water), soil subsidence could cause serious damages to people, street pavements and buildings. It is hence necessary to monitor soil status and movements and to be able to foresee them to intervene before the damage happens.

City administrators also need to have a constant and complete overview of the transport, energy and water networks, to make sure that they function, that all residents have access to them, and that new developments are based on a holistic approach.

Satellites can help city managers making urban infrastructure and buildings more sustainable: satellite imagery allows for the precise mapping of buildings and infrastructure. It is also widely used to monitor soil subsidence and the risk of slopes with great accuracy, allowing city administrators to implement works where it is most urgent, with no need for costly field surveys.

Furthermore, satellite images, combined with ground sensors, allow for the identification and monitoring of urban heat islands, and for the identification of correlations between soil sealing, building materials and temperature. City planners can hence recommend construction materials according to the specific needs of the areas in which new buildings are created. Satellite imagery can be used to test different traffic and construction scenarios, and their impact on air quality, and to design new urban infrastructure accordingly. Satellite navigation is also useful in mapping buildings and infrastructure, and it is already employed to verify the accuracy of new buried optic fibre, gas and electric lines.

22.7 Conclusion

To address the challenges of urbanisation, it is today necessary to develop new integral approaches to city management, leading to sustainable urbanisation. A city is sustainable when decisions are taken by considering their effects with a holistic approach, taking into account different areas, such as transport, health, environment, infrastructure and education, and their correlations.

Satellites alone cannot fight global warming or inequalities, but they offer data and signals that can be used to improve the life of those inhabiting urban areas.

Satellite imagery allows for an integrated view of land uses and infrastructures and it is already employed by city managers, for example to target soil and infrastructure maintenance works where they are most needed.

Satellite navigation is a precious tool to improve city management, in particular to monitor and optimise public and private transport. Indeed, satellite navigation has today a crucial role in providing real-time information on public transport and in the implementation of intermodal transport systems in cities and their hinterlands. Numerous apps use satellite navigation signals, e.g. to help persons with disabilities in their daily movements or to enable residents to access information about public services and provide feedback to their local authorities.

Satellite communication is also used in cities, to connect rescue teams when other connections are down, or to perform health checks in public spaces, among others.

These are just a few of the many existing applications of satellite data and signals which can help building safe, inclusive and resilient cities, as envisaged by the United Nations Sustainable Development Goal 11.[15] Seizing the impacts of satellite applications on urban areas is just a first step to envisage the full exploitation of these tools to build sustainable cities.

The new constellations of satellites and services deployed by single governments and by the European Union (Galileo, Copernicus and EGNOS) will provide us with additional tools to enhance economic, social and environmental well-being in cities. It is hence fundamental that cities are prepared to profit from these resources and that their needs are carefully assessed and considered when developing new products and services based on satellite applications.

[15]Sustainable Development Knowledge Platform (2018) Sustainable Development Goal 11 Make cities and human settlements inclusive, safe, resilient and sustainable. https://sustainabledevelopment.un.org/sdg11. Accessed 12 Mar 2018.

Chapter 23
Future Space Technologies for Sustainability on Earth

Stefano Ferretti, Barbara Imhof and Werner Balogh

Abstract The United Nations 2030 Agenda for Sustainable Development is based on 17 Sustainable Development Goals with 169 targets and indicators. Space science, technology and its applications can provide a wide range of solutions to help achieve the Sustainable Development Goals and thus contribute to achieving economic, social and environmental sustainable development. This paper addresses the contributions of space activities from three different perspectives: policy, strategy and technology. It focuses on a sub-set of challenges linked to the Sustainable Development Goals, namely global health, water, energy and urban development. The proposed strategy perspective considers elements, such as interdisciplinarity, spin-in and spin-off transfers, open innovation processes, and sustainability at large. Space exploration programmes are usually conceived around space mission requirements and technologies with a maturity level that allows for their implementation into planned technology roadmaps. Therefore, it is discussed how such roadmaps could better integrate policy and strategic aspects linked to sustainability on Earth. What if a system is efficiently designed to operate in space and at the same time allows for sustainable development on Earth? The further integration of key enabling technologies (big data, artificial intelligence systems, advanced robotics) are opening a new era in the exploration of other planets where autonomy is an essential requirement. At the same time such developments can become an integral part of the future developments on Earth, providing smart solutions to the citizens of tomorrow and opening up new business sectors associated to these spin-offs. For example, the development of Additive Layer Manufacturing technology will simplify the production of mechanical components and their logistic chain. However, it will also open up new ways of thinking on a large scale, for instance in the construction of buildings and structures using local materials, such as Moon regolith or Earth sand. Other examples are In

S. Ferretti (✉)
European Space Policy Institute (ESPI), Vienna, Austria
e-mail: stefano.ferretti@esa.int

B. Imhof
LIQUIFER Systems Group, Vienna, Austria
e-mail: barbara.imhof@liquifer.com

W. Balogh
World Meteorological Organization (WMO), Geneva, Switzerland

S. Ferretti (ed.), *Space Capacity Building in the XXI Century*, Studies in Space Policy 22,
https://doi.org/10.1007/978-3-030-21938-3_23

Situ Resource Utilisation and viewing a building or a city as a spaceship system, which will not only allow space habitats to include self-regenerative functions, but also allow smart cities on Earth to become greener and more sustainable, especially in view of the expansion of population, the resulting densification and increase of urban areas. The aim of the strategy proposed in this paper is to approach sustainability for development from a holistic perspective looking not only at a product of space technology and how one could transfer this into a terrestrial application but also at strategies for achieving spin-offs in compliance with the most urging topics of this century.

23.1 Introduction

Since the beginning of the space age in the last century we have developed technologies that allow us to explore outer space, to live permanently on board of the International Space Station and to even safely land humans on the Moon.

The technological developments of humankind have been staggering, but at the same time, we are reaching the limits of many of Earth's planetary boundaries, depriving the options available to future generations and limiting their possibilities to live on our planet having the same resources available to them as their ancestors did.

Our home planet is subject to increasing pressures on environment and resources mainly due to the growth of the population, which is projected to increase from currently 7.5 billion to reach 8.5 billion by 2030, 9.7 billion by 2050 and 11.2 billion in the year 2100.[1]

To address these challenges, we cannot take a "business-as-usual" development approach, but we will require the help of new technologies in moving towards a trajectory for sustainable development. It is therefore important and timely to start changing the way we live. One possibility is to create a link between new space exploration ventures and sustainable living here on Earth.

Space is entering a new age, which in Europe is now called Space 4.0. At the same time, the world delegations at the United Nations, just approved the Agenda 2030 for sustainable development.[2]

The Agenda is a call and plan for action for people, planet, prosperity and peace to be achieved in partnership with no one left behind. Taking into account the lessons learned in the implementation of the Millennium Development Goals (MDGs) in

[1]United Nations. Department of Economic and Social Affairs, Population Division. World population prospects: 2015 revision. https://esa.un.org/unpd/wpp/. Accessed September 7, 2016.

[2]Resolution adopted by the General Assembly on 25 September 2016. Transforming our world: The 2030 Agenda for Sustainable Development. A/RES/70/1, 21 October 2015.

Fig. 23.1 SDGs and Agenda 2030

the 2000–2015 period,[3] the 2030 Agenda encompasses the three dimensions of sustainable development, namely, economic, social and environmental development.

This agenda is based on 17 Sustainable Development Goals with 169 targets and indicators (see Fig. 23.1).[4]

The 2030 Agenda is applicable to countries at all levels of development. Its successful implementation will require all stakeholders to contribute, national and international institutions and organizations as well as the individual citizens of this World. Achieving the SDGs will be essential for the future of our planet and its inhabitants.

Space science, technology and its applications can provide a wide range of solutions to help achieve these Sustainable Development Goals and will be essential for successfully implementing the 2030 Agenda. Several space-related organizations have published documents or undertaken studies that assess how space applications can contribute to achieving the SDGs. Among them are the European Space Agency (ESA),[5] the European Space Policy Institute (ESPI),[6] the Group on Earth Observations (GEO),[7] the Committee on Earth Observation Satellites

[3]United Nations. Millennium development goals. https://www.un.org/millenniumgoals/. Accessed September 7, 2016.

[4]United Nations. Sustainable development goals. https://sustainabledevelopment.un.org/sdgs. Accessed September 7, 2016.

[5]European Space Agency and the Sustainable Development Goals. http://www.esa.int/Our_Activities/Preparing_for_the_Future/Space_for_Earth/ESA_and_the_Sustainable_Development_Goals. Accessed September 7, 2016.

[6]European Space Policy Institute. (2016, June) *Space for Sustainable Development.* ESPI Report 59.

[7]Group on Earth Observations and the 2030 Sustainable Development Agenda. http://www.earthobservations.org/geo_sdgs.php. Accessed September 7, 2016.

(CEOS)[8] and DigitalGlobe, in collaboration with UNOOSA, GEO and the United Nations Committee of Experts on Global Geospatial Information Management (UN-GG IM).[9]

Information and Communications Technology (ICTs) is also making use of space technology, satellite telecommunications and space-based positioning, navigation and timing services.[10]

Space technology can support the 2030 Agenda implementation in two ways:

(a) By providing data, information and services that directly or indirectly contribute to achieving particular SDGs.
(b) By providing data and information on particular SDG indicators that allow us to assess and measure the status of the implementation progress.

It is, therefore, key that a constructive dialogue is built between space and terrestrial sustainable development actors, in order to exploit the full potential of space by providing technologies and services that address the needs in the field.[11]

The United Nations are well placed at this interface and the following paragraph will address the UN perspective.

23.2 The Larger Perspective Leading to UNISPACE+50 and the United Nations Plans for Future Space Applications

The United Nations have been involved in space activities since the beginning of the space age in the 1950s. As a result of the launch of the first Earth-orbiting artificial satellite, the Member States of the United Nations agreed to establish the United Nations Committee on the Peaceful Uses of Outer Space (COPUOS), which remains the only Committee of the General Assembly exclusively concerned with international cooperation in the exploration and peaceful uses of outer space.

The Committee is supported by the United Nations Office for Outer Space Affairs of the Secretariat (UNOOSA), located at the United Nations Office at Vienna. Among its wide range of responsibilities, UNOOSA is implementing the United Nations Programme on Space Applications, the United Nations Platform for Space-based Information for Disaster Management and Emergency Response (UNSPIDER) and

---

[8]Committee on Earth Observation Satellites. http://ceos.org/ourwork/other-ceos-activities/sustainable-development-goals/. Accessed September 7, 2016.

[9]DigitalGlobe. (2016). *Transforming our world—Geospatial information key to achieving the 2030 agenda for sustainable development.*

[10]Sustainable Development Solutions Network. (2016, May) *The Earth Institute Columbia University and Ericsson. ICT & SDGs—How information and communications technology can accelerate action on the sustainable development goals.* Final Report.

[11]Ferretti, S., Feustel-Büechl, J., Gibson, R., Hulsroj, P., Papp, A., & Veit, E. (2016, June) *Space for sustainable, development.* ESPI Report 59.

acting as the Secretariat for UN-Space, the inter-agency mechanism for the coordination of space-related activities in the United Nations, and for the International Committee on Global Navigation Satellite Systems (ICG).

The mandate of the Programme on Space Applications is to assist Member States, in particular the developing countries, in using space science, technology and its applications.[12]

The Programme focusses its activities on three initiatives (Basic Space Science Initiative, Basic Space Technology Initiative, Human Space Technology Initiative) and several thematic priorities (Biodiversity and Ecosystems, Climate Change, Disaster Management, Environmental Monitoring and Natural Resource Management, Global Health).

The Programme is implemented through the organization of workshops, training courses, technical assistance missions, long-term fellowship programmes and through activities of the Regional Centres for Space Science and Technology Education, affiliated to the United Nations.

Presently the Office, in consultation with Member States and COPUOS, is considering its future strategy towards assisting Member States with implementing the 2030 Agenda for Sustainable Development. Linked to this agenda is the Sendai Framework for Disaster Risk Reduction and the Paris Climate Change Agreement.

COPUOS is addressing a wide-range of issues linked to the applications of space technology in addressing global issues, which is reflected in the items on the agenda of the annual sessions of its main committee, which include “Space and sustainable development”, “Spin-off benefits of space technology: review of current status”, “Space and water”, “Space and climate change”, and “Use of space technology in the United Nations system”.[13]

In this context the Office and the Committee held in 2018 the fourth United Nations Conference on the Exploration and Peaceful Uses of Outer Space (UNISPACE), entitled UNISPACE+50, celebrating the 50th anniversary of the first UNISPACE Conference, held in 1968.[14,15,16]

UNISPACE+50 charted the future role of the Committee on the Peaceful Uses of Outer Space, its subsidiary bodies and the Office of Outer Space Affairs at a time of an evolving and increasingly complex space agenda, when a growing number

---

[12]United Nations. (2012, September). United Nations Programme on Space Applications. ST/SPACE/52/Rev 1, V.12-55442.

[13]Committee on the Peaceful Uses of Outer Space: 2016. http://www.unoosa.org/oosa/en/ourwork/copuos/2016/index.html. Accessed September 7, 2016.

[14]United Nations, General Assembly, Official Records Twenty-Third Session, Agenda item 24, Report of the Committee on the Peaceful Uses of Outer Space, Annex II “Documentation on the United Nations Conference on the Exploration and Peaceful Uses of Outer Space”, A/7285, New York 1968.

[15]Resolution adopted by the General Assembly on 9 December 2015. (2015, December 15).*International cooperation in the peaceful uses of outer space*. A/RES/70/82, p. 14.

[16]UNSPACE+50 webpage. http://www.unoosa.org/oosa/en/ourwork/unispaceplus50/index.html. Accessed September 7, 2016.

of new actors, both governmental and nongovernmental, are getting involved with space activities.

It provides concrete deliverables of space activities for the development of nations under the four pillars space economy, space society, space accessibility and space diplomacy. It is built around seven thematic priorities:[17]

1. Global partnership in space exploration and innovation;
2. Legal regime of outer space and global space governance: current and future perspectives;
3. Enhanced information exchange on space objects and events;
4. International framework for space weather services;
5. Strengthened space cooperation for global health;
6. International cooperation towards low-emission and resilient societies; and
7. Capacity-building for the twenty-first century.

These thematic priorities are linked to the 2030 Agenda. There are various opportunities for development actors to contribute to the work of the Committee and the Office, either by participating in relevant discussions of COPUOS, its subcommittees, working and expert groups, by contributing to the activities of the United Nations Programme on Space Applications, UN-SPIDER, ICG and the open informal sessions of UN-Space and by participating in UNISPACE+50 and its High Level Fora.

23.3 Future Challenges and Space Technologies

Space activities can contribute to the Agenda 2030 from three different perspectives:

1. policy
2. strategy
3. technology.

The current major role played by space, concerns the majority of challenges linked to the Sustainable Development Goals, for example global health, water, energy and urban development.

Global health can be tackled making use of telemedicine, via satellite communications links, for primary and secondary care, tele-epidemiology tools based on Earth Observation (EO) data, spin-off technologies developed for human spaceflight missions.

But we should also keep in mind the policy tools offered by space, such as the Charter of Space and major disasters, which has been activated in case of major health crises around the world, providing support information to key stakeholders and decision makers.

[17]United Nations. Report of the Committee on the Peaceful Uses of Outer Space, Fifty-ninth session (8–17 June 2016), A/71/20, pp. 296–297.

Similarly, water management can be enhanced by the exploitation of EO data, but water can be purified by using ISS derivative technologies, as in the case of the Melissa technology transfer project in developing countries.

In the field of energy, space offers a number of effective solutions and efficient technologies, particularly for solar photovoltaic installations. But space can also offer tools to precisely estimate the solar energy potential of an area, the wind power or the assessment of geothermal reservoirs.

Infrastructure management and planning as well as smart city development can highly benefit of integrated space services, linking energy, mobility, clean air, waste management, education and e-health services in an inextricable nexus.[18]

The overall strategic perspective presented in this chapter, considers elements, such as interdisciplinarity, spin-in and spin-off transfers, open innovation processes, and sustainability at large, as key factors for success.

23.4 From Policy to Strategy: How Top Level Policy Objectives Can Become Embedded in Roadmaps and Fit into a New Model of Thinking and Doing

In order to implement the policy objectives reflected in Agenda 2030, Space Agencies are developing strategies and roadmaps, which aim to integrate space and society to the maximum extent. The European approach consist in the full deployment of Space 4.0, which is based upon:

- Seizing the potential of growth and cooperation
- the dynamics of innovation
- full integration of digital technologies
- new business models leading to smart integrated services

This innovative approach includes further development of existing partnerships with non-traditional space actors and the creation of new ones, in order to enhance the exchange between the suppliers of (space) services and the demand side (e.g. civil society, non-governmental organizations, development actors).[19]

In this context space could become an enabler of economic growth, political and strategic alliances, particularly in the domain of sustainable development and environmental protection, where Europe can play a leading role in the global context. In order to achieve this enabling function at best, Space needs to move from its comfort zone of a technology push approach, towards a new model based on a technology demand pull model, which puts the end user at the centre. In this pull-model the end user's role would entail participating in the requirements definition and conceptual design phase including the operational activities and full services

[18] Aliberti, M., Ferretti, S., Hulsroj, P., Lahcen, A. (2016, January) Europe in the Future and the Contributions of Space, ESPI, Report 55.

[19] See Footnote 11.

provision. This can be well achieved through the implementation of open innovation and open service innovation models.[20]

In order to open up space to a co-creation culture, as advocated in the open innovation models, the key is interdisciplinarity. Space has been quite interdisciplinary so far, involving all sorts of knowledge areas: from science and engineering to social sciences, from business and management to economy and finance, from history and art to law and ethics. However, since the 1980s there has been a dominance of engineers over the other disciplines so a new era of Space 4.0 in a co-creation environment would give each player and discipline a more democratic importance.

It is important that all these professional fields receive equal chances of active contribution so that the full potential of Space 4.0 can be released for the benefit of future generations. This concept can be taken further, as intended by the European Space Agency, which is developing the Space 4.0i, "Innovate! Inform! Inspire! Interact!" (see Fig. 23.2).

In this context, it is worth considering how new roadmaps could better integrate policy and strategic aspects linked to sustainability on Earth.

In the specific case of space exploration programmes, they are usually conceived around space mission requirements and technologies with a maturity level that allows for their implementation, not only in space but also on Earth.

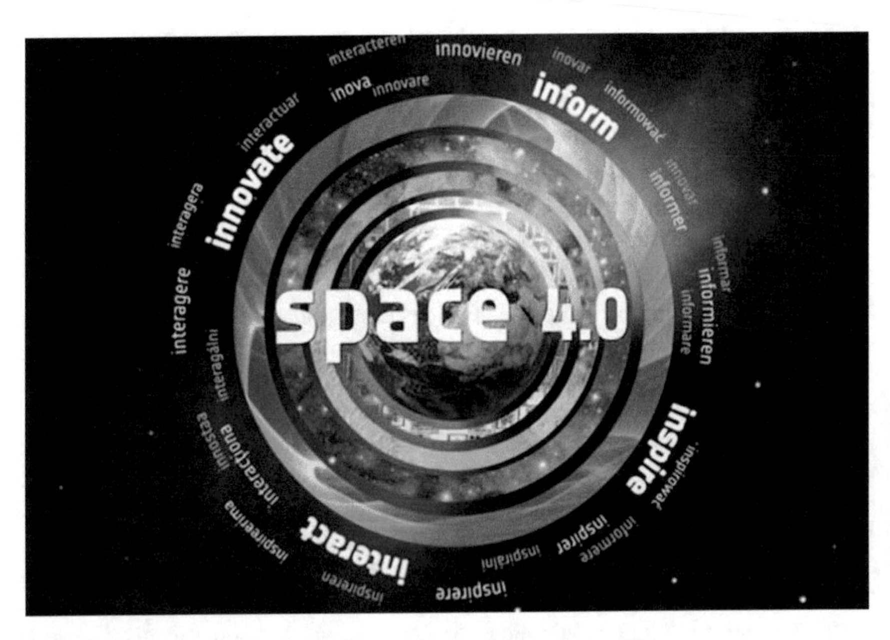

Fig. 23.2 Space 4.0, "Innovate! Inform! Inspire! Interact!". *Credits* ESA

[20]Chesbrough, H. (2003). Open Innovation: The new imperative for creating and profiting from technology. Harvard Business School Press.

MELiSSA (Micro-Ecological Life Support System Alternative) is a European Project to develop a closed life support system, aiming at producing food, water and oxygen for manned space missions. But in fact this project demonstrates that a system efficiently designed to operate in space, can in principle allow for sustainable development on Earth as well.

The further integration of key enabling technologies (big data, artificial intelligence systems, advanced robotics) are opening a new era in the exploration of other planets where autonomy is an essential requirement.

At the same time such developments can become an integral part of future developments on Earth, providing smart solutions to a future generation and opening up new business sectors associated to these spin-offs.

For example, the development of Additive Layer Manufacturing technology will simplify the production of mechanical components and their logistic chain, offering new options for space exploration missions.

In this context, the examples of companies such as "D-Shape", "Made in Space" and "Field Ready" demonstrate that developing 3D printing solutions for space create know-how that can be transferred to the logistic and manufacturing sectors on Earth.

On a larger scale and broader horizon, it can be noted that developing concepts for 3D printing constructions blocks for the "Moon Village" out of lunar regolith, could also take us a step further in 3D printing from sand on terrestrial grounds.

An example for this is the spin-off of the company "D-Shape" who works in the field of construction of structures using local materials, such as Earth sand.

Other examples of In Situ Resource Utilisation are related to viewing a building or a city as a spaceship system. This idea is thought around five parameters or ecologies which are inherent to a spaceship and should also be defining buildings on our spaceship earth.

23.5 The Paradigm of the City as a Spaceship (CAAS) as Inspiration Towards a Sustainable Use of Space Technologies

As first change of paradigm we have to establish Buckminster Fuller‘s notion of the fact that we are in space, whether we live on earth or on ISS. We are astronauts living on spaceship earth.[21]

As a human species we have exploited our home base since the dawn of humankind and today, with a constant increase of the world's population and the fast advancement of technology we have become extremely efficient in milking the earth's resources in a dangerous way.

Fast climate change can be observed, a surge of disasters and destruction of natural environments and eco-systems.

---

[21]Buckminster Fuller, R. (1968). Operating Manual For Spaceship Earth.

Humans refer to the countermeasures as a need to protect and guard the environment. But it is not so much about the environment than our species we need to protect. The environment will survive once we are long gone, earth, too—but humans are in danger.

With this premise in mind, it seems useful to start efforts to flee our spaceship to another, maybe the moon or Mars. Will humans then and on foreign terrains be able to find a sustainable way of living there? It seems advisable to start here and now and not delay the protection of humankind to another time or another place.

However, mainly current launch technologies and their costs keeps people on earth. Reusable rockets, cheaper launches, more efficiency with propellant and optimised rocket structures to reduce the weight, will support more affordable lift-offs with greater mass.

If a space station or spaceship is conceived from these constraining parameters of a launch system then the requirements include light-weight materials, limited habitation space and closed-loop life support systems to recycle every ounce of scarce resource and sustainable energy sources. This leads to primary spaceship parameters defined in.[22] "City as a Spaceship (CAAS) inspires technological human innovation by positioning the spaceship as an analogy of the modern densely built urban space with its complex structures and technologically-intelligent infrastructure".[23]

Currently, there is only one permanently inhabited spaceship, the International Space Station. Therefore, CAAS takes the ISS as reference and with this example the authors derived five parameters which they called ecologies. In a space terminology they could be interpreted as sub-systems.

Shelter as Transformable: a space that allows for multi-functionality, which defines interior and exterior and distinguishes private from public. The ISS is in an orbit 350 km above the earth's surface, in micro-gravity with an outside temperature range of −250 to +150 Degrees Celsius. In addition it orbits our planet in a near vacuum. As in a city every m3 of space must be used in many ways, it might be transformed during missions and needs to offer a lot of storage space. The main themes whether we live in space or on terrestrial grounds are: transformability of spaces, multi-functionality of furniture and systems including mobility, important to design and planning in all scales.

Energy as Renewable: with the advancement of technology and the increase of equal distribution amongst humankind the energy demand will rise especially in urban areas. Fossil fuel resources are limited and will be dry in the near future or too expensive to harvest. Three strategies can therefore be adopted: technologies requiring less energy, being more careful in using energy, energy sources from clean and renewable sources. A spaceship energy ecology implies to use clean and renewable energy sources, as exemplified on the International Space Station where all energy is harvested from the sun.

[22]CITY AS A SPACESHIP (CAAS), Fairburn, S., Mohanty, S., Imhof, B. (2014). 65th International Astronautical Congress, Toronto, Canada. IAC-14-E4.2.8,#20927.

[23]See Footnote 22.

Fig. 23.3 ISS Canadarm2 grapples the SpaceX Dragon. *Credits* NASA

Technology, Automation, and Infrastructure: the ISS was assembled solely through support of the robotic device, namely the 'Canadarm' (see Fig. 23.3) and through automated docking processes. Further, the whole space of ISS resembles a technologized envelope in which the crew lives in. Space Station can be viewed as precursor to the development of future smart homes on our planet.

All on-board ISS life support systems are automatic. If needed astronauts can manually override the system and control it. A diagnostic tool has been installed that allows detailed on-orbit monitoring and logging of all avionics bus messages; the nerve system of the Space Station. CAAS renders the view of spaceflight parameters feeding into future smart city designs.

Construction, communication, health, traffic and infrastructure support and maintenance will be controlled by intelligent systems.

Inhabitants: the ISS crews come like citizens in conurbations from different professional and cultural backgrounds. They come from the US, Canada, Russia, European countries, and Japan and are medical doctors, engineers, and scientists. Living together, in these tight spaces in extreme environments bears its' challenges. On ISS they are mitigated through cultural training and extensive work time together preparing for the mission. Cities are faced with the challenges of integration of very different groups of population especially in times of migration and war refugees. Cultural training is nothing superfluous but essential and the costs will pay off in the future.

Life support systems: water and air, essential to life are only available in a very limited way in closed environments. These and all other resources need to be treated carefully, responsibly, and sustainably through recycling. On Space Station astronauts drink purified water recycled from their urine. When going for long duration missions on extra-terrestrial surfaces we will need a sustainable food system that securely produces nutrition for the crew. In the loop of air, water and food we also need to integrate the waste management that is up to now still "messy" and not treated as part of a closed-loop system. Intensification of harvest in limited space, reducing the amount of waste and recovering waste into useful material will become essential factors to address.

We will not be able to ship food to Mars once we have settled there. Cities on our home planet are market places, however, fresh goods are better taken from the close environment.

When cities grow into mega-structures, space will be sparse like on a spaceship so we need functioning greenhouse systems. Spaceflight could help to visualize the concept and even technologies and knowledge originally developed for the extreme requirements of space could be implemented on our home spaceship.

These parameters were established as part of the project CAAS with the intention to show the sustainability of space exploration. These five ecologies as presented above can become a conceptual role model for how humans live together, cities are built and available resources used (see Figs. 23.4 and 23.5).

They can also be set against Sustainable Development Goals (SDGs) as defined in[24] or put in reference to UN's 17 Goals for Sustainable Development.

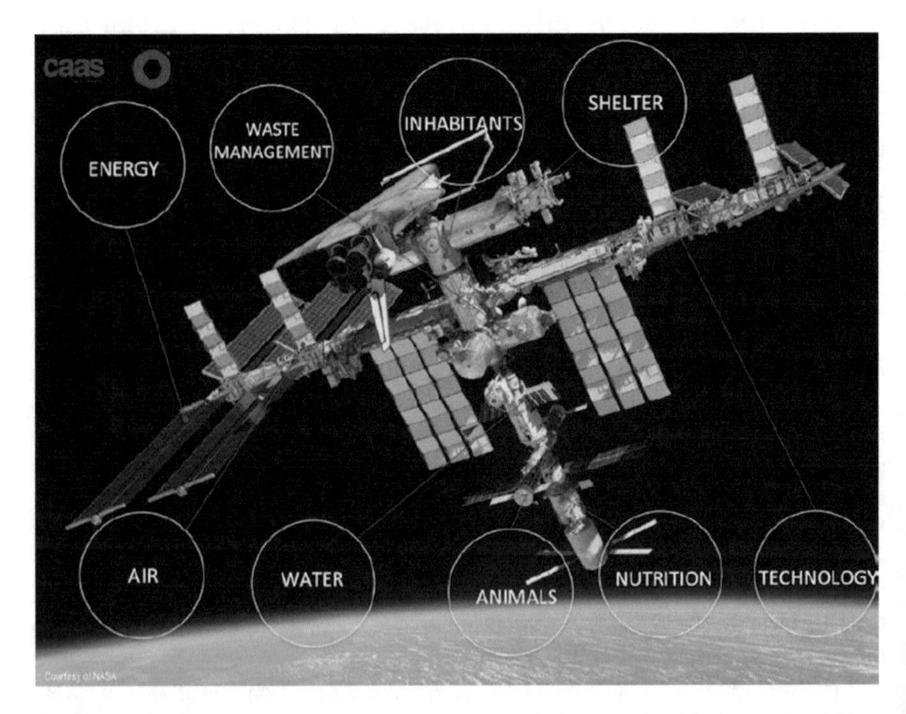

Fig. 23.4 Spaceship parameters (ecologies) displayed with the example of the International Space Station, background photo. *Credit* NASA

[24]See Footnote 11.

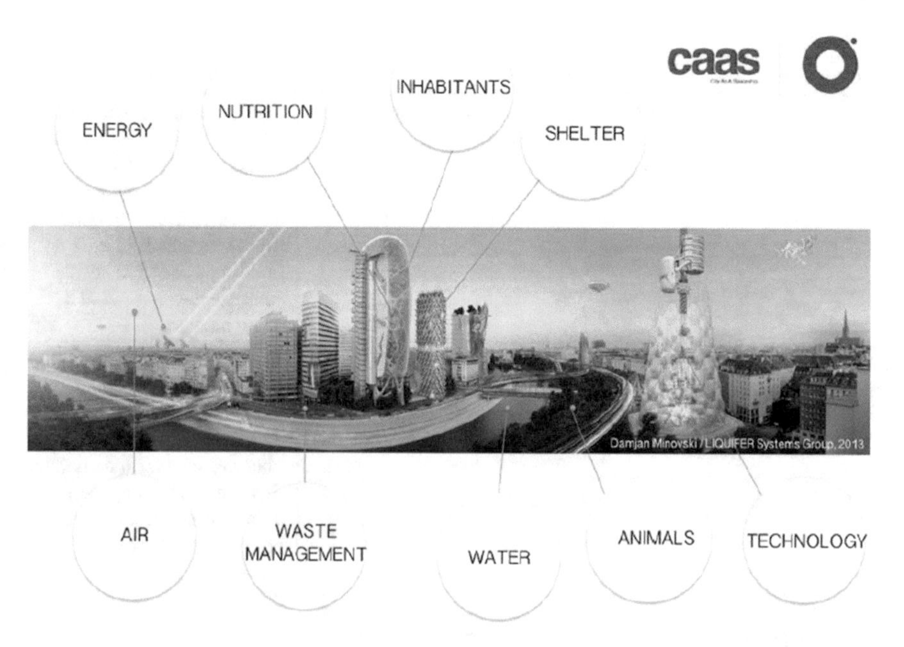

Fig. 23.5 Spaceship parameters (ecologies) overlaid on a future vision of a city. *Credit* Damjan Minovski/LIQUIFER Systems Group, 2013

The table below shows how the five CAAS ecologies can be related to SDGs under the condition that the SDGs are expanded in their meaning and implication, interpreted in a broader sense and are not only limited to the utilization of Earth Observation Data.

CAAS ecologies	SDG	Rationale
Shelter as transformable	SDG 11: Sustainable cities and communities	Mega-cities in Asia. South America and Africa have only limited land available, a comfortable living in small and crowed spaces through a careful planning as one would do for a spaceship levitates the stresses of dense population
Energy as renewable	SDG 7: Affordable and clean energy	The paradigm of ISS to harvest energy from a renewable source (the sun) can serve as role model for clean and affordable energy

(continued)

(continued)

CAAS ecologies	SDG	Rationale
Technology. automation, and infrastructure	SDG 9: Industry, innovation and infrastructure	The fact that the whole ISS was constructed via robotic arms and automated processes can show an alternative model to building processes that after a building's life time are also easily to deconstruct in a clean manner
Inhabitants	SDG 17: Partnerships SDG 16: Peace and justice—strong institutions	The ISS is model for international cooperation and many different nations, cultures and work attitudes play together On a daily basis this can be challenge but cross-cultural training the necessity to work together make people overcome differences, learn and thrive
Life support systems	SDG 6: Clean water and sanitation SDG 2: Zero hunger	From anticipated closed-loop life supporting systems we can leam to re-use water, cleanse air. and optimise food production within the city that can help solving issues of large populated areas also where they occur

23.6 Conclusion

This chapter has shown how current and future Space technology is interconnected with Sustainable Development, and it provides examples to look forward towards a wider horizon that encompasses a holistic view, underlining the importance of interdisciplinarity in looking at the sustainable development theme and space.

The aim of the strategy proposed is to approach sustainability for development from a holistic perspective looking not only at a product of space technology and how one could transfer this into a terrestrial application but also at strategies for achieving spin-offs in compliance with the most urgent topics of this century.

Space 4.0 and the Agenda 2030 will be inextricably linked in defining our global future! New innovation models are increasingly spread across sectors and disciplines, including Space, which is becoming an integral part of many societal activities (incl. telecoms, weather, climate change and environmental monitoring, civil protection, infrastructures, transportation and navigation, healthcare and education).

It can be argued that innovation drivers are definitive enablers of new functions, and Space 4.0 may certainly be essential in the successful evolution of Europe and the full implementation of the Agenda 2030 worldwide.

It is clear that it is now the time to look at how the wider perspectives and strategies can be implemented in actionable programmes in the coming years.

The future steps will be to review the outcomes of the ESA Council at Ministerial Level and the EC Space Strategy for Europe, and construct a roadmap, involving key stakeholders and civil society, in order to map out the available options and summarize the ideal programmatic conditions for successful implementation of the Agenda 2030, presented at the UNISPACE+50 conference in 2018.

These conditions may include future innovative frameworks and collaborations around Sustainable Development, to be further explored and proposed in addition to the already existing ones (e.g. ESA-World Bank agreement; the United Nations Framework Convention on Climate Change (UNFCCC) and the IAA joint Space Agencies declaration on Climate Change; the 2016 CNES-ISRO Delhi Declaration underlining contributions by the space sector in support of the COP21 outcomes; ESA-EPB agreement).

The second recommendation is to analyse how Space Agencies could improve the dialogue with NGOs and civil society in order to make them aware of the potentialities of space, and how Space actors may listen to the field and collect its needs, in order to fully implement and exploit future space programmes.

For example, designing new Copernicus services around citizen´s needs, while targeting sustainable development, may open up opportunities for Europe to play a new role worldwide.

Considering that, in the coming years, Galileo will see its full implementation, this system may be used beyond its original intent, addressing new needs and services (e.g. new vehicles like drones and self-driving cars).

SATCOMS are also on the verge of a revolution, which will impact our societies: high flying drones will be able to provide connectivity to underserved countries, satellite operators will develop new connectivity concepts and business cases (e.g. e-health and tele-education), mobile applications and innovative services will be made available to the users at an increasing pace.

The United Nations, through its Committee on the Peaceful Uses of Outer Space, provides a framework for addressing these issues at the global level. Making full use of the opportunities provided by UNISPACE+50, the Office for Outer Space Affairs, as the Secretariat of COPUOS, stands ready to assist Member States with developing the future of space governance and with implementing the capacity building actions necessary to address the global challenges of our rapidly changing world in the XXI century.[25]

Finally, it can be argued that Space is already becoming the link among systems of systems, and its enabling function may therefore represent the new element that

---

[25]United Nations General Assembly Seventieth session, Transforming our world: the 2030 Agenda for Sustainable Development, UN A/RES/70/1 (2015).

will help to bring our societies towards the goal of a sustainable living on this planet for all.

The views expressed herein are those of the authors and do not necessarily reflect the views of the United Nations. This paper was presented at the 67th International Astronautical Congress, 26–30 September 2016, Guadalajara, Mexico, https://www.iafastro.org

Chapter 24
Building Capacity and Resilience Against Diseases Transmitted via Water Under Climate Perturbations and Extreme Weather Stress

Shubha Sathyendranath, Anas Abdulaziz, Nandini Menon, Grinson George, Hayley Evers-King, Gemma Kulk, Rita Colwell, Antarpreet Jutla and Trevor Platt

Abstract It is now generally accepted that climate variability and change, occurrences of extreme weather events, urbanisation and human pressures on the environment, and high mobility of human populations, all contribute to the spread of pathogens and to outbreaks of water-borne and vector-borne diseases such as cholera and malaria. The threats are heightened by natural disasters such as floods, droughts, earth-quakes that disrupt sanitation facilities. Aligned against these risks are the laudable Sustainable Development Goals of the United Nations dealing with health, climate, life below water, and reduced inequalities. Rising to the challenges posed by these goals requires an integrated approach bringing together various scientific disciplines that deal with parts of the problem, and also the various stakeholders including the populations at risk, local governing bodies, health workers, medical professionals, international organisations, charities, and non-governmental organisations. Satellite-based instruments capable of monitoring various properties of the aquatic ecosystems and the environs have important contributions to make in this context. In this chapter, we present two case studies—the Ganga Delta region and the Vembanad Lake region in south-western India—to illustrate some of the benefits that remote sensing can bring to address the problem of global health, and use

S. Sathyendranath (✉) · H. Evers-King · G. Kulk · T. Platt
Plymouth Marine Laboratory, Plymouth, UK
e-mail: ssat@pml.ac.uk

A. Abdulaziz
National Institute of Oceanography, Kochi, India

N. Menon
Nansen Environmental Research Centre, Kochi, India

G. George
Central Marine Fisheries Research Institute, Kochi, India

R. Colwell
University of Maryland, College Park, USA

A. Jutla
West Virginia University, Morgantown, USA

S. Ferretti (ed.), *Space Capacity Building in the XXI Century*, Studies in Space Policy 22,
https://doi.org/10.1007/978-3-030-21938-3_24

these examples to identify the capacity building that is essential to maximise the exploitation of the remote sensing potential in this context.

24.1 Introduction

The human population is under threat from a suite of environmental perturbations, including global warming; shifts in global-scale distributions of precipitation and drought; population growth and the resulting increased stress on the environment; and increased threats from diseases transmitted via water (either fresh, brackish or salt). Against these multiple-stresses, there arises an imperative for strengthening human resilience to minimise risk. Building resilience is a priority to mitigate environmental damage that will be exacerbated by population pressure and poor sanitation conditions. In this chapter, we provide a perspective on demonstrated and potential capabilities of remote sensing to build capacity and resilience against water-borne and vector-borne diseases. Although the role of remote sensing in this context has been well demonstrated in some regions, it is as yet unexplored as a global public health tool. Earth observation data from satellites are increasingly available at the global scale in real time, with free, open and timely access. This resource enhances the potential for developing cost-effective methods to build a capacity to monitor risks to human health from water-borne diseases. Indeed, the problem is global, but its manifestations may be sporadic and regional in scope. With global warming and changing environmental conditions, geographic regions under threat are evolving and expanding, and so the need for both capacity and awareness building is not limited to developing countries in the tropical belt where such threats are currently more acute.

This chapter addresses effects of water-borne diseases impacting humans at the coastal fringe. A significant proportion of the world's population is at risk from such diseases, and the potential effects are severe, encapsulated in the United Nations Sustainable Development Goal 3 (Health) Target 3.3 "*By 2030, end the epidemic…. of malaria and neglected tropical diseases and combat hepatitis, water-borne disease and other communicable disease*", and in Target 3.d to "*Strengthen the capacity of all countries, in particular developing countries, for early warning, risk reduction and management of national and global health risks*". Given the influence of a changing climate on distribution and intensity of outbreaks of water-associated diseases, it is relevant to SDG13 (climate), Target 13.3 to "*improve education, awareness-raising and human institutional capacity on climate change mitigation, adaptation, impact reduction and early warning*". In the decade of the 2020s, incidence of waterborne diseases will, increasingly, be a major threat to low and middle-income countries (LMIC), notably India, where a major shortage of drinking water is predicted by 2030. Approximately 15% of the population of India lives within a 100 km periphery of a 7500 km long coastal line, and therefore is significantly vulnerable to water-associated diseases. Several diarrheal diseases (e.g., cholera, shigella, rotavirus) are frequently reported to occur in coastal regions of the developing world (Fig. 24.1).

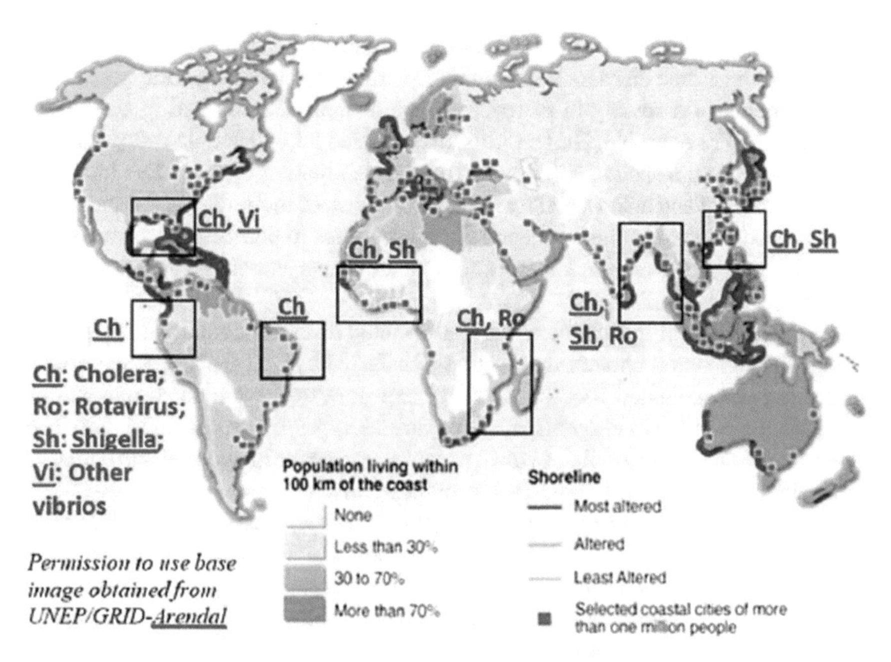

Fig. 24.1 Population distribution and water-borne disease outbreak. Rectangles represent endemic diarrheal disease regions, *Source* UNEP/GRID Arendal

The chapter is also relevant to SDG 14 (Life below water). Coastal and maritime tourism rely on healthy marine ecosystems. Tourism development must be a part of Integrated Coastal Zone Management to help conserve and preserve fragile marine ecosystems and to serve as a vehicle to promote a blue economy, contributing to the sustainable use of marine resources. This perspective also highlights the relevance of this chapter to SDG 10 (Reduced inequalities). Tourism can be a powerful tool for reducing inequalities if it engages local populations and all key stakeholders in its development. Tourism can contribute to urban renewal and rural development by giving people the opportunity to prosper in their place of origin. Furthermore, tourism serves as an effective mean for economic integration and diversification.

Specific, recent metrics demonstrating a need in the context of water-borne diseases include 216 million cases of malaria reported worldwide, and 870,000 deaths associated with lack of safe drinking water, unsafe sanitation, and lack of hygiene in 2016; they were caused mainly by diarrhoeal diseases, but also by malnutrition and intestinal nematode infections. According to the World Health Organisation, some two billion people use faecally-contaminated drinking water, rendering them vulnerable to death or chronic poor health from water-borne infectious diseases such as cholera, dysentery, typhoid and polio. Provision of safe drinking water is hostage to the influence of extreme weather, including drought and flooding, the frequency of which is increasing, as a consequence of climate change.

Apart from fatalities, the effect of a chronic burden of low-grade infections associated with water-borne diseases is antagonistic to maintenance of a healthy work force and well-being of a society in general, as well as being detrimental to sustainable development. For example, cholera kills an estimated 95,000 people every year, but causes 2.9 million to become seriously ill with debilitating disease. The incidence of diarrhoea has been acknowledged as a major cause of morbidity in coastal states, hence a need to address the resilience of communities to perturbation of their supply of safe drinking water under extreme weather events associated with a changing climate.

For the purpose of this chapter, water-associated diseases include those diseases, such as diarrhoea and cholera that can be transferred by oral-faecal routes through consuming contaminated food or water, or exposure to contaminated recreational beaches. The responsible agents may be various types of pathogenic bacteria or viruses. If we take the example of *Vibrio* bacteria,[1] the genus includes *Vibrio cholerae*, responsible for cholera (of some 200 serotypes *Vibrio cholerae*, only two—O1 and O139—are associated with cholera epidemics); *V. alginolyticus* that causes infections of wounds and ears exposed to contaminated water; *V. parahaemolyticus*, associated with food poisoning from consumption of seafood, causing acute gastroenteritis, with symptoms including cramps, diarrhoea, headache, fever and chills; and *V. vulnificus*, a food-borne pathogen known to cause wound infections, primary septicaemia and food-poisoning that can be fatal. Diseases carried by vectors, such as mosquitoes, which require (stagnant) water to complete parts of their life cycle, are also included in the family of water-associated diseases. Mosquitoes are known vectors for a host of diseases, including malaria, dengue fever, chikungunya and Zika. Fish and seafood can be sources of food-borne pathogens such as *Salmonella*, *Listeria*, *Escherichia coli*, and *Campylobacter*.

Here we focus largely on two contrasting case studies from coastal regions of the Indian subcontinent that are highly vulnerable to water-associated diseases: the Ganga Delta where seasonal outbreaks of cholera are routine, and which has been the subject of a number of studies demonstrating the applications of remote sensing; and Kerala in South India, also an area where cholera is endemic, but which presents quite different patterns of disease outbreaks. Mortality rates are comparatively low in Kerala owing probably, and at least in part, to the high standard of personal hygiene and health facilities. Global implications of regional problems are also touched upon.

Dealing with water-associated diseases requires a whole-ecosystem approach to management, including: monitoring the pathogen load in the environment; identifying its reservoirs, routes of transmission, and conditions for proliferation; optimization and validation of tools such as remote sensing; ecosystem models for early warning of disease incidence; and designing strategies to involve people in understanding the causes of disease occurrence and mitigate them.

---

[1] Baker-Austin, C., JoaquinTrinanes, J., Gonzalez-Escalona, N., & Martinez-Urtaza, J. (2017). Non-cholera vibrios: The microbial barometer of climate change. *Trends in Microbiology*, *25*(1), 76–84. https://doi.org/10.1016/j.tim.2016.09.008.

Within the armoury that might be mobilised against water-associated diseases, remote sensing is proving to be of increasing utility. To characterise the ocean, space-borne sensors use radiometric signals from various parts of the electromagnetic spectrum (visible, infra-red and microwave) to yield data at regional to global scales with fine local resolution (one kilometre or better) and temporal resolution (nominally one day) on a variety of environmental properties, including sea surface temperature, phytoplankton abundance indexed as concentration of chlorophyll, and sea surface height.

For illustration of the concepts and for identifying the priorities in future work, we turn next to two case studies.

24.2 The Ganga Delta Region

The Ganga Delta region, straddling India and Bangladesh, covers an area of over 100,000 km^2, making it the largest delta in the world. It is the drainage basin for many rivers, the most important being the Ganga and the Brahmaputra. It is a highly fertile region, and an important fishing zone. Over a 100 million people live in the delta. It is exposed periodically to flooding during the monsoon season, and to droughts in summer. Akanda et al.[2] classify cholera as epidemic, mixed-mode, and endemic. All three types are reported to occur in the Ganga Delta, and typically there are two seasonal outbreaks of cholera in the region every year. Several authors have reported the use of environmental data in general and satellite data in particular, to understand and predict outbreaks of cholera in the region. There is a wealth of clinical and environmental data related to cholera from the Ganga Delta region; it has been the subject of various studies on the relationships between environmental conditions and disease outbreaks; and various methods have been proposed to forecast risks of disease outbreaks in the region (Table 24.1).

Mutreja et al.[3] employed molecular techniques to study the bacteria responsible for the seventh pandemic of cholera and concluded that the pandemic occurred globally over the years, from a source in the Bay of Bengal area, primarily through human-to-human transmission. However, it has been established that the cholera bacterium, *Vibrio cholerae* is autochthonous to aquatic environment.[4] Thus, when exploring regional disease outbreaks, it is important not to overlook the global distribution of this bacterium.

---

[2]Akanda, A. S., Aziz, S., Jutla, A., Huq, A., Alam, M., Ahsan, G. U., et al. (2018) Satellites and cell phones form a cholera early-warning system. *Eos*, 99. https://doi.org/10.1029/2018EO094839. Published on March 27, 2018.

[3]Mutreja, A., Kim, D. W., Thomson, N. R., Connor, T. R., Lee, J. H., Kariuki, S., et al. (2011). Evidence for several waves of global transmission in the seventh cholera pandemic. *Nature*, *477*, 462–465. https://doi.org/10.1038/nature10392.

[4]Huq, A., Sack, R. B., Nizam, A., Longini, I. M., Nair, G. B., Ali, A., et al. (2005). Critical factors influencing the occurrence of *Vibrio cholerae* in the environment of Bangladesh. *Applied and Environmental Microbiology, 71*(8), 4645–4654. 10.1128/AEM.71.8.4645–4654.2005.

Table 24.1 A few selected publications that have reported relationships between environmental conditions, environmental cholera and cholera outbreaks in the Ganga Delta region

Environmental variables studied	Remarks	Reference
Landsat: NDVI, land use MODIS/MERIS: SST, ocean colour SWOT (planned-): river discharge SMAP: soil moisture GRACE: water storage, river discharge TRMM/GPM: Precipitation TOPEX/Poseidon/JASON: Sea surface height AVHRR: SST	A consortium of satellite sensors used to predict cholera, with models and data used varying according to whether epidemic, mixed-mode endemic or endemic cholera is being studied. Reports on the use of cell phones to survey the local population and to disseminate advisories	Akanda et al. (2018)[a]
GRACE: terrestrial water storage	Region of study: the Bengal Delta. GRACE data used to study anomalies in terrestrial water storage in relation to cholera outbreaks. Predictive model	Jutla et al. (2015)[b]
Chlorophyll, river discharge	Regions of study: Bangladesh, Mozambique. Chlorophyll concentration is used as a proxy for plankton abundance, since cholera bacteria are known to attach themselves to plankton	Jutla et al. (2010)[c]
Ocean-colour reflectances at 412 and 555 nm	Predictive model for Bengal Delta and Mozambique. Proposes a “satellite water marker” based on the ocean colour signal at two wavelengths, which is then used in predictive model for cholera incidence	Jutla et al. (2013)[d]
Climate variability (El Niño southern oscillation) and flooding	A multidimensional inhomogeneous Markov chain Model used. Predictive model with 11 month lead time	Reiner et al. (2012)[e]
SST, SSH and chlorophyll	Region of study: Bangladesh. Demonstrates link between cholera epidemics and regional climate, using time series of satellite data	Lobitz et al. (2000)[f]
Chlorophyll concentration, Rainfall, SST	Regions of study: Kolkatta (India) and Matlab (Bangladesh). Predictive models developed showed differences between the two sites, highlighting the importance of local information	De Magny et al. (2008)[g]

(continued)

Table 24.1 (continued)

Environmental variables studied	Remarks	Reference
Water temperature, water depth, rainfall, conductivity, and copepod counts	Four isolated sites in Bangladesh studied. Predictive model proposed using water temperature (lag, 6 weeks), ctx gene probe count (lag, 0 weeks), conductivity (lag, 0 weeks), and rainfall (lag, 8 weeks). "Occurrence of cholera correlates with biological parameters, e.g., plankton population blooms"	Huq et al. (2005)[h]
pH, temperature, iron, salinity, copepods, diatoms, dinoflagellates	Region of focus: Bangladesh. One of the earliest papers to demonstrate the use of satellite data to illustrate the link between cholera and climate	Colwell (1996)[i]

When satellite data are used, the names of the satellites or sensors, and the products are noted. A list of the acronyms used and their expansions are provided at the beginning of the chapter. Note that ocean colour (remote sensing reflectances at various spectral bands in the visible domain) can be used to retrieve chlorophyll concentration and indicators of water clarity (diffuse attenuation coefficient)

[a]Akanda, A. S., Aziz, S., Jutla, A., Huq, A., Alam, M., Ahsan, G. U., et al. (2018) Satellites and cell phones form a cholera early-warning system. *Eos*, 99. https://doi.org/10.1029/2018EO094839. Published on March 27, 2018.

[b]Jutla, A., Akanda, A., Unnikrishnan, A., Huq, A., & Colwell, R. (2015). Predictive time series analysis linking Bengal cholera with terrestrial water storage measured from gravity recovery and climate experimental sensors. *American Journal of Tropical Medicine and Hygiene, 93*(6), 1179–1186.

[c]Jutla, A., Akanda, A. S., & Islan, S. (2010). Tracking cholera in coastal regions using satellite observations. *JAWRA Journal of the American Water Resources Association, 45*(4), 651–662. https://doi.org/10.1111/j.1752-1688.2010.00448.x.

[d]Jutla, A., Akanda, A. S., Huq, A., Faruque, A. S. G., Colwell, R., & Islam, S. (2013). A water marker monitored by satellites to predict seasonal endemic cholera. *Remote Sensing Letters, 4*(8), 822–831. https://doi.org/10.1080/2150704X.2013.802097.

[e]Reiner, R. C, Jr., King, A. A., Emch, M., Yunus, M., Faruque, A. S. G., & Pascual, M. (2012). Highly localized sensitivity to climate forcing drives endemic cholera in a megacity. *PNAS, 109*(6), 2033–2036.

[f]Lobitz, B., Beck, L., Huq, A., Wood, B., Fuchs, G., Faruque, A. S. G., et al. (2000). Climate and infectious disease: Use of remote sensing for detection of *Vibrio cholerae* by indirect measurement. *Proceedings of the National Academy of Sciences, 97*(4), 1438–1443.

[g]de Magny, G. C., Murtugudde, R., Sapiano, M. R. P., Nizam, A., Brown, C. W., Busalacchi, A. J., et al. (2008). Environmental signatures associated with cholera epidemics. *Proceedings of the National Academy of Science, 105*(46), 17676–17681. www.pnas.orgcgidoi10.1073pnas.0809654105.

[h]Huq, A., Sack, R. B., Nizam, A., Longini, I. M., Nair, G. B., Ali, A., et al. (2005). Critical factors influencing the occurrence of *Vibrio cholerae* in the environment of Bangladesh. *Applied and Environmental Microbiology, 71*(8), 4645–4654. 10.1128/AEM.71.8.4645–4654.2005.

[i]Colwell, R. R. (1996). Global climate and infectious disease: The cholera paradigm. *Science, 274,* 2025–2031.

24.3 Vembanad Lake in Kerala, India

Vembanad Lake, along with its neighbouring wetlands and coastal waters of Kerala in India provides a contrasting study area, although from a hydrological perspective, it mimics many aspects of the Ganga Delta, but at a much smaller scale. The lake, which straddles the districts of Alappuzha, Kottayam and Kochi in Kerala (India), is the largest lake in South India. With a total length of approximately 100 km, it is also the longest lake in all India. It forms part of the Vembanad-Kol wetland system, recognized internationally under the Ramsar Convention for conservation and sustainable use of wetlands. Some ten rivers flow into the lake. Its drainage area is estimated to be 40% of the entire state of Kerala, and 1.6 million people live on its shores. The lives of the people are intimately linked to the lake: it is used for bathing, washing, and cooking, and is an important means of local transport. Kuttanad region on the southern part of the lake, often referred to as the rice bowl of Kerala, is one of the few places in the world where rice is cultivated below sea level (1–3 m below mean sea level). Because of its "biosaline farming", the Kuttanad farming system is recognized by FAO as a Globally Important Agricultural Heritage System.

A cultural icon, and a location of exceptional natural beauty, Vembanad Lake is considered a "safe" tourist destination that brings much needed revenue to the region: according to the 2017 Kerala tourist statistics, the income to Kerala from tourism in 2017 was INR 33,383.68 crores (or approximately 3.8 billion GBP), an increase of 13% over the previous year, and was responsible for 23% of the total employment in the state. Vembanad Lake is central to the industry, with thousands of house boats plying the waters of the lake during the tourist season.

Yet, all is not well in "God's own country" (the tag line of Kerala Tourism). The lake is now highly polluted: pressure from the high population density around the lake, growing prosperity of the state with associated construction work on the banks of the river, increasing tourism, all of which contribute to the overall problem. Various industrial units on the estuarine stretches of Vembanad Lake utilise water from the lake for routine operations and discharge effluent back to the lake. Biodiversity and fisheries are on the decline, and invasive species, notably water hyacinth (*Eichhornia crassipes*), and water moss (*Salvinia molesta*) are particularly problematic. From a health perspective, the incidence of water-associated diseases such as chikungunya and dengue fever, once unheard of in the region, are now on the increase. Malaria, previously eradicated, is making a comeback (but Kerala is still considered low risk for malaria). The lake and the surrounding low-lying regions are subject to frequent flooding, especially during the monsoon season. The Kerala floods of 2018 were exceptionally catastrophic, with up to a million people displaced, a reported death toll close to 400, and widespread damage to public infrastructure, private homes and natural ecosystems. The region around Vembanad Lake felt the brunt of the devastation.

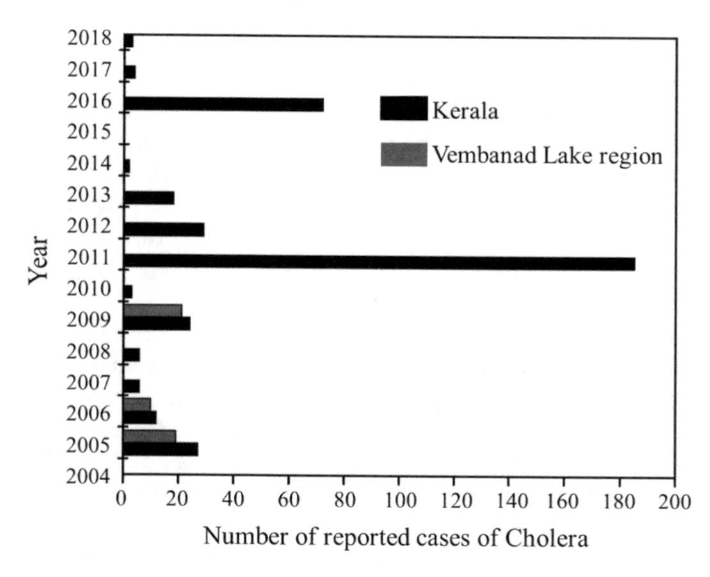

Fig. 24.2 Reported cases of Cholera in recent years from the Vembanad Lake area (study area) and all Kerala

Cholera (for which the causative agent is the bacterium *Vibrio cholerae*) is endemic to the region,[5] but the number of reported incidence is low (237 cases reported in Kerala during 2010–2015, compared with 5914 in West Bengal). According to Ali et al.,[6] there is a strong negative correlation between literacy rate and risk of cholera, and Kerala is a state claiming a literacy rate of 100%, which could be a contributing factor to the lower incidence of the disease in Kerala. Outbreaks of cholera in Vembanad Lake study area and in the whole state of Kerala are shown here in Fig. 24.2 (Kerala Directorate of Health Services Report on "Epidemiological Situation of Communicable Diseases in Kerala, 2006–2010"). Kanungo et al. (WHO)[7] synthesise the data on cholera in India for the period 1997 to 2006. They report a total of 1463 confirmed cases of cholera in Kerala during the period, and emphasise that this figure does not cover all clinical cases: "even if outbreaks were identified, the number of individuals affected was often underestimated". Often, health departments tend to report probable cases of cholera simply as dysentery, fever and diarrhoea. In addition to *Vibrio cholerae*, other *Vibrio* species occurring in Vembanad Lake cause debilitating illness involving many lost work days. A study organized by Cochin University of Science and Technology India, in collaboration with Northumbria University, reported the

[5]Ali, M., Sen Gupta, S., Arora, N., Khasnobis, P., Venkatesh, S., Sur, D., et al. (2017). Identification of burden hotspots and risk factors for cholera in India: An observational study. *PLoS ONE*, *12*(8), e0183100. https://doi.org/10.1371/journal.pone.0183100.

[6]See Footnote 5.

[7]Kanungo, S., Sah, B. K., Lopez, A. L., Sung, J. S., Paisley, A. M., & Sur, D. (2010). Cholera in India: An analysis of reports, 1997–2006. *Bulletin of the World Health Organization, 88,* 185–191. https://doi.org/10.2471/blt.09.073460.

prevalence of morbidity among rural populations of Alappuzha (part of study area) due to *Vibrio cholerae*. The report also ranked diarrheal disease as the second most important cause of working days lost through illness. Kerala is a relatively wealthy state, attracting migrant workers from the rest of India, and some of the reported cases of cholera in 2017 were traced to a worker from outside the state. Antibiotic resistance is a serious issue. Previous studies from NIO Kochi[8] have also shown that antibiotic and metal resistance are prevalent among bacteria isolated from the study area. Hence there is an urgency to find a solution to the problem of water-borne diseases in the area, particularly cholera.

The study area presents paradoxes begging explanations. The data available from the Integrated Disease Surveillance Program (IDSP) of Government of India on cholerae incidence in Kerala during the last decade, show that outbreaks occur in both dry and wet seasons. However, the region remained stubbornly free of water-borne diseases after the devastating floods of 2018, even though our own sampling showed wide-spread contamination of well-water and lake water by *E. coli*, and the presence of pathogenic *Vibrio cholerae* bacteria in the northern part of the lake (Fig. 24.3), in the vicinity of Kochi Municipal Corporation, the most densely-populated locality on the banks of the lake system.

A partial explanation for the absence of outbreaks of water-borne diseases in 2018 is perhaps the literate and well-informed inhabitants of the area, who are prepared to listen to scientists, especially in times of crisis. During the 2018 floods, scientists and volunteers worked with government ministers, health departments, municipal corporations, NGOs and international organisations to test water quality and bring awareness of water contamination to the fore. Some of the authors of this chapter (AA, NM and GG) took the lead to put in place "water clinics" and reported to the authorities on the level of contamination of well water in various localities around the Lake, and classified the wells according to distance from septic tanks and levels of water clarity. Government advisories exhorted the population to use only boiled water for drinking and cooking, demonstrating that the population and the authorities are predisposed to the concepts enshrined in Water, Sanitation and Hygiene (WASH).

Vembanad Lake and the surrounding areas constitute a natural laboratory of manageable size to study the epidemiology and ecology of cholera bacteria and the factors that influence their distribution: it is an aquatic system that contains elements of fresh lake water, brackish estuarine water and saline coastal water that can be treated as a prototypical tropical system with a densely populated shoreline and a community vulnerable to periodic flooding and consequent deterioration of the local sanitary conditions, and hence to exposure to water-borne diseases. Strong seasonality, punctuated by dry summers followed by heavy rains during monsoons, leads to strong seasonal variability in flushing rates, salinity levels, temperature and pH. River discharges into the lake also show strong seasonality: sluggish in summer, they become torrents during the rainy season. The Pampa river, which flows past a famous

[8]Vijayan, V., Abdulaziz, A., Sneha, K. G., Chandran, A., Jasmin, C., & Nair, S. (2015). Multiple antibiotic resistances among Vibrio cholerae isolated from Cochin Estuary, Southwest coast of India. In *Proceedings, 4th National Conference of Ocean Society of India*.

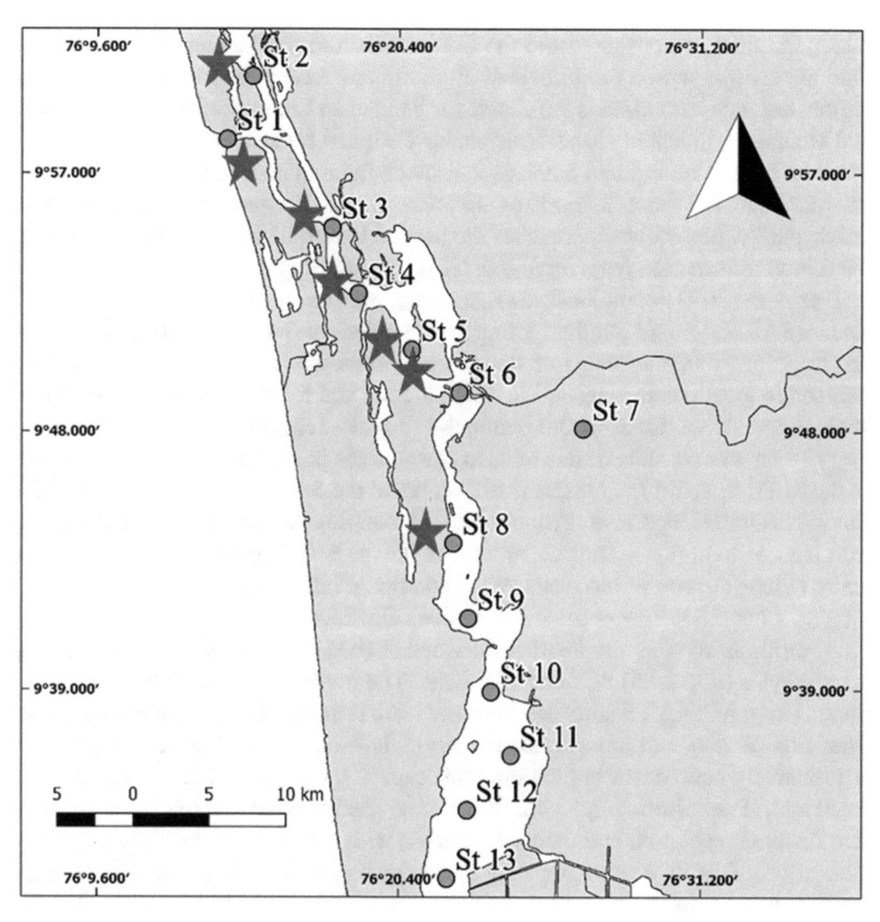

Fig. 24.3 Map of Vembanad Lake showing sampling stations as green dots. Red Asterisks indicate locations where pathogenic cholera bacteria were found at least once during the sampling period (April–November 2018). High correlation was observed between presence/absence of pathogenic cholera bacteria and *E. coli* in the samples

pilgrimage site in the neighbouring mountains of the Western Ghats, is suspected to be an important point source of bacterial pollution where the river enters the lake, but only during the pilgrimage season which follows the monsoons. In addition, the voluntary scientific work that was carried out by our team during these floods established excellent credentials, and high-level contacts in various departments of government, which, by positive feedback, will be particularly beneficial when the team addresses societal elements of the cholera problem.

No comprehensive studies have been undertaken to explore the use of remote sensing to predict and prevent outbreaks of infectious diseases in this region. Transposition of knowledge from the Ganga Delta region to the Vembanad Lake region is not always straightforward. Simple matters such as differences in scale complicate incorporation of algorithms from the Ganga region to the Vembanad Lake region. Satellite sensors with a footprint size of ~1 km are acceptable for the Ganga Delta region, but such resolution is too coarse for Vembanad Lake because of its small size and elongated, indented shape. Instruments designed for land applications, such as those on board Landsat and Sentinel-2, provide the required high spatial resolution of order 10 m, but those instruments are not optimised for quantitative estimation of water quality. Innovative approaches are needed to optimise algorithms that extract maximum information from these sensors for aquatic applications.

During the 2018 floods, local newspapers reported on the use of Synthetic Aperture Radar (SAR) on board Sentinel-1 to map the extent of floods. This is an important service, since dynamic maps of flood extent allow authorities to select the most vulnerable localities for targeted rescue missions and for prioritising health care to avoid outbreaks of diseases. But Sentinel-1 provides repeat temporal coverage only every 10 days or so, and it is desirable to consider the potential for other sensors such as the Multi-Spectral Instrument (MSI, on board the Sentinel-2 satellite, part of the European Union Copernicus Programme) for mapping the extent of water. Although MSI has no cloud-penetrating capability, unlike the SAR sensor, even partial images under cloud-free conditions can provide additional information on flood conditions on days when SAR images are not available (Fig. 24.4).

Algorithms such as the floating algal index (FAI) can be used to map floating macrophytes (Fig. 24.5) in Vembanad Lake. The image shows Sentinel-2 data processed using ACOLYTE software.[9] Further work is needed to separate floating algae from rice paddies and other forms of vegetation—an essential step, given that S2 is a relatively new sensor in providing this capacity. Once such algorithms are validated and, if necessary, improved, they can be used to study the dynamics of invasive floating vegetation and their relationship with proliferation of disease-bearing mosquitoes. Blocking of sunlight by these algae results in an increase in Biological Oxygen Demand caused by their decay and is thought to be responsible for decreasing fish yields. Further, anoxia in the saline stretches are often aggravated by organic-rich effluents released from urban settlements and industries. Some of the invasive species in Vembanad Lake are vulnerable to salinity, which leads to the question of whether flushing rates in the lake should be increased to destroy the species (the potential exists for changing the times when Thannermukkam Bund, which divides the northern and southern parts of the lake is opened or closed. The bund was built in 1974 to control incursion of salt water into the southern part of Vembanad Lake, and hence facilitate paddy cultivation).

More generally, satellites can serve a role in testing the effectiveness of remedial measures to eradicate invasive species in the lake, not an easy task according to previous attempts.

---

[9] Vanhellemont, Q. (2019). Adaptation of the dark spectrum fitting atmospheric correction for aquatic applications of the Landsat and Sentinel-2 archives. *Remote Sensing of Environment, 225*, 175–192. https://doi.org/10.1016/j.rse.2019.03.010.

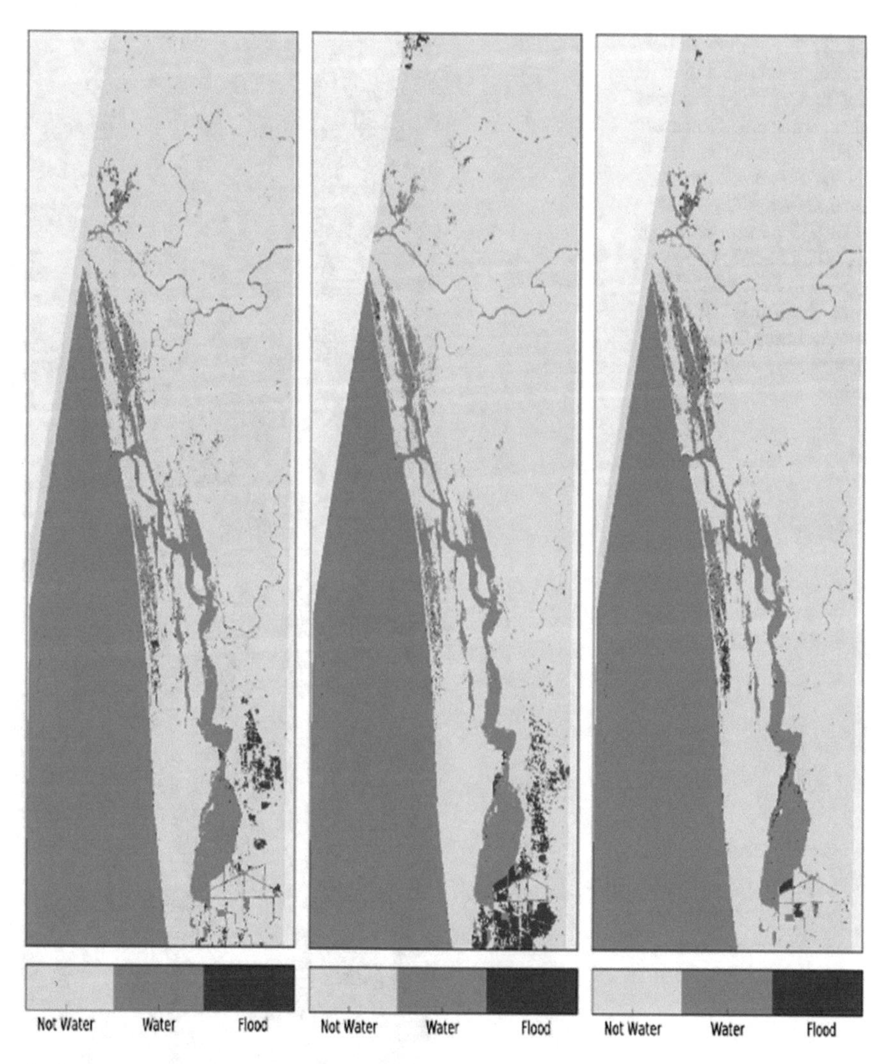

Fig. 24.4 Sentinel-2 images processed to show the extent of water in excess of a baseline, before, during, and after the 2018 floods in Kerala, India. Preliminary results. Note that "floods" as mapped by satellites may at times indicate paddy fields that are flooded as part of the cultivation cycle. Local information is essential to interpret and refine the results

24.4 Implications for Capacity Building

What can be learned from this brief comparison between the situation in the Bengal Delta Region and the state of Kerala? First, in building resilience against epidemics of water-associated diseases, there is a premium on education of all social classes, and strategies adopted to achieve this goal must include engagement with schools and community centres. Equally important is credibility of the scientific advice made

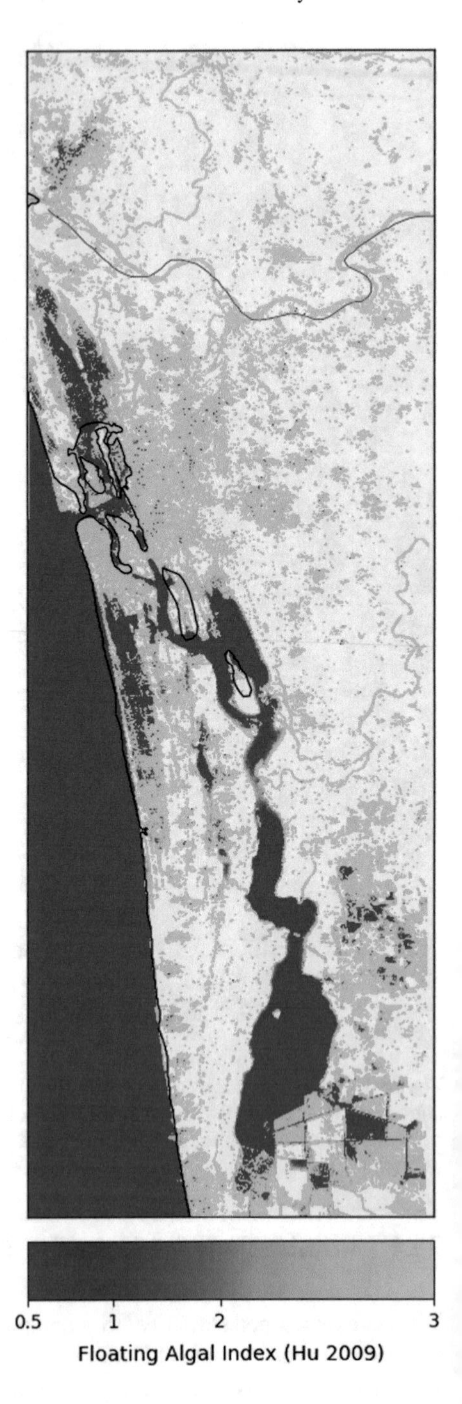

Fig. 24.5 Example of the floating algal index of Hu(Hu, C. (2009). A novel ocean color index to detect floating algae in the global oceans. *Remote Sensing of Environment, 113*, 2118–2129. https://doi.org/10.1016/j.rse.2009.05.012.) applied to Sentinel-2 data from Vembanad Lake area for 25 March, 2018. Preliminary, unvalidated results

available as a preventive measure and also the possible rapidly-evolving advice released during potential emergencies, for example during flooding. The public will heed advice given by relevant specialists, if the required credibility has been built up patiently in advance. If the entire community is engaged in, the foundation of resilience, already strong, will be further strengthened.

However, more can be done, especially engagement of citizens in regular scientific observations and in prompt dissemination of results will be invaluable in helping to identify possible development of conditions conducive to disease outbreaks. For example, epidemiological models exist to simulate the dynamics of bacterial pathogens, including the cholera Vibrio. But the models require parameters, which in general are not constant. Rather, they change with time as environmental conditions change. Consider the scheme depicted in Fig. 24.6. Here, citizen science can help in two ways. The first is to quantify ambient conditions in so far as they affect the growth rate of the cholera pathogen, for example water temperature in the lake. The next is to characterise the effect on local sanitary conditions that occur following damage to water infrastructure caused by flooding. Prompt collection and dissemination of data on a regular basis is vital to implement the epidemiological model and update local risk forecast models in a timely fashion. The models could even be implemented locally and regularly by the general public, using information collected on essential parameters. Typically, the models can be run on personal laptop computers. With the use of open-source web applications, one may anticipate that such models might be run using smart phones. In the case that the model results indicated risk of a disease outbreak, scientific authorities should be alerted for checking and further action, as required. Modern technologies, such as cell phones, can play a very helpful role in engaging communities and to collect and disseminate data.[10]

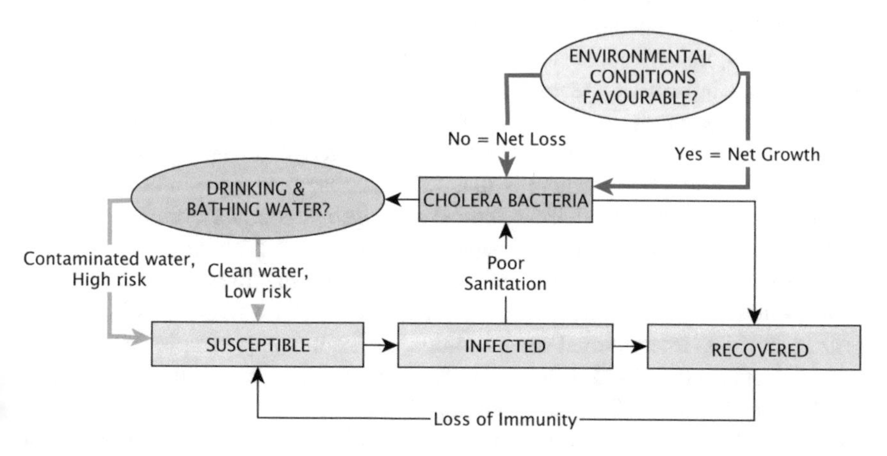

Fig. 24.6 Schematic diagram showing an epidemiological model developed for Vembanad Lake region

[10]See Footnote 2.

Finger et al.[11] used cell phone geo-location information to track people travelling from an area suffering from a cholera outbreak to unaffected regions, and used the information to predict the spread of the disease. We envisage that a cell-phone based approach could be developed to establish dynamic, regional sanitation maps that could facilitate targeted remedial action at governmental level in periods of high risk of disease outbreak, and could also serve as a communication tool to inform population at risk of changes in risk levels.

A central pillar of building resilience against possible epidemics is forecasting, early warning, and risk mapping of water-associated diseases using remote sensing, in situ observations, and modelling. A suitable vision would be to enable cost-effective, regularly updated, geo-referenced and timely warning for areas vulnerable to water-associated diseases, which in turn would enable preventive measures to be deployed in an optimal manner to minimise the probability of epidemics. The environmental data provided by remote sensing will have large spatial extent (regional) and high resolution on the ground (1 km or better). The long-term target would be to establish a system applicable in a variety of localities and for a variety of diseases. By combining information acquired from intensive in situ sampling with data from the regional scale (such as can be collected by remote sensing), it should be possible to address geo-referenced risk of exposure to water-associated diseases in a more complete and penetrating procedure. Because the environmental component of disease risk is subject to control by fluctuations in climate, the issue of resilience to climate change must also be considered, addressable by employing modelling, using results obtained for risk forecasting, together with expected changes in intensity and frequency of extreme weather events. In application of the results to real-life situations, capacity building would benefit from complementary citizen science programmes; available and utilisable operation manuals and sanitation brochures to explain the high relevance and value of maintaining hygiene and, at the same time, protecting the ecosystem; modern tools such as smart phones and social media could be used to disseminate information regarding risk status.

To date, efforts by scientists to address water-associated diseases remain fragmented, in the sense that efforts are often focussed on a particular region, to address a specific outbreak. The problem is that different scientific disciplines (microbiology, molecular genetics, modelling, remote sensing, epidemiology, clinical research, social scientists) explore different aspects of a complex problem such as waterborne diseases. Furthermore, different stakeholder communities, not just scientists but including health workers, medical practitioners, social workers, NGOs, government authorities, international organisations, the community of vulnerable people exposed to the threat, hold different parts of the puzzle. Networking and collaboration across scientific disciplines and across relevant communities is essential to ensure that the benefits of scientific advances reach the communities at risk, and

[11]Finger, F., Genolet, T., Mari, L., de Magny, G. C., Manga, N. M., Rinaldo, A., et al. (2016). Mobile phone data highlights the role of mass gatherings in the spreading of cholera outbreaks. *Proceedings of the National Academy of Sciences of the United States of America, 113*(23), 6421–6426. ISSN-8424. https://doi.org/10.1073/pnas.1522305113.

on a global scale. To promote networking, "open innovation" approaches, which exploit multiple external sources to drive innovation, could be invoked for stakeholder engagement and knowledge sharing (via open platforms such as wikiversity, for example).

There is ample evidence[12,13,14] that climate change and climate variability and associated changes conditions of the world oceans can alter geographic areas under high risk. Jutla et al.[15] suggested that ocean corridors might guide the transmission of cholera bacteria in tropical regions, and Martinez-Urtaza et al.[16] provided evidence to support the suggestion that the ocean corridors might be altered during El Niño Southern Oscillation (ENSO) events. Escobar et al.[17] used climate projections, along with an ecological niche model of cholera, to illustrate how oceanic conditions favourable for cholera bacteria might expand to higher latitudes, with changing climate. Nasr et al.[18,19] provided one of the first studies to link global climate change models with cholera in the Bengal Delta, The use of artificial intelligence (such as support vector and neural networks) can provide an unparalleled opportunity to understand emergence of water-related diseases under various climate change scenarios. This will aid in development of new models (and understanding) and facilitate decision making under deep uncertainty. There is no time to be lost before collective and innovative action be taken to build resilience in the human population worldwide, against the broad and extensive threats from diseases that will be exacerbated by altered environmental conditions.

---

[12]Escobar, L. E., Sadie, J., Ryan, S. J., Stewart-Ibarra, A. M., Finkelstein, J. L., King, C. A., et al. (2015). A global map of suitability for coastal *Vibrio cholerae* under current and future climate conditions. *Acta Tropica, 149*, 202–211.

[13]Jutla, A., Akanda, A., Unnikrishnan, A., Huq, A., & Colwell, R. (2015). Predictive time series analysis linking Bengal cholera with terrestrial water storage measured from gravity recovery and climate experimental sensors. *American Journal of Tropical Medicine and Hygiene, 93*(6), 1179–1186.

[14]Martinez-Urtaza, J., Trinanes, J., Gonzalez-Escalona, N., & Baker-Austin, C. (2016). Is El Niño a long-distance corridor for waterborne disease? *Nature Microbiology, 1*, 16018. https://doi.org/10.1038/NMICROBIOL.2016.18.

[15]See Footnote 13

[16]See Footnote 14.

[17]See Footnote 12.

[18]Nasr, F. A., Unnikrishnan, A., Akanda, A., Islam, S., Alam, M., Huq, A., et al. (2015). A framework for downscaling river discharge to access impacts of climate change on endemic cholera. *Climate Research, 64*, 257–274. https://doi.org/10.3354/cr01310.

[19]Nasr, F., Khan, R., Rahimikollu, J., Unnikrishnan, A., Akanda, A., Alam, A., et al. (2016). Hydroclimatic sustainability assessment of changing climate on cholera in the Ganges-Brahmaputra Basin. *Advances in Water Resources*. https://doi.org/10.1016/j.advwaters.2016.11.018.

24.5 Conclusion

Here, we have suggested a powerful role for remote sensing in building resilience against outbreaks of water-associated diseases. We illustrated the problem by comparing situations in two locations: the Bengal Delta Region (Bangladesh) and Vembabad Lake region in Kerala state in India. We envision a whole-ecosystem approach including monitoring pathogen load in the environment, identifying its reservoirs, routes of transmission, conditions for proliferation, optimization and validation of tools including remote sensing and ecosystem-simulation models for early warning of disease incidence and designing strategies to involve the world community so that it can understand disease occurrence and its causes, thereby being empowered to mitigate them. The global patterns of disease outbreaks are likely to be modified with climate change, adding to the imperative to understand better the underlying causes of disease outbreak, so we can be better prepared to deal with changing patterns of outbreaks. The importance of various communities working together to implement solutions cannot be exaggerated.

Acknowledgements This work is a contribution to the REVIVAL (REhabilitation of Vibrio Infested waters of VembanAd Lake: pollution and solution) Project, jointly funded by NERC (UK) and DST (India).

Part IV
Capacity Building for the XXI century

Chapter 25
ESA and Capacity Building for the XXI Century—The Catalogue of ESA Activities Supporting the UN SDGs

Isabelle Duvaux-Béchon

Abstract One of the roles of ESA is to increase support to its Member States and help them to achieve their goals and more widely the societal challenges faced by all in the world. This role was reinforced in December 2016 at a Council at Ministerial level. A specific emphasis was placed on sustainability goals and societal benefits. Capacity building for the XXI century is the prerequisite for growth in general and for achieving those goals. There are many ways to propose capacity building. It covers inter alia training and education (at all levels including tertiary and continuous education), access to data and technologies, creativity and innovation. The concept of a catalogue describing activities supporting the UN Sustainable Development Goals (SDG) and supporting capacity building via the use of space is explained, together with the steps leading to the first online version, as well as some elements for the next steps to be pursued.

25.1 Introduction

The United Nations Sustainable Development Goals (UN SDG)[1] were adopted on the 25th of September 2015 and are now widely known. These 17 goals have to be reached by all nations (developed and developing) by 2030, and this will require the efforts of all.

The adoption of those goals happened to be concomitant with the beginning of an ESA internal initiative at corporate level of assessing and promoting how the ESA programmes and projects could support sustainable development, both for its Member States and for all who could benefit from them.

[1]Sustainable Development Goals—17 Goals to Transform our World. http://www.un.org/sustainabledevelopment/sustainable-development-goals/. Accessed Mar 19, 2018.

I. Duvaux-Béchon (✉)
European Space Agency, Paris, France
e-mail: isabelle.duvaux-bechon@esa.int

S. Ferretti (ed.), *Space Capacity Building in the XXI Century*, Studies in Space Policy 22,
https://doi.org/10.1007/978-3-030-21938-3_25

Indeed, space programmes, projects, data, technologies or processes are among the tools that will help both measuring the level of achievement of the goals and reaching those goals.

25.2 Background

Many ESA activities, in all domains and since decades, are in support, directly or indirectly, of sustainable development on Earth. At the same time, it is felt that most potential users that could benefit from these activities have a limited understanding or knowledge of what space could bring, either to help them in assessing their development level (taking the "temperature" of their development) or in increasing it (providing "medicine" to improve the level). This is what led ESA to develop a catalogue presenting those activities, a first step of what is hoped to become a global "Coordination Platform".[2]

At the same time, when the initiative was started, the United Nations adopted the 17 Sustainable Development Goals (SDGs), defining what should be reached to achieve sustainable development and setting relevant targets. It was thus natural and easy to select those SDGs as the entry parameters of the catalogue.

The main aim was to increase the knowledge in the non-space world of the importance of the space activities and projects for sustainable development, and to increase efficiency and use of public-funded activities, enlarging it to all that could benefit. Indeed, ensuring coordination across the space sectors is compulsory to better serve and pass information to end users that might not have ideas on how space can support them.

A web page was also developed on the ESA website in the frame of the "Space for Earth" initiative (www.esa.int/spaceforearth)[3] to group all news items that are related to sustainable development (www.esa.int/SDG). News are coming from all ESA Directorates and these sites allow to find them all in one place.

25.3 The Draft Catalogue

To expand on what can be proposed on the web page, and following an action received by Member States, a review was started across ESA of those activities and projects that could support sustainable development. ESA decided to classify the activities according to their support of one or more of the UN Sustainable Development Goals (SDG), in order to simplify external actor's understanding of the listed activities.

[2]Duvaux-Béchon, I. (2016). Elaboration of a "coordination platform for the use of space capabilities to support sustainable development goals (including migration)". In IAC-16-D4.2.3, 67th International Astronautical Congress, Guadalajara, Mexico, September 26–30, 2016.

[3]Space for Earth. (2018). www.esa.int/spaceforearth. Accessed Mar 19, 2018.

Indeed the UN SDG categoriation is increasingly recognised and used in developing countries. Each of the activities identified can support the achievement of at least one SDG or the measurement of the progress of the goals (through the indicators). Performing this review has shown that, indeed, all types of programmes can provide some support, and hundreds of examples have been identified. Several Directorates have also specific projects targeting sustainable development.

Just to mention some examples, Earth observation data can help calculate many indicators (climate evolution, temperatures, etc.) assessing progress in the achievement of the goals, and it can also support the achievement of some goals, for example supporting the identification of good locations for the implementation of renewable energy, supporting agriculture or water management. Telecommunications satellites, in addition to lowering the digital divide, allow the development and provision of applications and services in remote areas. European navigation satellites provide independent, precise positioning for efficient navigation of all mobile telephones and can be combined with other data for improved services. Also human spaceflight activities are supporting sustainable development, be it for health and telemedicine, water recycling or nutrition. Many space technologies can also be transferred on Earth, for example for energy production, robotics, fast-deployment antennas or diagnosis, etc.

Selected examples could be found for each of the 17 goals and a first draft of the catalogue, with a selection of activities supporting each of the goals, was published at the end of November 2016 on https://www.esa.int/Enabling_Support/Preparing_for_the_Future/Space_for_Earth/ESA_and_the_Sustainable_Development_Goals. This allowed the validation of the concept and formed the seed of the online catalogue.

25.4 The Online Database

The draft catalogue supported the validation of the interest of the concept and the need for a further, exhaustive, identification of solutions that could support achieving and measuring sustainable development. Examples of applications or technologies are now available by hundreds and often go unnoticed by citizens and even, in many occasions, by the users themselves, be it a ministry, a region, a non-space industry, a local government or an NGO. An overview of all these applications is difficult to build and, for the potential "normal" user, it is often difficult, if not impossible, to easily find what kind of support can be obtained from space programmes.

The development of a full online database then started, with the aim to identify all relevant activities to make them easily accessible to potential users, and to serve as a basis for discussion with them to identify their needs and propose possible solutions.[4,5,6]

The intention was to include all elements considered as important for a potential user to identify interesting activities, and support her/his informed decision (short description, status of development or availability, conditions of access, link to web page, contact points, etc.). The difficulty was to set up a structure that could be adequate for all sorts of activities (e.g. images, data, projects, processes, technologies, etc.), contrary to the more traditional catalogues that usually present only one category of information (e.g. images, applications or technologies). It was also important that the database would be searchable along non-space parameters, thus SDG, SDG targets and keywords.

The online catalogue is now available since the 15th of March 2018 on https://sdg.esa.int.[7] It started with around 300 initial activities and is complemented over time (at least 200 more were identified and are being validated). The address sdg@esa.int is also available should there be questions that are not answered within or about the catalogue.

25.5 Next Steps

25.5.1 Support to UNOOSA

By the time the concept of the catalogue was presented outside ESA in conferences, it appeared that the United Nations Office of Outer Space Affairs (UNOOSA) was elaborating a similar concept called the "Space Solutions Compendium". This is planned to be the second element of a two-parts concept that includes a Space Development Profile (evaluating the actual capability of a State) and a Space Solutions Compendium (giving information on what exists that would help a State improving its profile). As this Compendium is quite close in its concept to the ESA catalogue, discussions started between ESA and UNOOSA and it was decided to explore the

[4]Duvaux-Béchon, I. (2017). The catalogue of ESA activities supporting the UN Sustainable Development Goals. In IAC-17-E5.2.2, 68th International Astronautical Congress, Adelaide, Australia, September 25–29, 2017.

[5]Duvaux-Béchon, I. (2017). ESA Catalogue of activities supporting capacity-building. Presented at the United Nations/Austria Symposium on "Access to Space: Holistic Capacity-Building for the 21st Century", Graz, Austria, September 3–7, 2017. http://www.unoosa.org/documents/pdf/psa/activities/2017/GrazSymposium/presentations/Tuesday/Presentation18.pdf. Accessed Mar 19, 2018.

[6]Duvaux-Béchon, I. (2017). Space-based systems for resilient and low-emission societies. Presented at the United Nations/Germany International Conference on International Cooperation Towards Low-Emission and Resilient Societies, Bonn, Germany, November 22–24, 2017. http://www.un-spider.org/sites/default/files/Session3-WG1-IsabelleDuvaux-Bechon-ESA_0.pdf. Accessed Mar 19, 2018.

[7]ESA Catalogue of activities supporting the UN SDGs. https://sdg.esa.int. Accessed Mar 19, 2018.

possibility that the elements in the ESA catalogue would form the seed of the Compendium. This was presented at various occasions in UN conferences preparing UNISPACE+50 (e.g. UN/Austria on capacity building in Graz, UN/Germany on low-emission and resilient societies in Bonn) and has been well received as there has been an overall recognition that a catalogue was indeed long overdue and expected. It has been recognised, on those occasions, that such a catalogue would be an important step in the development of space capabilities and use of space assets.

A joint statement was signed by the ESA Director General and the UNOOSA Director on the 24th of May 2019.

25.5.2 Promotion to Potential Users

In parallel to the identification of ESA activities that are and will be presented in the online catalogue, it has been recognised that there is a need to be proactive towards selected organisations, categories of users or potential European contributors, to provide them with information on the benefits they might achieve by starting to use or increasing their use of space assets, derived data, services and technology.

Information is targeted towards the main domains of expertise and action of those users, so that a fruitful dialogue can start for a better understanding of their needs. Such information, when it exists (and it will be provided via the catalogue from now onwards), is very often unknown.

Those discussions are enriching both sides by providing a better understanding of both the "offer" and the "needs". They are planned to lead to specific partnerships agreements to develop ESA cooperation with non-space entities. An internal database was developed to provide everyone at ESA with an overall view, and allowing expansion of existing partnerships to other areas when relevant. Being proactive is also a way to identify the types of challenges those categories of users are likely to face, that are not yet solved by space projects, and to be in a better position in the future to adapt and define ESA programmes and activities to the needs of the external world.

25.5.3 Opening to Other Providers

It was clear since the beginning of the initiative, that ESA was not the only space Agency in Europe or in the world, government entity or applications provider, supporting SDGs, and that it might be worth to mutualise the information provided in order to allow a "user" to find more easily the relevant programmes, wherever they were coming from. The purpose of the catalogue is then to be the core of what has been proposed back in 2016 as a global "Coordination Platform".

In parallel with the elaboration of the catalogue, discussions started with ESA Member States in order to propose to them to join the initiative and include their nationally developed technologies, applications and services. Such discussions will be pursued now that the catalogue is online, and hopefully it will soon become the European catalogue.

In addition, since the ultimate aim of the catalogue and the Coordination Platform is to become the world reference with inputs from all contributors, this will be pursued, in particular through the cooperation with UNOOSA and the Compendium,

25.5.4 Test Case on Africa

Following the request from several of its Member States, that wish to increase the role of space in national development projects, ESA is developing a specific case for Africa, the closest continent to Europe, as a way to see how in a specific region, the examples of the catalogue could be implemented and contribute to the development policy of those Member States. The ultimate aim is to combine all efforts from space actors to optimise the support space can bring to help achieving the UN Sustainable Development Goals, whether they are for the benefit of our own countries or the ones we are supporting or partnering with.

25.5.5 Socio-Economic Benefits

It is recognized that governments or companies financing space projects are looking for "results" or the proof and level of the socio-economic benefits they can obtain from the budget invested in the area. This occurs especially before starting a new activity, in order to have some decision tools. Even if studies performed up to now, on selected programmes, show multiplier effects several times the one of the investment, there are still too few studies and not on all types of activities.

Concerning those socio-economic aspects, it is our aim to obtain data or information for as many projects as possible. It is rather complex, as the activities selected are very diverse by nature, from images to services, from a technology transferred or an activity targeting end users to one targeting large companies or government services. There will not be one global indicator that could be proposed for all. Indicators can concern the number of users, the regions/countries where it is implemented, the savings obtained, the revenues reached, the progress made in the achievement of the goals, even qualitative benefits, or any other information that may be important for a potential user. The effect of "space" will also have to be, when possible, distinguished from the effect of better governance or non-space measures. This is being built and will be included in the catalogue, when relevant, as soon as the data will become available, in order to give some indications to the potential users on the interest of each proposed solution.

25.6 Conclusion

The initiative that started at ESA in September 2015 is now reaching full momentum, with a catalogue published online on the 15th of March 2018 following several intermediate steps. It answers many expectations for an easier access to information on how space is supporting and can support sustainable development on Earth, both by measuring the level of development and helping to improve it. Space is increasingly an indispensable tool at the disposal of those who can benefit from the data, services, projects, tools or technologies stemming from the programmes. This will also be a long lasting initiative as new activities are starting every month.

This catalogue was developed at a time when it matched other similar initiatives, in particular the one of the UNOOSA Compendium, allowing ESA to propose a comprehensive list of some 300 activities at the start, as potential seed information for the Compendium, to be complemented by equivalent information coming from other space agencies or service providers.

The whole process, initiated at ESA, is aimed at maximising the efficiency of public funding injected into ESA for the benefit of the citizens of the world and of a sustainable development of our planet, by facilitating the process of informing potential users about space applications, services or technologies they can use to solve the challenges they are facing.

Acknowledgements This catalogue could not have been prepared without the support and data inputs from all the ESA projects and activities and we are grateful for that. It was really a corporate initiative. Special thanks to Micky Elanga Yangongo who was in charge of the "physical" development of the catalogue and Matija Rencelj who took over.

Chapter 26
Leveraging Existing Societal Platforms to Enhance Capacity Building Activities at the United Nations Regional Centre in Nigeria

Funmilayo Erinfolami

Abstract Space Science has without doubt made life better and more comfortable in this century. A number of the advances that have been made in health care, disaster monitoring, navigation, and food security are spin-offs of Space Science. Space is also capable of making a difference in the approach to building capacity and this needs to be explored from an unconventional stand point. This, in essence, is what is required for the realities of the XXI century. The attainment of the Sustainable Development Goals (SDGs) calls for pragmatic, innovative and all-embracing approach. There is no doubt that space science has a crucial role to play in attaining these goals, hence the seven (7) thematic priorities for the implementation of UNISPACE+50, that were endorsed at the Fifty-ninth Session of the Committee on the Peaceful Uses of Outer Space (COPUOS) in June 2016. The thematic priorities are aimed at "**strengthening the contribution of space activities and space tools for the achievement of global agendas**" (A/AC.105/C.1/L364, Draft Resolution on Space as a driver for sustainable development.). In particular, thematic priority 7, which is focused on capacity building for the XXI century is the most cross-cutting aspect of the initiative.

Clearly, there are a lot of space actors with laudable initiatives, but these initiatives are not accessible or known to all. Hence, the recommendation of the 2016 United Nations/International Astronautical Federation Workshop on Space Technology for Socioeconomic Benefits: Integrated Space Technologies and Applications for a Better Society in Guadalajara, Mexico, that the UN Office for Outer Space Affairs (UNOOSA) should develop a space capacity index and a space solutions compendium.[1] This chapter considers ways in which the UN Regional Centre in

[1] A/AC.105/1128, 2016, Report on the United Nations/International Astronautical Federation Workshop on Space Technology for Socioeconomic Benefits: Integrated Space Technologies and Applications for a Better Society.

F. Erinfolami (✉)
ARCSSTE-E, Ile-Ife, Nigeria
e-mail: vieve4real@gmail.com

S. Ferretti (ed.), *Space Capacity Building in the XXI Century*, Studies in Space Policy 22,
https://doi.org/10.1007/978-3-030-21938-3_26

Nigeria[2] and others alike can enhance their existing roles to meet the capacity building needs of the XXI century, achieving an overall global effect, even with the existing constraints and limited funds. Funding is the common constraint of the UN Regional Centres and this limits the number of participants that can benefit from their educational programmes.[3] These Regional Centres can generally serve as information centres on space science and technology initiatives, just as they already do for the International Committee on GNSS (ICG).[4] These existing institutions should be supported and strengthened.

26.1 Establishment of United Nations Regional Centres for Space Science and Technology Education

The Second United Nations Conference on Exploration and Peaceful Uses of Outer Space (UNISPACE II), held in 1982 in Vienna, Austria, recommended that the United Nations Office for Outer Space Affairs (UNOOSA), through its Programme on Space Applications (PSA), should focus its attention, inter alia, on building indigenous capacities for the development and utilization of Space Science and Technology, particularly at the local level.[5] Consequently, this recommendation was endorsed by the United Nations General Assembly (GA) in its resolution 37/90 of on 10th December 1982.

Sequel to this, the United Nations General Assembly (GA), in its resolution 45/72 of 11 December 1990 endorsed the recommendation of the Scientific and Technical Subcommittee of the Committee on the Peaceful Uses of Outer Space (A/AC.105/456 of 1990) that:

> … the United Nations should lead, with the active support of its specialized agencies and other international organizations, an international effort to establish regional Centres for Space Science and Technology Education in existing national/regional educational institutions in the developing countries.[6]

Following the endorsement of these resolutions, the UN Office for Outer Space Affairs (UNOOSA) outlined a number of steps to translate the GA resolutions mentioned above into operational programmes [Ref. (A/AC.105/498 of 1992 and

---

[2]The full name of the UN regional centre in Nigeria is African Regional Centre for Space Science and Technology Education in English (ARCSSTE-E).

[3]P. Martinez/Is there a need for an African Space Agency? (2012) Elsevier Space Policy 28 (2012) 142-145.

[4]E. Offiong, Using the ISS for Capacity Building in Developing Countries (2011), JBIS, Vol. 64, pp. 275–279.

[5]ARCSSTEE, 2011, Achievements, Challenges and Future Projections.

[6]Regional Centres: Background, Mandate and Objectives. http://www.unoosa.org/oosa/en/ourwork/psa/regional-centres/background.html.

A/AC.105/534 of 1993)]. The established Regional Centres for Space Science and Technology Education are[7]:

Africa:	**Nigeria**—ARCSSTE-E: African Regional Centre for Space Science and Technology Education in English
	Morocco—CRASTE-LF: African Regional Centre for Space Science and Technology Education—in French language
Asia and the Pacific:	**India**—CSSTEAP: Centre for Space Science and Technology Education in Asia and the Pacific
	Jordan for Western Asia—RCSSTEWA: Regional Centre for Space Science and Technology Education for Western Asia
	China for South East Asia—RCSTEAP: Regional Centre for Space Science and Technology Education in Asia and the Pacific
Latin America and the Caribbean:	**Brazil/Mexico**—CRECTEALC: Regional Centre for Space Science and Technology Education for Latin America and the Caribbean (Fig. 26.1).

The main goal of each Centre is the "**development of the skills and knowledge of university educators and research and applications scientists, through rigorous**

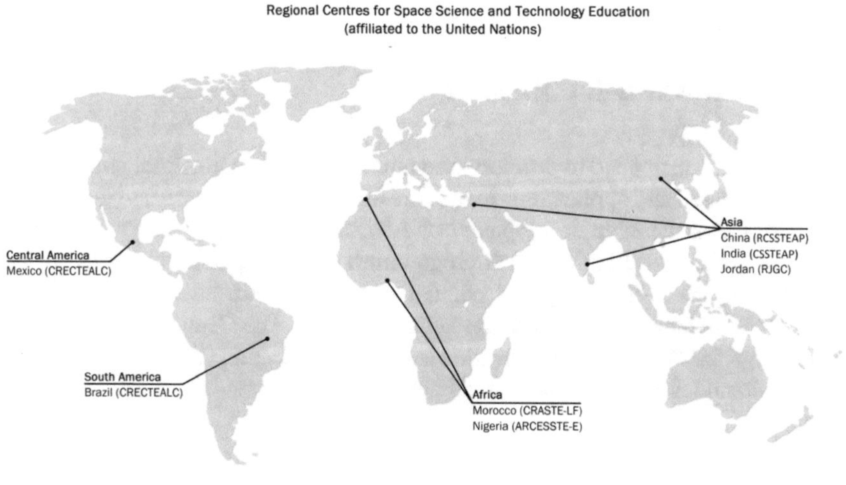

Fig. 26.1 United Nations Regional Centres

[7] A/AC.105/625, Dec. 1995, Report of United Nations Expert on Space Applications. www.unoosa.org/pdf/reports/ac105/AC105_625E.pdf.

theory, research, applications, field exercises, and pilot projects in those aspects of space science and technology that can contribute to sustainable development in each country."[8]

26.2 About ARCSSTE-E

The African Regional Centre for Space Science and Technology Education in English (ARCSSTE-E) was formally inaugurated on the 24th November, 1998, in Nigeria. The Centre is strategically located within the campus of the Obafemi Awolowo University (OAU) as stipulated by the United Nations that the regional Centres be established in an existing national/regional educational institution.[9] ARCSSTEE also serves as an activity Centre for the National Space Research and Development Agency (NASRDA)—an agency of the Federal Ministry of Science and Technology, Nigeria, to create awareness about space science and carry out space education outreach activities across all levels of education. Since inception, Member States made up of 25 English-speaking African countries are yet to make financial commitments to the running of the Centre's programme. The programme is still primarily funded by the Federal Government of Nigeria through the Ministry of Science and Technology while travel grants are provided for the foreign participants by UNOOSA.

The emphasis of the Centre is to concentrate on in-depth education, research and application programmes, execution of projects, continuing education/awareness programmes etc. Scholars and professionals who contribute to its educational programmes are drawn from both within and outside the region.

26.2.1 ARCSSTE-E's Mandate

ARCSSTE-E is mandated to develop, through in-depth education, indigenous capability for research and applications in the core areas of space science and technology (SST) for the sustainable development of Africa. The Centre is also mandated to undertake space education outreach programmes to create public awareness of the benefits of space to humanity and popularize space science education in African institutions, thereby demystifying the notion of space. The Centre fulfills its mandate through the Postgraduate Education and Space Education Outreach activities at all educational levels and to the general public.

---

[8]See Footnote 7.

[9]See Footnote 7: Annex 1, para 2.

26.3 Ongoing Capacity Building Programmes/Initiatives at ARCSSTE-E

ARCSSTEE fulfills its capacity building mandates through three main channels: Postgraduate Education, Space Education Outreach and International collaboration/Conferences/Workshops.

26.3.1 Postgraduate Programmes

The Centre runs two postgraduate programmes: a 9-month Postgraduate Diploma Programme; and an 18-month Masters of Technology (MTech) in Space Science and Technology Applications. To encourage students from other English speaking African countries to benefit from this initiative, UNOOSA offers scholarships (travel grants) for some non-Nigerian participants. Other needs (accommodation, monthly stipend, health care etc.) are provided by the Federal Government of Nigeria. However, it is pertinent to mention here that all Nigerian participants of the PGD and MTech programme do not get any scholarship, they are self-funded (e.g. for tuition, accommodation and healthcare).

26.3.1.1 Postgraduate Diploma Programme (PGD)

The United Nations Office for Outer Space Affairs' (UNOOSA) 9-month Postgraduate Diploma (PGD) Programme in Space Science and Technology (SST) application, run by the African Regional Centre for Space Science and Technology Education in English (ARCSSTE-E) creates the needed opportunity and platform for English speaking African countries, to acquire knowledge and skills in six specialized areas of SST: Remote Sensing/Geographic Information Systems, Satellite Communication, Basic Space and Atmospheric Science, Satellite Meteorology, Space Law and Global Navigation Satellite System (GNSS) for national and regional sustainable development. The PGD course curricula, used by the UN Centres, is developed by the Programme on Space Application (PSA). Since its inception and until today, four hundred and forty-six (446) participants from seventeen African countries (Botswana, Democratic Republic of Congo, Cameroon, Ghana, Ethiopia, Gambia, Kenya, Liberia, Malawi, Nigeria, South Africa, Sierra Leone, Sudan, Tanzania, Uganda, Zimbabwe and Zambia) have participated in the PGD programme.[10]

The Postgraduate programme is implemented in three phases:

(i) Two-month common course module
(ii) Four-month specialised module
(iii) 3-month project.

[10]2017 ARCSSTE-E Annual report.

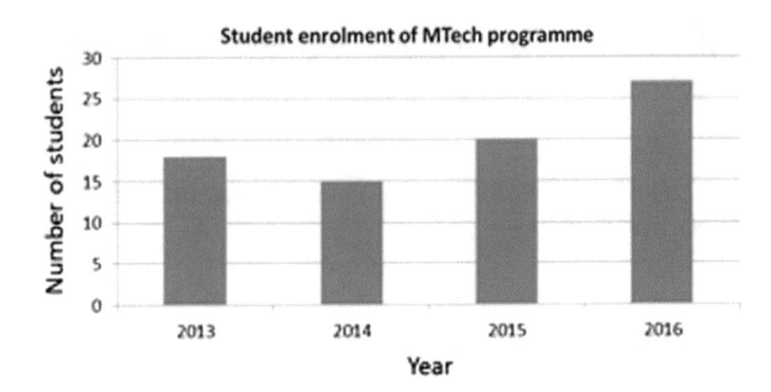

Fig. 26.2 Student enrollment in the MTech programme. *Source* ARCSSTEE

PGD Certificate is awarded to all participants that successfully and satisfactorily completed and fulfill the requirement for the diploma.

26.3.1.2 Master of Technology (MTech.)

The Masters programme is done in collaboration with the Federal University of Technology, Akure (FUTA), Nigeria. At the end of the programme, participants are awarded a Master of Technology (MTech.) degree in Space Science and Technology Applications. So far, over eighty (80) participants have benefitted since the inception of the programme in 2013[11] (Fig. 26.2). The programme runs for 18 months and it is implemented in two phases: course work and thesis.

Areas of Specialization are:

- Remote Sensing and GIS
- Basic Space and Atmospheric Science
- Satellite Communications
- Satellite Meteorology
- Global Navigation Satellite Systems
- Space Law
- Space Engineering (under development).

26.3.2 *Space Education Outreach Programme (SEOP)*

Using a systematic and targeted approach, the Centre creates awareness about space through its "catch-them-young" programmes. These programmes are designed for nursery, primary, secondary schools and tertiary institutions. The activities include;

[11] See Footnote 10.

Robotics, Microgravity Experiments, Workshops and Training of teachers, Competitions, Exhibition/tours and commemoration of World Space Week. Relevant activities are organized across all school grades and for the general public including appropriate ones for the law makers. Enlightening the law makers is particularly important as it would ensure the sustainability of the space programme with good budgetary allocation by the legislatures.

The outreach activities and programmes are:

- **Annual Space Science Schools' Workshop**: Schools' space education workshop is organized annually for Nursery/Primary and Secondary schools. It is usually a 2-day programme; one for nursery and primary schools, the other for secondary schools. The centre has been able to cover all the six geopolitical zones in Nigeria. Activities like hands-on, powerpoint presentation, amongst others, are packaged to give the students a learning experience in an informal way.
- **Competitions**: Quiz, debate, Poetry/nursery/rhymes/songs, Art/essay, water rocketry are organized to get a feedback of how much has been imparted on the students and the level of creativity that can be expressed. It is akin to giving them wings to fly and watch how high they can go!
  - **Rocketry**: Building Water rockets and amateur rocketry are part of the hands-on activity the students are encouraged to participate in. Water rocket competitions are sometimes organized and the level of complexity the students can attain through research is amazing and very encouraging.
  - **Quiz/debate competition**: These competitions are organized annually and the debate is usually targeted at thematic issues.
- **Zeronaut Programme**: The Zero-Gravity flight is a good example of one of the motivational activities organized under ARCSSTE-E's SEOP. This parabolic flight (an initiative of a Houston-based non- profit organization, the Space-week International Association in collaboration with the Zero-G Corporation of Florida, USA) is a two-hour mission onboard the G-FORCE ONE airplane from the Kennedy Space Centre in Orlando, Florida, by trained pilots maneuvering the aircraft between 24,000 and 32,000 feet altitude. The flight includes 2 Martian (at one-third of gravity), 3 Lunar (at one-sixth of gravity), and 10 weightless (at zero gravity) experiences. This flight simulates the weightlessness experienced by astronauts in space. The winner of a nation-wide schools' competition is usually sponsored to visit the Kennedy Space centre in USA to participate in the weightless flight and the participants are called **Zeronauts** (Fig. 26.3).
- **World Space Week (WSW)**
  Annual celebration of the WSW is an event that is organized for the law makers, general public, primary, secondary and tertiary institutions around a theme for the given year.
- **Teachers' Seminars**: This is the train-the-trainers exercise of the Centre aimed at building the capacity and capability of teachers. This way they can confidently propagate the knowledge gained in the traditional classroom.

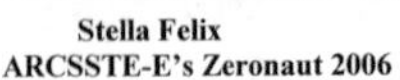
Stella Felix
ARCSSTE-E's Zeronaut 2006

Adeolu Akano
ARCSSTE-E's Zeronaut 2007

Omolola Ibrahim
ARCSSTE-E's Zeronaut 2008

Fig. 26.3 ARCSSTEE's Zeronauts. *Source* ARCSSTEE

- **Curriculum Development**: The curriculum development exercise aims to add more dedicated space science content in the primary and secondary education curricula in Nigeria. The phase one of this project has been successfully completed.
- **Launching Space Clubs**: Space clubs are launched in schools to promote continuity of space activities in schools. To date, over a thousand space clubs have been launched.
- **Educational Tour/Excursion**: Schools are also encouraged to visit the mini space museum at the Centre.
- **Science Project Exhibition by Schools**: Students are encouraged to design projects around a space-related theme that is displayed during the exhibition section of the schools' workshop. The projects are assessed and the best project is awarded a prize.
- **Cansats**: The Cansat programme was introduced to the space club students at the tertiary level. So far, the launched Cansats which were designed and built by the students, have been used to make atmospheric measurements and the data obtained to do basic atmospheric research.
- **Robotics**: This is another activity that the students find both interesting and inspiring. It enhances their understanding of the STEM subjects of the formal school curriculum. ARCSSTE-E—iLab OAU Robotic Education Programme is aimed at inspiring interest of young people in STEM and robotics. Students receive training on robot building and programming using Lego Mindstorm® kits. ARCSSTE-E supported Nigerian secondary school students to participate in the 2011 World Robot Olympiad, held in Abu Dhabi, from 18 to 20 November, 2011.

- **Zero Gravity Instrument Project (ZGIP)**: The United Nations Office for Outer Space Affairs, Human Space Technology Initiative (HSTI) launched the ZGIP on 1st February 2013, and distributed the microgravity simulation instruments (Clinostat) to qualified schools, universities, research centres and institutes to promote space education and research in microgravity. The Centre was one of the beneficiaries of the Clinostat and this has facilitated exposing students to do microgravity experiments and learn the basics for conducting research activities. The programme has been very successful so far.

26.3.3 International Collaboration/Conferences/Workshops

In addition to the above mentioned capacity building activities, ARCSSTE-E has fostered a number of important and strategic collaborations to achieve its goals. International collaboration is a crucial platform that has been identified as essential for capacity building and enhancing the capability of the workforce. Collaboration has been fostered with institutions like Beihang University and RCSSTEAP, China, through which the workforce capacity is being enhanced. Other existing collaborations are with Samara University, EUMETSAT, participating organization status with GEO (Group on Earth Observations), and serves as an information Centre for the International Committee on GNSS (ICG). A number of international training/workshops on GNSS have also been organized and hosted by the Centre. A few are listed below:

- **Global Navigation Satellite Systems (GNSS) Training Workshop**
  (a) A training was held from 4th–29th October, 2010 at ARCSSTE-E, Ile-Ife, Nigeria.
  (b) Another one was organized in collaboration with Beihang University, which took place in August 2016 at the Obafemi Awolowo University campus, Nigeria.

All of these programmes are, in essence, targeted at creating awareness, inspiring young people, building Space science capacity amongst Africans and within the workforce.

26.4 Enhancing/Leveraging on Other Societal Platforms for Capacity Building in the XXI Century

In view of the foregoing, there is no doubt that the regional Centre in Nigeria is forging ahead to achieve its institutional goals irrespective of whatever challenges it may be dealing with/facing. It is also non-contestable that building capacity in the XXI century has to be an all encompassing endeavor taking into consideration advances in technology, societal interactions and changes in societal needs in order to adequately bridge the necessary gap for a progressive society. However, in order to expand the reach of the capacity building efforts by leveraging on existing societal platforms, there is need to embrace the realities of the XXI century—highly technological, digitalised and fast paced- and adopt innovative and effective ways to achieve an overall sustainable capacity building and inevitably socio-economic development and growth.

To achieve the desired development and growth in the developing countries, time and time again, it has been proven that rather than have tailor-made solutions imported to developing countries, it is better and more efficient, to allow for indigenous engagement and efforts at tackling the needs of the society. In this way, collaboration geared towards consultative support is more beneficial for sustainable development and allows for capacity building. At times, solutions that work for the developed world may not necessarily work effectively in the developing nations, since there is the need to consider their peculiar challenges. In Africa, for example, there is a wide cultural and economic diversity, therefore the one solution fits all may not even suffice. The need to consider each country separately and factor in the uniqueness of each partnership should therefore be appreciated.

Some existing platforms in the society have been identified and leveraging on the strengths of these entities will help to strengthen capacity building in the XXI century and to build a "space" society, for an overall global impact, in the long run.

Existing Societal Platforms for Integrated capacity building:

(A) **Non-Governmental Organisations**—Their mandates and terms of reference makes it important for them to take on projects that are relevant to the benefits of civil society. The World Bank defines NGOs as "**private organizations that pursue activities to relieve suffering, promote the interests of the poor, protect the environment, provide basic social services or undertake community development**". They serve as a gap-bridging entity for the society. In this light, fostering collaboration with NGOs is beneficial, whereby the regional Centre offers space-related consultative support to them in carrying out projects they embark on. That is, providing space solutions that can help them deliver better results in instances where it is economically feasible. For example, NGOs that are focused on public health can be supported to utilize space assets from the research to the implementation stage, as long as it is cheaper and provides better impact on the citizens. In the process of creating this kind of work-synergy, while working together, skills are being developed and space technology gradually becomes more popular.

The regional Centre can organize a stakeholders' meetings or conference where representatives of accredited NGOs are enlightened and sensitized about space and the initiative to collaborate with them, whilst the range of possibilities that can be achieved by working together is emphasized and possibly demonstrated.

(B) **Partnering with Technology Hubs**

Engaging young entrepreneurs and aspiring entrepreneurs to facilitate innovations that can solve or alleviate pertinent societal challenges is one way to ensure sustainable development. Globally, the role of technology hubs in driving socio-economic transformation cannot be ignored. An example is the Silicon Valley in the United States of America. The Centre can play a supportive role for these technology and innovation hubs. Developing countries have a lot of basic challenges they are facing and that are best known to them. In view of this, it is particularly important for developing countries to develop solutions that meet their societal/regional needs by themselves and for themselves. In most developing countries, there is a crop of home-grown technology-savvy youths who are good with coding/programming, developing apps, and solutions for various sectors like banking, education to mention a few. Some of these solutions can be made more robust and successful by integrating space technology. There are a number of such hubs in Africa- JoziHub in South Africa, iHub in Kenya and in Nigeria, Co-creation hub (CcHub), iDEA hub, Roar Nigeria, Delta State Innovation hub, BD hub and Civic innovation lab and the GE Lagos garage skill training programme for hardware entrepreneurs.

This will invariably engender an entrepreneurial new-space ecosystem in developing countries which will allow for giant strides in the space sector without having to wait on the government to act, provided that the right policies are in place.

(C) **Mentoring**

The need for mentoring cannot be over-emphasized as the youths and young professionals out there needs someone or an institution that can bridge the knowledge gap by supporting and guiding them through their career paths. The Centre can effectively play this role by matching suitable mentors and mentees in the space industry.

(D) **Online and Social Media tools**

Embracing social media for propagating capacity building is an indication of moving with the times to meet the needs of the XXI century. Organising trainings, creating awareness about possibilities, providing information and networking—all of these can be successfully achieved and propagated to a wide range and larger number of audience at minimal overhead cost.

26.5 Conclusion

To meet the capacity building needs and goals of the 21st century, it is pertinent to employ the available and most effective tools of our time to meet the needs of the people.

The Regional Centres can readily serve as focal points for information dissemination and can bridge the knowledge gap by employing the available tools for the realization of their mandates and inevitably achieve the global goal of all the agendas like Space 2030, ESA 4.0, UNISPACE+50 and ultimately the Sustainable Development Goals.

Finally, it is recommended that the six (6) UN Regional Centres with the involvement of UNOOSA organize a joint conference every two years to facilitate interaction and exchange of ideas amongst students, the faculty and alumni.

Chapter 27
Capacity Building to Help Mobilise Space for Sustainable Development: A European Perspective

Servaas van Thiel

Abstract The Chapter provides a brief overview of the main European space policy objectives and actors and sets out a number of recommendations on how space technologies and applications can be mobilised to help realise the 2030 Agenda for Sustainable Development and its 17 Sustainable Development Goals and 169 targets.

27.1 Introduction and Preliminary Remarks

The 3 February 2018 Conference of the European Space Policy Institute (ESPI) on synergies for capacity building in the XXI Century,[1] took place at the right moment, because it took place in the middle of the 55th Session of the Scientific and Technical Subcommittee (STS 55) of the UN Committee on the Peaceful Use of Outer Space (COPUOS), which opened broad international consultations on the Draft General Assembly Resolution on 'Space as a Driver for Sustainable Development'.[2] On the occasion of these consultations it was a pleasure to be able to say a few words about capacity building to help mobilise space for the realisation of the 2030 Agenda and its 17 Sustainable Development Goals (SDG) and 169 targets.[3]

Before doing that, two preliminary remarks should give some background, respectively on European space policy objectives and on European space actors. As a first

[1]See: https://www.espi.or.at/images/Space2030_and_Space_4.0_synergies_for_capacity_building_in_the_XXI_century_-_Program_31_Jan_2018.pdf. Accessed on 11 February 2018.

[2]See latest version at: http://www.unoosa.org/oosa/oosadoc/data/documents/2018/aac.105c.12018crp/aac.105c.12018crp.6_0.html. Accessed on 11 February 2018.

[3]Transforming our world: the 2030 Agenda for Sustainable Development', UNGA Resolution 70/1 of 25 September 2016 (http://www.un.org/ga/search/view_doc.asp?symbol=A/RES/70/1&Lang=E). Accessed on 9 February 2018.

S. van Thiel (✉)
Delegation of the European Union to the International Organisations in Vienna, Vienna, Austria
e-mail: Servatius.VAN-THIEL@eeas.europa.eu

Free University Brussels, Brussels, Belgium

Vienna University of Economics and Business, Vienna, Austria

S. Ferretti (ed.), *Space Capacity Building in the XXI Century*,
Studies in Space Policy 22, https://doi.org/10.1007/978-3-030-21938-3_27

preliminary remark, it is useful to recall that the EU and its Member States draw up a European Space Policy to promote scientific and technical progress, industrial competitiveness and the implementation of its polices (Arts 4 and 189 TFEU).[4] European Space Policy is articulated more specifically by the EU Council of Ministers (in which all EU Member States are represented), which has been particularly active in 2017 by adopting four sets of Council Conclusions respectively on UNISPACE+50, on the mid-term evaluation of Galileo and Copernicus, and on the Space Strategy for Europe.[5]

From these Council Conclusions three main EU space policy objectives can be distilled. First, the EU wants an autonomous European space capacity, and for that reason it invests significant resources in innovative European space research (Horizon 2020), a competitive European space industry (Space Strategy for Europe),[6] a self-standing European launch capacity (ESA and national space agencies), European satellite navigation systems (Galileo and EGNOSS) and European Earth Observation services (Copernicus).[7]

Second, the EU wants this space capacity to be able to function in a secure and safe environment. For that reason it supports international cooperation on space hazards, such as space weather and Near Earth Object's, and it pursues international agreement on Guidelines on the Long term Sustainability of Outer Space, Transparency and Capacity Building Measures (TCBM), Principles Of Responsible Behaviour In

[4]The 'Treaty on the Functioning of the European Union' which is better known as the Lisbon Treaty, entered into force in December 2009 (together with the Treaty on the European Union and the Charter of Fundamental Rights), and is available at: http://eur-lex.europa.eu/legal-content/EN/TXT/?uri=celex%3A12012E%2FTXT. Accessed on 9 February 2018.

[5]Council Conclusions on UNISPACE+50 (http://www.consilium.europa.eu/media/32289/st15628en17.pdf) of 11 December 2017 (adopting Council Document: ST 13766 2017 INIT.); Council conclusions of 5 December 2017 on "The Mid-term Evaluation of the Galileo and EGNOS programmes and of the performance of the European GNSS Agency" (http://www.consilium.europa.eu/register/en/content/out?&typ=ENTRY&i=ADV&DOC_ID=ST-15435-2017-INIT); Council Conclusions of 1 December 2017 on the midterm review of the Copernicus programme (http://www.consilium.europa.eu/en/press/press-releases/2017/12/01/the-mid-term-evaluation-of-the-copernicus-programme-council-adopts-conclusions/); Council conclusions of 30 May 2017 on "A Space Strategy for Europe" (http://data.consilium.europa.eu/doc/document/ST-9817-2017-INIT/en/pdf); See also the 26 October 2016 European Commission Communication on: Space strategy for Europe (https://ec.europa.eu/transparency/regdoc/rep/1/2016/EN/COM-2016-705-F1-EN-MAIN.PDF). All accessed on 10 February 2018.

[6]The Commission will support R&D, innovation, entrepreneurship and new business opportunities. For that purpose it will use innovative procurement to stimulate demand for innovation; it will address the new skills requirements in the sector; it will step up finance to investors, entrepreneurs, start-ups and R&D.

[7]The EU, ESA and Members States together have a large public space budget. The EU alone will invest over EUR 12 billion in space activities in the 2014–2020 financial framework. The EU has autonomous access to space with state of the art space systems such as Copernicus for Earth observation, and EGNOS and Galileo for satellite navigation and geo-positioning. The EU is a large institutional customer for launch services in Europe with 22 satellites currently in orbit and over 30 launches planned in the next 10–15 years.

Outer Space (PORBUOS)[8] and the Prevention of an Arms Race in Outer Space (PAROS). Moreover, at European level, the EU ensures the protection and resilience of its critical space infrastructure, inter alia by its Space Surveillance and Tracking –SST- support framework (which started delivering operational services based on a pool of Member States' capacities), to address the issue of space debris and prevent collisions in space.[9] Furthermore, the EU is setting up a secure system of Governmental Satellite Communications, to serve its Embassies and its Common Foreign and Defence Policy.[10]

A third European space policy objective is to mobilise space for sustainable development. For that reason the EU will seek to ensure that space technology, infrastructure, data and applications contribute to the realisation of the 2030 Agenda and its 17 Sustainable Development Goals and 169 targets, both in Europe and world-wide. For the economic dimension of sustainable development this means mobilising space for the creation of jobs and economic growth (SDG 8) which may help reduce poverty, hunger and inequalities (SDG 1, 2, 5 and 10). For the environmental dimension this means mobilising space to improve resource efficiency (SDG 6 and 7 for water and energy and SDG 12 for sustainable patterns of production) and to reduce emissions and pollution (SDG 13, 14 and 15). For the social dimension this means mobilising space for improving health and education (SDG 3 and 4) and reducing conflict and crime (SDG 16). Also crucial will be to mobilise space for sustainable cities (SDG 11) since 60–70% of the world population is expected to live in cities in the second half of the 21st century.[11]

A second preliminary remark concerns the multitude of European space actors. We have European Universities, space industry and the national space agencies of the Member States, and the main actors at European level are the European Space Agency (ESA) and the EU, which work closely together on the basis of their 2004 Framework Agreement.[12] ESA is an international organisation with 22 Member States that pool their resources for space exploration missions since 1968 (2018 budget of 5.6 Billion EUR).[13] ESA also constructs the space infrastructure of the

---

[8]The EU is in favour of a non-binding instrument negotiated in the UN Framework to establish standards of responsible behaviour across the full range of space activity, strengthening commitments to non-interference in the peaceful exploration and use of outer space, facilitating equitable access to outer space and increasing transparency of space activities.

[9]Decision No 541/2014/EU of the European Parliament and of the Council of 16 April 2014 establishing a Framework for Space Surveillance and Tracking Support available at: http://eur-lex.europa.eu/legal-content/EN/TXT/HTML/?uri=CELEX:32014D0541&from=EN. See also: http://www.cesmamil.org/wordpress/wp-content/uploads/2017/05/9-_-Matarazzo-Brancati-_-Thales-Telespazio.pdf. Both accessed on 10 February 2018.

[10]See: https://www.eda.europa.eu/what-we-do/activities/activities-search/governmental-satellite-communications-(govsatcom). Accessed on 11 February 2018.

[11]See: https://www.thegef.org/topics/sustainable-cities. Accessed on 11 February 2018.

[12]OJ L 261 of 6.8.2004, p. 64. See also: http://www.esa.int/About_Us/Welcome_to_ESA/ESA_and_the_EU. Accessed on 11 February 2018.

[13]See www.esa.int. ESA elaborates and implements a long-term European space policy and activities and programmes in the space field. ESA also progressively integrates the national space programs of its 22 member states into a European space programme (in particular as regards

satellite navigation and earth observation programs (Galileo and Copernicus) of the EU, which will contribute an additional 1.5 billion Euro to the ESA budget in 2018 (1.315 Million from the EU budget and 221 Million from Eumetsat).[14]

The European Commission, DG Internal Market, is the second main European space actor, and it is responsible for overseeing the European research program Horizon 2020,[15] the European satellite navigation programmes (Galileo and EGNOS)[16] and the European earth observation program Copernicus.[17] Copernicus is a good example of intense cooperation between different European Space actors because it is funded and overseen by the EU Commission, the technical coordination is done by ESA, the Sentinel missions are operated by ESA and Eumetsat[18] and the different services are managed by specialised agencies including services in the fields of: atmosphere- and climate change monitoring (ECMWF),[19] marine monitoring (Mercator Ocean),[20] land-monitoring (JRC and EEA),[21] emergency management (JRC, EFAR and EFFIS)[22] and security (Frontex, EMSA and Satcen).[23]

---

the development of applications satellites) and recommends a coherent industrial policy to its Member States. See Article II of the 'Convention for the establishment of a European Space Agency' available at: http://download.esa.int/docs/LEX-L/ESA-Convention/20101200-SP-1317-EN_Extract_ESA-Convention.pdf (Accessed on 27 January 2018). The ESA Convention was signed in 1975 and entered into force in 1980 and is at present ratified by 22 States (Sweden and Switzerland ratified in 1976, Germany and Denmark in 1977, Italy, the UK and Belgium in 1978, Netherlands and Spain in 1979, France and Ireland in 1980, Austria and Norway in 1986, Finland in 1995, Portugal in 2000, Greece and Luxembourg in 2005, Czechia in 2008, Romania in 2011, Poland in 2012, Estonia and Hungary in 2015. Slovenia is an Associate Member. Canada takes part in some projects under a cooperation agreement. Bulgaria, Cyprus, Malta, Latvia, Lithuania and Slovakia have cooperation agreements with ESA. Discussions are under way with Croatia.

[14]See: http://www.esa.int/About_Us/Welcome_to_ESA/Funding. Accessed on 27 January 2018.

[15]The Programme for Research and Innovation, Horizon 2020 has a space research component of €1.4 billion which is implemented by the Research Executive Agency (REA). Accessed on 27 January 2018.

[16]The Programme for Satellite Navigation, (Galileo/EGNOS) (Accessed on 27 January 2018) has a budget of EUR 7 billion which is implemented by the European GNSS Agency (GSA) (Accessed on 27 January 2018) and has applications in agriculture, transport, location based services and mapping and surveying. EGNOS is also supported by Eurocontrol (the European air traffic control organisation).

[17]The European Programme for Global Earth Observation (Copernicus) (Accessed on 27 January 2018) has a budget of EUR 4.3 billion, and consists of six families of dedicated, EU-owned earth observation satellites and instruments, the so-called Sentinels,.

[18]https://www.eumetsat.int/website/home/index.html. Accessed on 27 January 2018.

[19]Sentinels 1 and 4 are managed by European Centre for Medium-Range Weather Forecasts (ECMWF) and EUMETSAT). Accessed on 27 January 2018.

[20]Sentinel 2 is managed by Mercator Ocean. Accessed on 27 January 2018.

[21]Sentinel 3 is managed by the Joint Research Centre (https://ec.europa.eu/jrc/en) and the European Environment Agency (EEA). Accessed on 27 January 2018.

[22]Sentinel 5 is managed by the Joint Research Centre, supported by the European Flood Awareness System (EFAR) and the European Forest Fire Information System (EFFIS) (for more information see http://emergency.copernicus.eu/. Accessed on 27 January 2018.

[23]Sentinel 6 is managed by different agencies. The border control element is managed by the Agency for the Management of Operational Cooperation at the External Borders (FRONTEX), the

27.2 Space Related Assistance and Capacity Building for Sustainable Development

Availability of Data

If you want to mobilise space data, technology and applications for sustainable development, you have to ensure that space data are actually available to users. The good news is that both Galileo/Egnos and Copernicus are operational and that European satellite navigation and earth observation data are available free of charge. These data can thus be, and are in fact already, used to help realise the sustainable development goals.

Galileo's advanced navigation and positioning services, for instance, are very accurate and they can be used to make (air, rail, road, maritime) transport safer and more efficient which improves health (SDG 3), reduces fuel consumption (SDG 7: energy efficiency) and reduces emissions (SDG 13). Copernicus facilitates the formulation of numerous policies e.g. on: climate and environment (SDG 13, 14 and 15), on maritime safety and security (SDG 14), on agriculture (SDG 2), on disaster management, and on urban planning and infrastructure (SDG 11). Its practical value is illustrated, for instance, by the Copernicus Emergency Management and Land Monitoring Services, which allow for Early Warning for floods and forest fires and provide reliable maps derived from satellite images. These maps help key actors to assess the impact and define the response to natural and man-made disasters all over the world. Copernicus thus has become invaluable to crisis managers, civil protection authorities, humanitarian aid actors, and those involved in preparedness and recovery activities. These activities are carried out as much as possible in coordination with international partners, including UN-SPIDER.[24]

In fact European space-based applications and services support many activities including agriculture and fisheries (SDG 2),[25] health (SDG 3),[26] telecommunication

maritime element by the European Maritime Safety Agency (EMSA) and the CFDP element by the European Union Satellite Centre (SATCEN - https://www.satcen.europa.eu/). In addition the European Defence Agency (EDA) implements parts of the European Parliament Pilot Project on Common Security Defence Policy-related Research, the European Space Agency (ESA) implements the main parts of the Galileo and Copernicus programmes including the construction of space infrastructure and the European Investment Fund (EIF) implements part of the COSME programme called 'Financial Instruments'. All accessed on 27 January 2018.

[24]http://www.un-spider.org/. Accessed on 27 January 2018.

[25]Satellite-enabled applications, for instance, improve the mapping of cropland in need of irrigation, harvest forecasts, and fisheries control. This guarantees better food quality and security while safeguarding the environment.

[26]Space-based applications can significantly improve healthcare and the health education of patients through remote medical support. They also help in preventing or mitigating the outbreak of disease and delivering medicines (drones).

and connectivity of remote communities (SDG 4 and 10), transport and fuel efficiency (SDG 7 and 13),[27] crisis response,[28] protection of the environment (SDG 13, 14 and 15),[29] and the reduction of conflict and crime (SDG 16).[30] An interesting new development is the integrated use of Galileo and Copernicus data, which has great potential to support realisation of the SDG. One example is precisions agriculture, which makes use of earth-observation data to monitor land use, crop growth and health and soil humidity, and subsequently uses precise navigation data for the targeted delivery of water, fertilisers and pesticides. But there are many more as illustrated by a 2018 study by the European GNNS Agency and UNOOSA.[31]

Facilitate Access to and Use of Data

The availability of space data is a good step in the right direction, but data availability does not automatically mean that end users will have access to those data and will know how to use them for the realisation of the SDGs. This is where traditional European capacity building comes in. The EU and its member states are the largest providers of development assistance[32] and their projects and programs include assistance to allow users in developing countries to have access to European space data and to develop their capacity to use those data.

For example the 'European Geostationary Navigation Overlay Service (EGNOS) in Africa Support Programme' (EUR 4.7 Million EUR for 2015–2017) supports the implementation of EGNOS (which improves GPS and Galileo satellite navigation data) in Africa to the benefit of air transportation. It creates an African project management office and seeks to modernise air navigation systems across the continent by deploying satellite navigation systems, installing and operating ground stations, and training flight controllers, with the objective of increasing flight safety and efficiency (SDG 3, SDG 7) and lowering carbon emissions (SDG 13). The project was preceded by the 2013 SAFIR (Satellite navigation services for the African Region)

[27] When combined with enhanced communication capabilities, highly accurate satellite positioning contributes to a modern and reliable transport sector for cars, planes, and ships. It optimises fleet management, vessel traceability, collision prevention, speed control, assistance for ship manoeuvres, etc.

[28] Satellite services help shorten response times in emergencies. Damage images and assessment maps contribute to more efficient planning of rescue and relief efforts.

[29] Environment monitoring provides crucial information on vegetation, ocean currents, water quality, natural resources, atmospheric pollutants, greenhouse gases, and the ozone layer.

[30] Satellite positioning, satellite communications, and Earth observation help detect illegal immigration, prevent organised crime, and combat piracy at sea.

[31] The study: "EGNOS and COPERNICUS: Supporting the Sustainable Development Goals. Building blocks towards the 2030 Agenda" (prepared by the European GNNS Agency and UNOOSA) is available at: http://www.unoosa.org/res/oosadoc/data/documents/2018/stspace/stspace71_0_html/st_space_71E.pdf. Accessed on 27 January 2018.

[32] The OECD Development Assistance Committee provides statistics on Official Development Assistance and it 2016 Table 'DAC Member's net ODA in 2016' shows that the EU and its Member States together provided around 100 billion USD of ODA in 2016. The 30 DAC members together provided around 145 Billion of ODA in 2016. See: http://www.oecd.org/dac/financing-sustainable-development/development-finance-data/. Accessed on 27 January 2018.

project which set up a Joint Program Office to build capacity in ACP countries to use Galileo/EGNOS data.[33] These projects can in the future be extended to other sectors like maritime (costal) navigation, agriculture, mining, movement of goods (containers) and any other application benefiting from the provision of precise navigation data/land survey.[34]

A second example of a traditional European space related capacity building project is the 'Global Monitoring for Environment and Security & Africa Support Program (EUR 32 million for 2014–2017).[35] The GMES & Africa project builds on previous projects that helped prepare Africa to use Meteosat data (the 2001–2006, 11 Million EUR PUMA or Preparation for use of Meteosat project), to monitor the environment (the 2007–2013, 21 Million AMESD African monitoring of the environment for sustainable development project) and the intra-ACP'Monitoring of Environment and Security in Africa' (the 2013–2017, 37 million, MESA project). The GMES project aims to improve capacities of African policy-makers and planners to use Earth observation and Copernicus data so as to better understand long term environmental trends and to better design and implement national, regional and continental environmental responses/policies. The project has a broad potential scope with 9 priority areas defined by the 2007 *Action Plan on GMES and Africa*', but there is an initial focus on 3 areas: (1) long term management of natural resources, (2) marine and coastal areas, and (3) water resource management.[36]

Again the project has very concrete deliverables. It will install receiving stations and it will support African institutions to design and develop new monitoring applications, and five African universities or specialised training centres to train policymakers, companies and non-governmental organisations, so as to build concrete capacity on Earth Observation. The project will notably create two specialised African services, an African *Water & Natural Resources* Service and an African *Marine & Coastal* Service, in order to receive and process EO and geo-data; elaborate and disseminate EO-based information; build African capacities to fully benefit from these services to allow to better assist policy makers; and increase the impact of EO derived information on the decision-making process. The marine and coastal service will also monitor oceanic variables (physical, biological and fisheries), coastal areas (vulnerability and ecosystems), ship traffic and pollution (oil spills) and marine weather forecast (3 days). The Water and Natural Resources Service will monitor

---

[33] See http://gpsworld.com/egnos-africa-joint-programme-office-launched/ and http://www.aviation-africa.eu/safir. Accessed on 27 January 2018.

[34] See http://www.aviation-africa.eu/jpo. Accessed on 27 January 2018.

[35] https://europa.eu/capacity4dev/africa-eu-part.gmes/minisite/gmes-and-africa. Accessed on 27 January 2018.

[36] Next three priority areas to be validated are: climate change adaptation, disasters risk reduction and, rural development and food security, and after that the project can cover infrastructure, conflicts and political crisis, and health management.

various environmental topics from crop monitoring to forest degradation assessment, including inland water quality and quantity measurements, protected areas and wetlands mapping, flood, drought and wildfire monitoring and land degradation.[37]

The EU space related capacity building activities are not limited to Africa. In September 2017, for instance, the European Commission organised two TAIEX workshops on Space Applications in Chile and Bolivia. The objective of these workshops was to promote European initiatives on Space applications to officials in Latin America, and also to transmit information about opportunities for Latin American stakeholders to use Copernicus and Galileo for their own needs. In addition, the activity helped to identify opportunities for cooperation with European stakeholders (e.g. academia, government and industry).[38]

27.3 Suggestions for the Future of Capacity Building Activities to Mobilise Space for Sustainable Development

Making data available and assisting potential users to actually make use of those data, are very good steps in the right direction and they should be pursued. In a more future oriented perspective, however, the question arises what other steps could be undertaken to make sure that the potential contribution of space data and applications to the sustainable development goals is fully realised. In other words, what steps could be envisaged to scale up the positive impact of space data and applications on sustainable development?

One suggestion would be to make sure that there is a central online overview of all existing 'space for sustainable development' technologies and applications, with a simple indication on how and for what particular purpose these can be used in a concrete way and where those technologies and applications would be available. This would allow end users anywhere in the world to have real time access to the latest available applications and to the respective suppliers. In this respect, the initiative of ESA to build a catalogue of space applications deserves praise, as does the fact that they have published a first version of their catalogue online,[39] and that they are discussing with UNOOSA how that catalogue can contribute to the Space Solutions Compendium which the international space community will most probably

[37]See: http://www.copernicus.eu/news/african-union-commission-launches-gmes-africa-call-proposals. Accessed on 27 January 2018.

[38]For the workshops in Chile and Bolivia, please find the link to the programme and all presentations at: http://www.copernicus.eu/library/detail/2479 and http://www.copernicus.eu/library/detail/2509. Accessed on 27 January 2018.

[39]See: http://esamultimedia.esa.int/docs/spaceforearth/SDG_Catalogue_October2017.pdf. See also: *www.esa.int/SDG*. Accessed on 27 January 2018.

be invited to support by the UNGA Resolution on Space as a driver for sustainable development.[40]

A second suggestion would be to promote the development of new applications and in this respect the Commission's intention to actively support downstream industries and service providers is very good news.[41] Very helpful in this respect could be the global space partnership which is proposed by UNOOSA to analyse the relevance of space applications for sustainable development and to identify gaps and action priorities on the basis of which work on new apps can be undertaken. Also important is UNOOSA's idea to establish a global compact for space which would serve as an entry point for the private sector (industry and academia) into the space solutions compendium.

A third suggestion could be to encourage traditional capacity builders at European and international level to mainstream space applications (and other innovative technologies) into their traditional capacity building activities. One example of creating such synergies in the Vienna context could be to combine the integrated earth-observation and satellite navigation space technology for precision agriculture, of which UNOOSA is aware, with the nuclear technology for crop control and precision dripping, of which the IAEA is aware, and integrate these technologies in the EU funded projects that are implemented by UNIDO and that seek to boost agricultural production in Africa.

A fourth suggestion could be to make an extra effort to ensure that the numerous space related capacity building activities of the many space actors are to a maximum extent coordinated so that synergies can be created and duplication of work can be avoided. UNOOSA could play a crucial clearing house role here, and a capacity building compendium could easily be added to the proposed global compact and space solutions compendium, so that demand for and supply of capacity building activities is transparent.

A fifth suggestion, would be to really bring space technology and applications to the individual end-users, so that the farmer can see his field and crop through space, the doctor and nurse can see their patients and deliver their diagnosis and medicine through space, and the businessman can create economies of scale and therefore jobs and economic growth through space. The way to do that is to make space applications for sustainable development available on mobile phones, which have become the central information tool of the XXI century.

Finally, a good way forwards would be to reach agreement as soon as possible on a Resolution that supports proposals to create a global space partnership and a global compact for space, to do more work on space and health, to set up the Space Climate Observatory, to strengthen UN Spider and international collaboration on space weather, to promote the space and women- and open university initiatives, to

---

[40]See: http://www.unoosa.org/oosa/en/oosadoc/data/documents/2018/aac.105c.1l/aac.105c.1l.364_0.html. Accessed on 27 January 2018.

[41]See the 26 October 2016 European Commission Communication on: Space strategy for Europe (https://ec.europa.eu/transparency/regdoc/rep/1/2016/EN/COM-2016-705-F1-EN-MAIN.PDF). Accessed on 27 January 2018.

create a space for development profile and a space solutions compendium and to coordinate work on research and achieve progress on the Guidelines on Long Term Sustainability.

27.4 Conclusion

In conclusion, European actors invest significant resources in an autonomous European space capacity which they want to be able to operate in a safe and secure environment and to help realise sustainable development, both in Europe and globally. Different European Actors that help achieve these objectives include universities, space industry, national space agencies and, at European level, in particular the European Space Agency and the European Commission (DG Internal Market), which work closely together in particular on Galileo and Copernicus, which are operational and provide data and services through a multitude of implementing agencies.

There are at present well developed European assistance and capacity building activities to help mobilise space for sustainable development. First of all, European space data (Galileo, Copernicus) are available free of charge and are already used to serve numerous SDGs. This is a good start, but in itself not enough, because not everybody has access to, or knows how to use those freely available data. That is why EU programs also help build capacity in developing countries to use space based data and applications for the realisation of the SDGs. One example is the EGNOS for Africa project which makes precision navigation available to African air traffic management to increase flight safety and reduce fuel consumption and emissions. A second example is a series of projects that extend Earth Observation capacity to Africa, to African decision makers and planners to make better informed decisions on national, regional and continental environmental policies.

But even though having the capacity to make use of space data free of charge is a very good beginning, further steps can be taken to ensure that the potential contribution of space data and applications to the SDGs if fully realised. Important for instance is to increase awareness on existing space applications and their suppliers, for which the online ESA catalogue of space applications and the ESA and UNOOSA Space Solutions Compendium could be instrumental. Important also is the development of new applications, including in particular also for mobile phones, which, in Europe, is actively supported by the Commission. Mainstreaming space data and applications into traditional development assistance budgets and projects/programs, could also be important together with a good coordination between different actors, so as to ensure synergies and avoid duplication of work.

Finally, the international Community should consider the post UNISPACE+50 negotiations as a unique opportunity to raise political awareness on the potential contribution of space to sustainable development and to strengthen international cooperation and coordination through better and strengthened mandates of COPUOS and UNOOSA.

Acknowledgements The Author thanks his colleagues from the EEAS, DG GROW and DG DEVCO for their valuable input. The observations made in this contribution are purely personal and cannot be attributed to the organisations mentioned.

Chapter 28
Capacity Building and Space Applications

Roberta Mugellesi

Abstract The European Space Agency is working towards the development of integrated applications, using satellite data, connectivity, positioning and other space technologies that can concur to the solution of societal problems. Space Applications are solutions that make simultaneous use of different space services and technologies. They combine different space systems (Earth observation, navigation, telecommunications, etc.) with airborne and ground-based systems to deliver solutions to local, national and global needs. The chapter discusses the different actors within ESA that are in charge of the building blocks for space applications, and the different programs that external firms in ESA member states can leverage to support the feasibility study, development and market application of integrated applications.

28.1 Introduction

The concept of digitizing and connecting everything forms the basis of how the Fourth Industrial Revolution, Industry 4.0, is influencing and impacting the world. Emerging technologies, such as Machine learning, Artificial Intelligence, Internet of Things, and other advanced technologies are rapidly revolutionizing and reshaping infrastructure and global-local economies. Leveraging these new transformations and understanding their disruption potential with respect to technology, shifting demographics and global connectivity is essential for the space technologies.

The ability of satellite technology to provide ubiquitous and increasingly fast connectivity to billions of people globally is at the core of the Fourth Industrial Revolution. Connectivity is not the only element in the Fourth Industrial Revolution that can be harnessed by the satellite industry. Innovative technologies will open the door to new opportunities incorporating multiple disciplines and industries to create new markets and growth. New business models (e.g. the impact of AI on satellite data processing) and the evolving economic/trade landscape, for example related to the autonomous technologies, will lower barriers to entrepreneurs with new ideas to

R. Mugellesi (✉)
European Space Agency, Harwell, UK
e-mail: Roberta.mugellesi@esa.int

S. Ferretti (ed.), *Space Capacity Building in the XXI Century*, Studies in Space Policy 22,
https://doi.org/10.1007/978-3-030-21938-3_28

access the markets. Space systems are more and more involved in the delivery of global utilitarian services to end-users. The concept of Space Applications encompasses the simultaneous use of basic space services and technologies. Space applications combine different space systems (Earth observation, navigation, telecommunications, etc.) with airborne and ground-based systems to deliver solutions to local, national and global needs.

28.2 Space Applications in the Global Context

The European Space Agency is involved in the international arena of space applications with several activities since more than a decade, but it is the Symposium of the Integrated Applications and its related Committee at the International Astronautical Congress, organized by IAF, IAA and ISL, that collects every year distinguished professionals globally to discuss future strategies and trends in the area of space applications. The symposium addresses various aspects of integrated applications and exploits the synergies between different data sources to provide the right information at the right time to the right user in a cost-effective manner and to deliver the data to users in a readily usable form. The goal of the symposium is to enable the development of end-to-end solutions by connecting the user communities that are driving toward end-to-end solutions with those that are developing enabling technologies for integrated applications.

In May 2018, the Global Space Applications Conference (GLAC) chaired by ESA was held in Montevideo with the purpose of enabling the development of end-to-end solutions by connecting the user communities that are driving space solutions with those that are developing enabling technologies for integrated applications. This was the first time ever that IAF took one of its Global Conferences to South America and also the first time that it organised an event in Uruguay. GLAC 2018 was co-organised together with the Centro de Investigacion y Difuson Aeronautico-Espacial (CIDA-E), a member of the IAF since 1985 and one of ten IAF members from South America.

This conference was a follow up of GLAC 2014 which was organised in Paris, France. During the 4 years since the previous GLAC, the international satellite-based applications community significantly moved forward with their respective planning and developments and it was therefore timely to take stock of the progress and undertake an outlook to the future of space applications on a global scale, with a specific focus also on developing space nations, such as Uruguay. GLAC 2018 encouraged the sharing of programmatic, technical and policy information, as well as collaborative solutions, challenges, lessons learnt, and paths forward among all nations with the desire to improve space applications and their usage. The conference provided an excellent opportunity to review the state of the art of satellite-based applications, with a focus on several sectors, as farming and fishing, integrated risk management, climate, natural resources, mapping and legal aspects.

28.3 Space Applications at ESA

At the European Space Agency, Space Applications are developed across several directorates. The Earth Observation (EO) directorate focuses on both the space and ground segments, to support forefront EO applications related development activities, including, for example, the Thematic Exploitation Platforms. The Navigation Directorate supports European industry in succeeding in the highly competitive and rapidly-evolving global market of Satellite Navigation while supporting Member States in enhancing their national objectives and capabilities in the sector. The aim of the Telecommunications and Integrated Applications Directorate is to reinforce and extend the impact of the space sector through development of wider application of commercial products and services using space-derived data with a key element of the implementation being the identification and involvement of Users/Stakeholders and particularly Customers.

These high-level objectives for Earth Observation, Navigation and Telecommunications are entirely complementary and clearly drive the focus of the activities carried out in each ESA Directorate. It is noted that some data products/services are intended for commercial exploitation, while others may be intended for scientific exploitation or for free-of-charge utilization by large user communities. For example, many space applications depend on by-properties of the atmosphere which are subjects of the investigations of space science. Weather forecasting and the prediction of climatic trends depend on a knowledge of atmospheric models. This requires knowing how long certain circulation patterns may be expected to persist, the ways in which energy exchanges are likely to occur within the atmosphere and between the atmosphere and the land and sea, and how all these are influenced by the continuous input of energy from the sun. As a consequence, meteorology assumes a dual aspect, the practical one of forecasting weather and climate, impacting economic activities and closely related to commercial space applications, and the scientific aspect of research on the atmosphere.

At the European Space Agency, in the frame of the *ARTES (Advanced Research in Telecommunication Systems)* programme, the Intergrated Applications element is dedicated to develop and validate downstream applications and services based on the use of space based technologies. With more than 400 activities placed so far, this part of the programme, called as Business Applications, has the final objective to facilitate the creation of novel business opportunities leading to sustainable use of space technologies for the everyday life, and indirectly help the uptake of the space domain as a whole. The ESA Business Applications programme was introduced more than 10 years ago as Integrated Applications Promotion (IAP) programme aiming to pursue an holistic approach to space applications, enabling cross-cutting functionalities among the relevant space capabilities. In doing so it builds upon the rich competences developed within ESA, to address the utilisation of space capabilities and to create value along the entire satellite applications value chain.

Business Applications activities are leveraging the utilisation of Satellite Communications, Satellite Navigation, Earth Observation and the use of technologies coming from Human Space Flight, used as stand-alone or integrated together with terrestrial based systems, leading to user driven solutions in many different domains. Specific focus is given to the need to contextualise the utilisation of the downstream application in a pre-operational environment, where pilot user communities will be using the results of the development to validate the added value brought in their business context.

The programme has developed a range of new applications by utilising and integrating different space assets, resulting in new or improved services for the citizens of Europe on a regional, European and even global scale. Intrinsic to these new applications are the added value of space as a facilitating capability and the long-term sustainability of the resulting services. The programme does not push any particular technology but instead explores and responds to users' needs, addressing a variety of thematic areas, such as Aviation, Health, Infrastructure & Smart Cities, Transport & Logistics, Safety & Security, Energy, Food & Agriculture, and others.

Specific focus is given to the involvement of users and the service demonstration in a pre-operational environment, where pilot users validate the added value brought in their business context. A successful activity does not require only the identification of promising opportunities, but also the development of complete business cases, combining technical aspects with commercial, financial, legal and regulatory considerations. The focus of the Business Applications programme is the application of space capabilities into a non-space business context with focus on promoting the utilization of space technology or space-derived data in a multiplicity of economic market sectors identifying opportunities for utilization of space capabilities to the benefits of suppliers, users and the wider ecosystems.

User involvement is a pre-requisite for the development and implementation of the application.

This is because the programme does not push any particular technology but instead explores and responds to users' needs. The user-driven approach of the programme ensures that the user needs are satisfied by providing solutions that are technically, economically viable and cost effective.

Activities proposed within the Business Applications (BA) programme are either implemented as feasibility studies or demonstration projects. Feasibility studies provide the preparatory framework to identify, analyse and define new potentially sustainable applications and services. The study is required to provide definitions of the technical solutions that can be envisaged, and an assessment of the investment and of the running costs providing a first evaluation of the sustainability of an operational system. The study is also required to identify technological adaptations and gaps in current systems which need to be filled for the service to be operational and sustainable. The applications and/or services covered by the proposed feasibility studies have to be customer/user driven, benefit from the integrated use of space assets, and aim to evolve the targeted applications and services to marketability and operational roll-out, potentially through a BA demonstration project after successful completion of the feasibility study.

Kick-start studies are a subcategory of feasibility studies to be initiated by industry/institutions following thematic calls for ideas or opportunities initiated by the

Agency. Kick-start activities aim at exploring the viability of new service/application concepts while fostering innovation and reaching out to new players. This instrument has been successfully targeted at new entrants to the Space Economy and provides an accessible starting point for smaller companies to work with ESA. Following an analysis on the awarded contracts for kick-start studies it emerged that more than 80% were conducted by SMEs.

Demonstration projects are dedicated to the implementation and demonstration of pre-operational services, they have to be customer/user driven (including user involvement and contribution) and benefit from the integrated use of at least one space asset, with clear potential to become sustainable in the post-project phase. Demonstration projects can be initiated following a successful feasibility study or as a spontaneous proposal by users or industry. The engagement of prospective customers (either public or private) in the activity is considered essential to mitigate the commercial risk and ensure that customers' feedback is adequately considered in the consolidation of the service offer and preparation for the commercialization phase. Proposals for a feasibility study or a demonstration projects can be provided based on the permanent open call which explains the bidding process as well as the tender condition and contract conditions. Before engaging in more detailed elaboration of a potential outline or full proposal, an Activity Pitch Questionnaire allows the Tenderer in a reduced, standardised pitch form to present his initial idea to ESA, and it allows ESA to quickly take an informed decision on the next steps pointing the Tenderer to the most appropriate activity stream and providing support on the further implementation steps.

28.4 Users and Stakeholders

ESA Business Applications has established many successful partnerships over the years with European institutional organizations or industry associations as well non-profit organisations. These partnership were concluded to federate user needs addressing specific user communities (e.g. maritime) and support pre-operational demonstrations. Partnerships have been also set up with organizations outside of the members states. The partnership with the UK investor and business support community has generate referrals of a rising number of promising companies offering also additional finance and resources for the companies with ESA contracts. Further formal and informal partnerships are continuously explored, extending the approach on partnerships benefitting the entire programme's participating Member States and favouring the ones with global commercial reach.

In the context of new opportunities arising in the space sector and addressing emerging needs, the Atlantic has acquired a position of key priority since its resources are recognized as being essential for addressing the multiple challenges that will be faced in the decades to come, as it is a key source of food, energy, minerals, jobs and transport. The challenge is promoting sustainable development of Atlantic economic sectors with high growth potential as the marine aquaculture, the ocean renewable energy exploiting the potential of tides and waves, the seabed mining, the hydrocarbon exploration and production, and the bio-economy. Emerging Atlantic industries are in particular characterised by the key role played in their operations by cutting-edge science and technologies, moving increasingly to high level of automation and benefiting from satellite technology, tracking and imaging. The ultimate goal is to promote and enable the emergence of new markets and the growth of skilled employment in different areas and contribute to the successful achievement of the UN Sustainable Development Goals.

The focus is on the application of space data for the development of services within emerging ocean industries while managing the Atlantic in a responsible and sustainable way. These applications will demonstrate the value of satellite data in the development of services integrating new and existing navigational tools for enhancing the safety of ships at sea through better organization and exchange of data between ships and shore. The use of Integrated Atlantic multi-use platforms as joint location of offshore industries will allow to capture the synergies offered by the use of different ocean-based technologies, renewable energy (wind, wave, etc.), marine aquaculture, maritime transport and logistics, marine research, biotechnology deployed on the same site. AI techniques can contribute in important ways to a more effective enforcement and conservation of Atlantic ecosystem by making possible to extract valuable information from the huge data volumes and allowing to monitor, model, and manage the environmental systems.

28.5 Conclusion

Based on the experiences collected at ESA as well in the international community of IAF, we can conclude that space applications can indeed make a major contribution to global policies, technological infrastructures, economies, along with social and cultural development. A growing number of space applications such as precision farming, land and sea monitoring, mobility, can potentially provide enormous opportunities to reduce social and economic inequalities; support sustainable rural wealth creation by overcoming barriers of geographic isolation, along with providing access to information and in communication services at affordable costs. Moreover, the benefits from Space applications that address crop and soil management, water and costal resources, and disaster monitoring and mitigation are instrumental to the support and implementation of the Sustainable Development Goals.

Satellites' ability to connect and communicate is essential for developing countries, especially in rural areas, for example for communication in disaster areas, obtaining information about weather for use in agriculture and fishing and sharing medical information. By improving these areas, satellite technology can be used as one solution for many SDGs. For example, by using information gathered about agriculture and fisheries, satellites are assisting in the development of economies. In some countries, illegal fishing is a problem that satellite technology can help reduce, which improves the livelihoods and security of people in the fishing industry. Communication about healthcare (for general care or in disaster areas) and education improves people's safety and gives them access to education.

The experience in Space Applications allows to draw important recommendations. As first, the importance of integrating all data sources that are available, from the space data to the terrestrial ones and making use of advanced technological breakthroughs. It is the integration of the data and technologies that enables the exploitation of the full capability of the application. Fundamental for success, is the involvement of the users/stakeholders in all the phases of the development, testing and validation. The participation of the users throughout the development of the application is beneficial to guarantee that what is developed fits their needs, and is also beneficial to the developing contractor that can use the time to prepare for the full exploitation of the service.

Satellite technology is also assisting developing countries to address environmental monitoring and natural resources management issues and there are many experiences that can be reported successfully across ESA. For example, in Mexico, ESA is developing a project aimed at improving the management of the fishing industry, which is important to the Mexican economy, jointly with the National Space Agency. The project uses satellite-based tools on fishing boats in order for the government to collect data that can reduce illegal fishing practices, while also being a tool to relay relevant information to fishermen that are out at sea.

In Nigeria, ESA has launched another project that is benefiting healthcare systems. Satellite technology can be used in health centres for data collection to determine what actions need to be taken. The health centres use satellite technology for communication, disease surveillance and video-based training for their staff. By helping improve communication and data collection, satellites are a beneficial investment for the sustainable development of countries. The improvements that can be made to their economies as well as their healthcare systems can ultimately improve people's security and save lives. The lessons learned with these projects have allowed ESA to identify the key elements that enabled their success and could be used as key attributes to a model that can be applied in other developing countries to build Space capacity in the XXI century.

Chapter 29
SGAC: A Network for the Young Generation

Clémentine Decoopman and Antonio Stark

Abstract Over the recent years, the Space Generation Advisory Council (SGAC) in Support of the United Nations Programme on Space Applications has seen an increase in the organisation's membership base, as well as growth in its role and influence in critical international space organisations. This growth trajectory is set to increase as SGAC plans to further expand its reach on a global scale. Through projects, events, scholarships and dissemination of information, SpaceGen aims to be the premier network for young space professionals and students. For this reason, SGAC provides an environment for growth, engagement and professional development for our members from around the world to gain skills that can prove essential to their future careers. As an organisation run predominately by volunteers, without a membership fee, SGAC has proven itself to be the platform for the future space workforce.

29.1 Introduction

Along with the Space Generation Advisory Council, many other entities in the space sector have been developing capacity building initiatives to holistically provide and guarantee either a physical access to space or an access to the infrastructure necessary to contribute to the sector. The capacity building activities are also supported at the United Nations level by the UN Regional Centres for space science and technology education. Altogether, the different actors ensure that future capacity building activities contribute to sustainable development, in particular to the successful implementation of the UN 2030 Agenda.

Individually, each and every one of these capacity builders have a different model and structure coming with its own assets and strengths. In order not to waste a great deal of efforts in creating initiatives that already exist and work, there is a need to build a database and a network of all the entities who do capacity building globally (or regionally) and *coordinate the efforts*. Among others, the Space Generation

C. Decoopman (✉) · A. Stark
Space Generation Advisory Council (SGAC), Vienna, Austria
e-mail: clementine.decoopman@spacegeneration.org

S. Ferretti (ed.), *Space Capacity Building in the XXI Century*, Studies in Space Policy 22,
https://doi.org/10.1007/978-3-030-21938-3_29

Advisory Council platform structure constitutes a model which could be applied to other entities.

29.2 The Space Generation Advisory Council

The Space Generation Advisory Council is a global non-governmental, non-profit (US 501(c)3) organisation and network which aims to represent university students and young space professionals ages 18–35 to the United Nations, space agencies, industry, and academia.

Headquartered in Vienna, Austria, the SGAC network of members, volunteers and alumni has grown to more than 13 000 members and alumni representing the 6 United Nations regions and more than 150 countries.

29.2.1 SGAC History

SpaceGen was conceived at the Third United Nations Conference on the Exploration and Peaceful Uses of Outer Space (UNISPACE III) in 1999, whereby states resolved, as part of the Vienna Declaration, "To create a council to support the United Nations Committee on the Peaceful Uses of Outer Space, through raising awareness and exchange of fresh ideas by youth. The vision is to employ the creativity and vigour of youth in advancing humanity through the peaceful uses of space". SGAC holds Permanent Observer status at the United Nations Committee on the Peaceful Uses of Outer Space (UN COPUOS) and regularly takes part in the annual meeting, as well as its Legal and Scientific and Technical Subcommittees. SGAC holds consultative status at the United Nations Economic and Social Council (UN ECOSOC), contributing to discussions on the role of space in achieving the UN Sustainable Development Goals.

29.2.2 SGAC's Role as Part of UNISPACE + 50

As a product of UNISPACE III, SGAC has been expected to play an important role in fostering and shaping the UNISPACE + 50 thematic priorities, bringing into the process the views of the future generation of space leaders and their long-term visions for space.

Over recent years, SGAC has been focusing on different initiatives around Capacity Building, in line with Priority Item 7 of the Sustainable Development Goals (SDGs). The organisation has dedicated discussion groups at its international conferences, providing the next generation of space professionals a platform to express their views and insights to each of the UNISPACE + 50 thematic areas. Moreover,

SGAC is working with sponsors and partners from agency, industry and academia to raise awareness on and promote capacity building mechanisms that can be easily and efficiently implemented across the globe. The objective is to articulate and frame a new long-term vision for space that aligns with—to the greatest extent practicable and in a manner consistent with the UNISPACE + 50 principles—the views of the future generation of space leaders.

Capacity building has been flagged as a priority for SGAC which needs a *common vision and shared actions*. Since its inception in 1999, SGAC has been taking multiple initiatives to become a global platform for Capacity Building for the young generation of future space leaders.

29.2.3 SGAC's Impact

With members from over 150 countries, SGAC is in a unique position to provide a cumulative and comprehensive vision of the current and future industry. As a result of our global membership base, the organisation is able to tackle important and often sensitive topics in the industry, drawing on member's experiences, background and national interests. SGAC's 'umbrella view' of the global space industry, allows the organisation to provide direct feedback on mission plans and policies, advocating vigorously for plans that the organisation and its members deem valuable for the future of the space industry. Over recent years, SGAC has seen an increase in the uptake of the recommendations put forward by the organisation, focusing on topics such as inspiring the future through exploration, promoting space access through global partnerships or space resource governance among others. SGAC continues to rapidly progress on its growth trajectory—evidenced by the draw of SGAC's industry partners who make a point of using the organisation as a 'sounding board' for new ideas.

SpaceGen participation and contribution at the United Nations level is vital for the organisation because it constitutes a great platform for our members to collect and present insights from the next generation on key issues of the global space industry. The outcomes of each SGAC activities ranging from its 8 working groups, to its global, regional and local event, and the activities of its Regional Coordinators and National Point of Contacts in their respective regions or countries are turned into reports. These report are then presented to SGAC's partners, sponsors, as well as at global conferences and during the UN COPUOS subcommittees' sessions.

29.3 The Space Generation Advisory Council Network

The organisation's network is culturally and geographically diverse, and includes members from government, military, non-profit and private organisations. SGAC's network has more than 13,000 members and alumni in over 150 countries around the world with educational backgrounds that include engineering, management, law, international relations, economics, and more (Fig. 29.1).

The network of SpaceGen is not limited to its members; in fact, a large number of professionals and experts contribute to the success of this organisation. These senior members follow the events and project groups as mentors, and are a source of inspiration for their younger colleagues, enriching the network with their heritage and knowledge. SGAC's Advisory and Honorary Boards reflect the diversity of our advisors, who come from different regions and activity sectors. SGAC network represents university students and young space professionals ages 18–35 to the United Nations, space agencies, industry, and academia. Alumni SGAC members often become advisors for SGAC team, mentors or enter the Advisory Board of SGAC.

The organisation brings together not only individuals but also institutions, industry and academia from all regions as partners. Each partner brings unique views about space topics, providing SGAC members with valuable insight and skills which are unreachable through traditional learning paths. Overall, while SGAC's mandate is to focus on young professionals ages 18–35, it goes beyond that by engaging experts and alumni. The diverse background of participants, the geographic diversity of our network, and the organic involvement of women in SGAC activities showcase some of the strengths that international collaboration could bring to space activities.

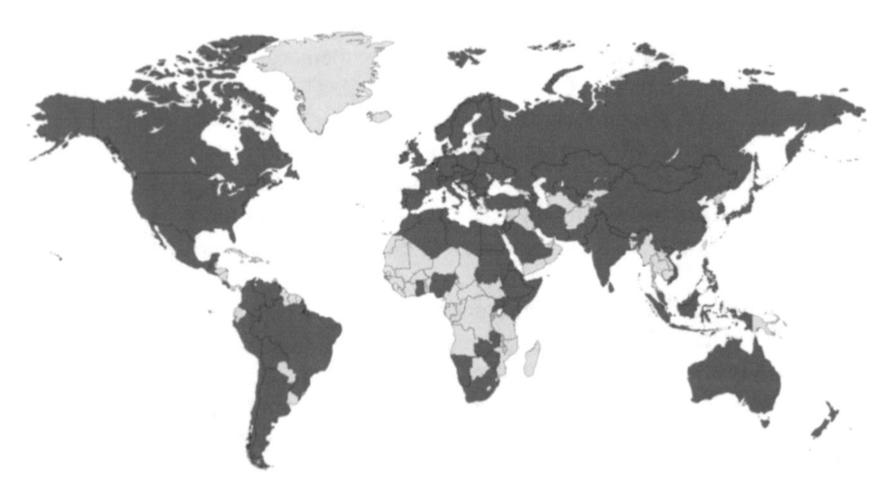

Fig. 29.1 SGAC Network and represented countries. *Source* Space Generation Advisory Council

29.4 The Space Generation Advisory Council Capacity Building Blocks

SGAC works diligently to raise awareness among the next generation of space professionals on a global scale working together with the United Nations Office for Outer Space Affairs (UN OOSA) in promoting UN workshops and activities, and in supporting SGAC members to attend space conferences around the world. By hosting international, regional and local events, SGAC provides its members with opportunities to expand their knowledge of international space policy issues, think creatively about the future direction of humanity's use of space and engage with current leaders from space agencies, industry and academia.

SGAC stewards the views and opinions of students and young professionals to ensure their creativity and vigour is employed for the advancement of humanity through the peaceful uses of outer space. Year-round project groups enable our members to further develop their thoughts on key topics of relevance to international space policy often resulting in technical papers, policy briefs and recommendations.

As a non-governmental, non-profit organisation, SGAC relies on the sponsorship and support of governmental, non-governmental, and industry partners as well as private individuals. This support is administered to fund activities. In addition, Space-Gen runs scholarships with its partners to enable participation of SGAC members in various events around the world.

Operation of SGAC relies on a global volunteer base. As a volunteer-run organisation, SGAC believes in empowering its members and providing them with opportunities for professional development through roles in the SGAC team. The highest governing body of the organisation is the SGAC Executive Committee; a body comprised of elected and appointed members supported by industry leaders and young professionals per the below organisation structure.

29.5 The Space Generation Advisory Council's Model for Capacity Building

SGAC's mission is to work at the international, regional and local level to link together university students and young professionals to think creatively about current space policy challenges and provide recommendations from the new generations' point of view. SGAC members are proactive members of the space community, who are the future workforce that will shape, develop and implement environmental legislation and policies, or take critical decisions from the experience and the information provided.

The organisation's structure is based on a decentralised model. According to the SGAC Bylaws, the highest governing body of the organisation is the SGAC Executive Committee, an elected body comprised of two co-Chairs and 12 Regional Coordinators (2 from each SGAC region). While voting on matters is restricted to elected Executive Committee members, Coordinators of different SpaceGen teams, and the SGAC Executive Director are members of the SGAC Executive Office, and also take part in the meetings of the Executive Committee.

In addition to the Regional Coordinators, SpaceGen is also represented at a national level through its National Points of Contact (2 from each SGAC represented country).

29.5.1 SGAC's Capacity Building Initiatives at a Global Level

29.5.1.1 Through events

By hosting international, regional and local events, SGAC provides its members with opportunities to expand their knowledge of international space policy issues, think creatively about the future direction of humanity's use of space and engage with current leaders from space agencies, industry and academia.

Every year, SGAC host three global events being the Space Generation Congress (SGC), the Space Generation Fusion Forum (SGFF) and the SGx event. Each event has a different format and a different purpose. Typically, those events welcome up to 160 participants from up to 40 different countries and at least 40% of women participants. The delegates are students (1/3) and young professionals (2/3) and comes from different backgrounds from science to engineering, business or policy.

The Space Generation Congress (SGC)

The Space Generation Congress (SGC) held in conjunction with the International Astronautical Congress (IAC), is SGAC's primary conference which annually brings together top young minds from around the world to focus on key space topics, providing them the chance to interact and engage with the incoming generation of space professionals from all over the world. This event usually welcomes up to 160 delegates from more than 40 different countries. The 3-day event is based on six working groups on different topics of relevance to the Space Agenda. Typically, the working groups range from Space Exploration to Space Transportation, Diplomacy, Law, Technologies and Innovation.

The Space Generation Fusion Forum (SGFF)

SGAC also organises the Space Generation Fusion Forum (SGFF) in Colorado Springs in conjunction with the Space Symposium. This global space event highlights international thinking and gathers a selected group of up to 60 top young adults from various areas of the space sector—government, industry, and academia. These intense, interactive discussions, moderated by today's international space sector leaders, gather the perspectives of tomorrow's space leaders on today's key space issues.

The SGx Event

Finally, SGAC also organises the SGx Event in conjunction with Satellite in Washington DC and in collaboration with the Future Space Leaders Foundation. SGx is a one-day thought-leadership and networking event that will bring together young professionals, industry experts and government leaders to discuss pressing issues that impact the global space community in an innovative way. The presentations are fast-paced and seek to inspire the next generation of space leaders to solve global challenges through the application of space technologies.

29.5.1.2 Through year-round project groups

SGAC is proud to act as the forum for the next generation of space sector leaders to discuss and debate current topics in international space policy. SGAC has eight year-round project groups as follow:

- Commercial Space Project Group
- Near Earth Object (NEO) Project Group
- Space Exploration Project Group
- Space Law & Policy Project Group
- Space Safety and Sustainability Project Group
- Small Satellite Project Group
- Space Technologies for Disaster Management Project Group
- Youth promoting Global Navigation Satellite System Project Group.

SGAC Project Groups produce papers with input from a broad sample of our members and embodies SGAC's purpose as envisioned from our beginnings at the United Nations. From perspectives on space situational awareness and space debris to thoughts on exploration and space workforce issues, the members of the young space community have opinions to share.

The SGAC Project Groups play a vital role in allowing our members to gain hands-on experience within specific technical areas, while also sharing their insights with leaders in the space sector. Led by students and young professionals, our project groups have more than 400 members distributed in all UN geographic regions.

Each project group is constituted of about 50 active volunteers who worked remotely on a large variety of activities such as publications, long-term projects and recommendations. The members use modern communication tools to coordinate internally among themselves but also externally to communicate about their activities. As a concrete example, in 2017, the Space Exploration Project Group organised its first Mars Analogue Simulation in Torun, Poland. This was a long-term project which required more than two years of preparation. The team represented members from more than 10 different nationalities who worked remotely on this project.

The structure of the SGAC working group could definitely be applied to other entities in the space sector and even potentially at the United Nations level through the United Nations Regional Centres.

29.5.2 SGAC's Capacity Building Initiatives at a Regional and Local Level

Because SGAC is based on a decentralised model and is represented regionally, through its Regional Coordinators, and locally, through its National Points of Contact, the organisation is able to create capacity building at a regional and a local level. Those volunteer members act as "ambassador" for SGAC and represents the organisation to their own community. Along with SGAC's mandate, they create capacity building for their region or country through events and outreach activities.

Since 2014, SGAC has initiated a series of regional events, dubbed the Space Generation Workshops (SGWs), which aim to bring together SGAC members and regional space leaders to discuss topics of relevance to the respective region.

The objectives of those events are to:

- To strengthen the regional network of the students and young professionals in the region;
- To examine and consider key questions in the region that the regional space community is facing and to provide inputs from the next generation of space professionals;
- To allow tomorrow's space sector leaders in the region to have the opportunity to interact with today's space leaders in the region.

For example, in 2017, the organisation hosted its first African Space Generation Workshop. The delegates were assigned to six different working groups on: socio-economic impacts of regional navigational satellite system, space diplomacy: bridging the divide, CubeSat as an enabler of space technology, ecosystem for young space entrepreneurs, regulatory hurdles and space policy and lunar exploration.

Another initiative started in 2015 with the SG[Country] series of events and local events, providing an opportunity for local members in each country to engage in dialogue about local space issues.

Finally, SGAC representatives at a regional and local level also engage their own community through outreach activities such as inspirational talks in local schools, mentoring events and gatherings.

29.6 The Importance of Using Modern Communication Tools

Because of the nature of SGAC's network including a community of 13 000 members and alumni, the use of modern communication tools internally and externally is vital. As described in the title "Space 2030 and Space 4.0: synergies for capacity building in the XXI century", there is a need to accordingly use the communication tools of the XXI century to create capacity building.

29.6.1 The Use of Modern Communication Tools as a Means to Function as an Organisation

The Space Generation Advisory Council literally uses modern communication tools as a vital mean to function as an organisation. The active volunteers of SpaceGen work *remotely* on various projects and activities ranging from the organisation of a global, regional or local event to the publication of a report on a specific topic.

Taking a concrete example, this year, SGAC will host an event in June in Vienna, Austria. The organising team for the event represents 10 different countries (Mexico, Korea, Italy, China…). Yet, the members are working together on putting together this event and so regularly communicate through social media messaging platforms, video conference tools among others.

Modern communication tools are also used at SGAC for training purposes. Because SGAC members are located worldwide, SpaceGen uses virtual training videos to "onboard" its new active members and provide them with the resources they need to develop capacity building activities for their own community. The videos are usually accompanied by an online and downloadable training pack.

29.6.2 The Use of Modern Communication Tools to Communicate About Its Activities to Civil Society

We now enter into a new era of Capacity Building where outreach activities, online platforms and tools as well as medias will come into play.

Although modern societies are heavily reliant on space systems, the vast majority of people are still not aware of how, and how much, outer space impacts our daily lives. Thus, story-telling among other means is seen as an essential tool to inspire the young generation to get involved in the space sector. The use of modern tool for education are also more and more popular with the development of Massive Open Online Courses (MOOCs) for instance.

The organisation also uses modern communication tools to communicate about its activities to its network of members and alumni as well as the civil society.

29.6.2.1 Through Social Media Platforms

SGAC heavily rely on social media to communicate about its activities. The organisation uses different platforms such as Facebook, Twitter, LinkedIn, Instagram and YouTube to share stories with its community. Publications topics varies from opportunities, announcement of new active members, call for applications for a position or a scholarship, event-related publications, story-telling from members etc.

One example that SpaceGen uses to inspire the young generation to get involved with SGAC and within the space sector in general is the *Human of SGAC* initiative. The objective is to share stories about and testimonials from our members, alumni or partners on social media. The format usually varies from videos to articles.

SpaceGen also regularly posts about the activities of its partners on its own social media platform and share the external opportunities which could be interesting for its members.

The organisation often takes videos during its events and share them on YouTube. For example, SGAC SGx event and its TEDTalk like environment is perfectly suitable for this media activity. During the event, the speakers share stories with the delegates about their career, their inspirations, their visions and perspectives of the sector. SGAC then share the video on its social media platform.

29.6.2.2 Through Virtual Platforms

The organisation also uses virtual platform to organise live session during some of its events. For example, the Space Exploration Project Group hold a Mars analogue simulation in Torun, Poland called the Poland Mars Analogue Simulation (PMAS) in 2017. It was the first combined Moon-Mars analogue mission in Poland. In this two-week mission, a crew of six analogue astronauts conducted scientific research and recorded their results while confined to a newly constructed habitat, LUNARES provided by the Space Garden Company—to simulate real planetary exploration conditions. Fully isolated from the rest of the world, the astronauts' only communication with the outside world will be through a 15-minute time-delayed link (in Mars mode) with the 30 international students and young professionals making up the Flight Support Team (FST), in the Mission Support Centre (MSC), located 89 miles away in Torun, Poland.

As another way to increase the interactions and awareness during its event, SpaceGen also regularly uses live polling platforms and live surveys platform to interact with the audience.

29.6.3 *The Potential of Modern Communication Tools for Conducting Projects*

In the modern era, many organisations use digital tools to communicate with teams, coordinate meetings, and distribute the workload. The adoption of digital tools has brought specific benefits: communication between team members can happen instantaneously; meetings can happen as long as members had a stable internet connection, and work distribution could be visualised and checked with ease.

The capacity of these digital tools to increase efficiency has led many organisations to deprioritise workflow innovation. As a result, many organisations continue to use the Standard Operating Procedures (SOPs) developed decades earlier, modified only by substituting some components with digital counterparts. This form of growth poses a limit as an organisation aims to become more diversified in geographic distribution and organisational complexity.

The potential of modern communication tools can only be harnessed when matched with a workflow that is designed around digital capabilities, as opposed to digital tools being used to support traditional workflows.

29.6.3.1 Multicultural Communication Protocols

Many Western organisations build communication channels that assume cultural homogeneity, especially Western culture. Multiple research shows that different cultures have different perceptions of time and punctuality. This difference is not only apparent in showing "on time" for meetings, but also on time allocation to tasks. In some cultures, people work as many or as few hours they require as long as they finished the allotted amount of work. In other cultures, like Japan, people will work strict hours, emphasising the role of managers in allocating the amount of work per teammate.

Different cultures have different levels of prioritising work. In U.S. and Europe, there is a tendency for members to prioritise personal life as much as work, leading people to be more vocal about extenuating circumstances or personal vacations. In many cultures in Asia, personal lives are dealt with more privacy, with members being less vocal when personal situations incur deadline extensions or leaves of absences.

Current communication channels are effective only when teams are homogeneous or when there are sub-teams from the same culture. Many organisations fail to create effective multicultural teams because they fail to recognise the diversity in communication methods. The shortcomings are woefully evident in online teams where there are no opportunities for members to interact outside of meeting hours.

Effective communication protocols not only create functional teams with clear expectations but allow organisations to harness experiential diversity that is very needed in the XXI century. In my career, I have created and maintained four different international online accelerator programs. Startup companies that established clear communication protocols, continuously outperformed traditional startups. Effective multicultural teams had more diverse ideas, innovated more quickly, found new patterns in client behaviour, and analysed the target base in more holistic ways.

Achievement of such performances is possible when digital platforms are equipped with an onboarding curriculum. Instead of "handing out" a work manual, managers should meet up with the team or individual members to discuss work expectations. The curriculum should be more interactive than didactic. The curriculum should allow different expectations to be reviewed and adopted rather than corrected and dismissed. In my accelerator programs, teams that revised project goals and success metrics according to the expectation of diversity performed far superior to teams that worked on existing, Western-centric guidelines.

29.6.3.2 Meta Digital Workspaces

Many organisations have lengthy disputes over which digital platform to use for which function. Emails, Asana, Slack, the Google Suite (including the Google Calendar, Hangouts, Drive) are but a few contenders. Supplementary services such as Skype, Appear.in, Zoom, Teamviewer, GoToMeeting, and Doodle are frequently seen in organisations that host multiple meetings. Data-centric organisations are often plagued with additional platforms such as AWS and Salesforce that also provide its communication methods.

The debate around the optimal all-in-one digital platform assumes the idea of a static organisation whose future requirements are already well known. A traditional organisation would set up an "ideal" persona for a job and recruit the person with the necessary skillsets. A modern organisation, while still placing necessary requirements, emphasises personality and willingness to adapt. These organisations bring capabilities previously unseen in the industry to the market.

The "expecting for the unexpected" philosophy opts not for a single digital platform, but for a meta workspace that utilises different digital platforms at will. The meta workspace can take on different forms. The simplest meta workspace I witnessed was just an editable document that catalogued the operations within the organisations, different teams, and a log of what digital platforms they used. The more complex meta workspace I came across was a board of team leaders who met bi-weekly to discuss the strategic goals of the organisation and the pros/cons of the current digital platforms being used. That organisation handled a variety of materials from social media data analysis results to 3D simulation models, making it optimal for different teams to use different platforms that best hosted the resources needed. The platforms were also continuously revised depending on team needs: some teams would have members working mostly on desktops, hence opting for a platform that had greater bandwidth and video conferencing options. Other teams had managers working in the field on time-sensitive articles. These teams opted for a platform that had a streamlined app interface and a robust communication service that was optimised for complex organisational groups.

29.6.3.3 The Digital Organisation

Organisations such as SGAC are increasingly growing more complex. Its teams are more likely to have a diverse geographic representation, and its tasks require capabilities such as instant literature review, 3D simulations, and data analytics. By structuring the organization around digital capabilities, SGAC can achieve more significant innovation by incorporating workflows from previously underrepresented cultures and become more adaptive to newly developed resources.

29.7 The Challenges Associated with Consistent Messaging Across the Organisation and Its Membership Base

Keeping the message consistent across an organisation like SGAC can be a real challenge. Although, if successfully maintained, having a consistent messaging can be used as a tool to unite the members around the same mission. It is an essential vector to federate, mobilise and motivate the members.

29.7.1 The Challenges

Given the scale and scope of the organisation, ensuring a consistent message across the SGAC membership base comes along with numerous challenges.

The network of SGAC is extremely large and is constituted of about 13,000 members and alumni. Among those volunteers, 33 members are active in the Executive Committee, the organisation is also represented by six Regional Coordinators and 90 National Points of Contacts. Due to the diversity of its network, each member has its own culture, perspectives and visions about the organisation and its activities. Additionally, each team member's term lasts between one and two years with a quick turnaround between active members of the Executive Committee and Office. In other words, it is essential for the organisation to keep the knowledge internal and it is necessary to ensure a proper on-boarding of each new active member.

29.7.2 The Solutions

To ensure consistent messaging across the SGAC membership, the organisation is based on a knowledge transfer model. SpaceGen developed a series of tools to ensure consistent messaging.

The aim is twofold: ensure that any new active member of the organisation starts with the same basic knowledge about the organisation's history, activities, structure and impact and ensuring a consistent external communication.

29.7.2.1 Ensuring Consistent Training Within the SGAC Team

The organisation implemented a system of training packs for each SGAC team. For example, the scholarship team who is in charge of the scholarship activities or the legal team in charge of the legal matters for SGAC prepared their own training pack including all the important information needed when a new team member integrates their teams. This is also a means for the organisation to keep the knowledge internal to SGAC and document on that activities and processes.

Another tool that the organisation is starting to put in place is virtual training sessions. Experienced members would "tag-up" with new members and teach them the processes of their respective activities remotely and online through conference platforms. The members also have access to their one member's page on the SGAC website where they can download templates as well as numerous online resources which are made available to them. SpaceGen also initiated its "team office hours" to give new SGAC members the opportunity to ask questions to the leadership and to the different team leads at fixed time slots during the week.

When possible, the organisation is committed to ensure that there is an overlap of the position terms between the experienced member and the new member. Usually, the overlap period ranges from four months to six months. The outgoing member can then become an advisor of the team and provide support to the new team members in their activities.

29.7.2.2 Ensuring Consistent External Communication

To ensure consistent messaging from the SGAC members when communicating about SGAC's activities, the organisation implemented its SGAC core message. It is a simple, structured and one-page short document which describes SGAC and its activities. It is structured as follows around four main questions:

- What is SGAC?
- What does SGAC do?
- How does SGAC work?
- What are the impact of SGAC?

Thanks to this core message, the organisation makes sure that its members know exactly how to communicate about SGAC with an external audience. SpaceGen also has an official SGAC presentation which is regularly updated and available to its members when they have to present SGAC at an event for instance.

29.8 The Importance of Global Partnerships

SGAC wouldn't exist without its partners and supporters. SpaceGen is a non-profit organisation which relies solely on the generous contribution and support of its partners and sponsors. The organisation collaborates with more than 100 partners worldwide in the sector.

SpaceGen collaborates with the private sector (Blue Origin, Lockheed Martin…), with Academia (ISU, Embry Riddle…), with 10 Space Agencies (DLR, CNES, ESA, JAXA, KARI, NASA…, with foundations (Secure World Foundation, Space Foundation…) and NGOs (Austrian Space Forum…).

The contribution constitutes a "win-win" relationship for SGAC and its partners. Most of the time, the partner would support SGAC for its events through:

- financial support directed to the organisation of the event
- financial support to SGAC members to attend the events
- the participation of keynote speakers and Subject Matter Experts from the partners' organisation to SGAC events to guide the delegates
- the support of a SGAC working group.

In return, the partners get access to the SGAC's large network of skilled members and alumni and can communicate about their organisation and their activities. The supporters also get to interact with the members and get the perspectives from the young generation on the topics which are relevant to their mission.

In this new era of capacity building, it is necessary to involve everyone in the discussion and take into account the perspectives from all the relevant actors of the sector (private sector, NGOs, the United Nations, Space Agencies…). Global partnership in particular is essential to remove existing barriers and solve issues that prevent countries to make progress towards and finally achieve the SDGs. Enhancing cooperation also helps to bridge the space divide and encourage a global effort with the leading players in each sector being involved to take responsibility in the Sustainable Development Goals (SDGs).

29.9 Conclusion

SpaceGen is one example among others of a model which could be applied to other entities in the sector and even at the United Nations level for some aspects, globally or regionally.

In particular, the way SGAC is structured and represented around the world as well as its capacity building initiatives and the processes attached to them constitute a great platform for capacity building globally and regionally.

Because there are many initiatives coming from all relevant actors in the space sector, there is a true need to build a database of all the entities who do capacity building globally or regionally. It would be helpful to even have a capacity building network including the capacity builders globally or regionally to encourage global partnership among them and to provide them a platform for cooperation.

Story-telling and mentorship are essential to inspire the young generation to get involved into the space sector. Having a network of mentors globally and regionally would provide the young generation with an opportunity to receive support from experienced space professionals and to learn from them. In a sense it also ensures a certain sustainability in the space sector as senior space leaders pass on the knowledge to the younger ones.

Young actors in the space sector also have different needs including:

- Belong to a network
- Meet with people who share the same passion
- Learn about space and gain skills
- Access to valuable opportunities (attend global events, networking, etc.)
- Enough support and visibility to launch their career.

And more than anything else, they have the desire to have an impact in the space sector. Thus, SGAC was created to give students and young professionals a voice in the space sector and to represent them to the United Nations, Space Agencies, Industry and Academia. Empowerment and opportunities are key to reach this goal. Additionally, there is a need to share with and promote to the young generation the different opportunities in the space sector in term of career and provide them with a career pathfinder tool.

Public outreach will also prove beneficial to raise awareness amongst governmental and commercial actors, and the general public alike, of the benefits and opportunities that space brings. Also, in this new era of capacity building there is a need to communicate differently through modern communication tools about the benefits of space activities and to multiply the outreach and story-telling initiatives.

Chapter 30
Concurrent Design for Innovative Capacity Building

Massimo Bandecchi

Abstract This article describes the Concurrent Design methodology adopted in the ESA CDF, the experience gained during the activities performed in the last 20 years and highlights the benefits of three specific aspects and applications of the concurrent design approach that can become assets for innovative capacity building: **Education, Interoperability, System of Systems (SoS) architecting**.

30.1 Introduction

In 1998, the European Space Agency (ESA) started investigating the use of Concurrent Engineering (CE) to improve its design processes and in particular to evaluate the benefits of Concurrent Design (CD), when applied to the definition and assessment of potential future space missions.[1]

ESA performs every year a number of pre-Phase-A/level 0 internal studies as part of the definition of future space missions and in order to assess their feasibility from the technical, programmatic and financial points of view. An experimental design facility, called Concurrent Design Facility (CDF, see Fig. 30.1) was created in the ESA Research and Technology Centre (ESTEC, see Fig. 30.2), with the scope to set up a mission design environment to be applied to internal pre-Phase A/system level studies and to improve the processes for the assessment and design of future space missions.

A space mission and the associated spacecraft are complex systems which require the contribution of several experts, in complementary disciplines, in order to be conceived and designed. Figure 30.3 ("Design" phase), illustrating the exploded view of MarsExpress, the first ESA mission to Mars, highlights that a number of disciplines have to be involved in the design of the spacecraft, e.g. in the mechanical, electrical,

[1] Bandecchi, M., Melton, B., Gardini, B., & Ongaro, F. (2000). The ESA/ESTEC concurrent design facility—Proceedings of EuSEC 2000, pp. 329–336.

M. Bandecchi (✉)
European Space Agency, Noordwijk, The Netherlands
e-mail: Massimo.bandecchi@esa.int

S. Ferretti (ed.), *Space Capacity Building in the XXI Century*, Studies in Space Policy 22,
https://doi.org/10.1007/978-3-030-21938-3_30

Fig. 30.1 The evolution of the ESA CDF infrastructure; left: the first experimental facility created in Nov. 1998; right: the current state-of-the-art design centre, operational at ESTEC since 2008 Courtesy of ESA

Fig. 30.2 A panoramic view of the ESA Research and Technology Centre (ESTEC) established in Noordwijk—The Netherlands, on April 3, 1968 (50 years ago) Courtesy of ESA

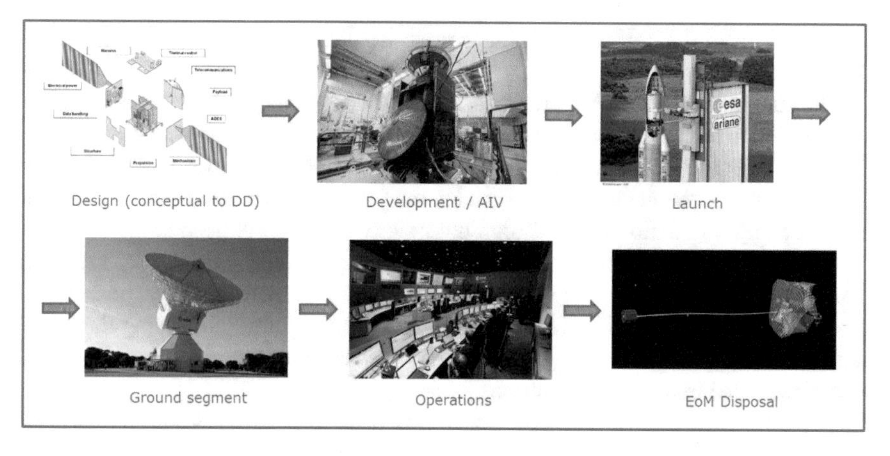

Fig. 30.3 The phases of a space mission life-cycle Courtesy of ESA

instrumentation areas. Each discipline, and expert (team) in charge, is responsible for the design of a "sub-system". However the design of each sub-system depends, to a certain extent, on the decision taken by other disciplines and on the requirements and constraints of other sub-systems. Furthermore the system design depends on, and has to take into account, the subsequent phases of the project life-cycle as illustrated in Fig. 30.3, e.g. development, launcher interfaces, communication with ground stations, etc. This means that the design of one sub-system cannot progress independently and disregarding the other disciplines, unless hypothesis or assumptions are made about the limitations posed by other sub-systems and by later phases. A strictly sequential design approach, where the communication and exchange of information among the experts is poor or absent, and the experts are loosely coupled, can lead to a divergent and very inefficient process, which implies mistakes, inconsistencies, need for corrections and re-work.

An alternative to the classical approach is offered by Concurrent Engineering (CE), which provides a more performant design method by taking full advantage of modern Information Technology (IT). The enabling factor for the CE approach has been the evolution of IT.

Co-location of experts from various disciplines to elucidate a preliminary design concept is not a novelty in itself. However, in the past the process and personal interactions were limited to very basic brainstorming sessions, because any numerical analysis required the use of tools which could not easily be co-located, nor interconnected nor could provide results in real time. Nowadays, with perhaps few exceptions, most of the analysis needed for a pre-Phase-A study can be performed in real time on a personal computer (PC) or on a laptop.

The simultaneous participation of all the specialists reduces the risk of incorrect or conflicting design assumptions, because each major decision is debated and agreed collectively. In this way the design progresses in parallel and allows those disciplines that were traditionally involved at a later stage of the process, the opportunity to participate from the beginning and to correct trends that might later invalidate the design.

The customer is invited to participate in all sessions along with other specialists of his/her choice (e.g. study scientist, project controller), so that they can contribute to the formulation of the study assumptions, answer questions from the team and follow the evolution of the design. This includes the possibility to discuss and correct in real-time any orientation of the design not in line with their expectations.

The first design session starts with the customer presenting the mission requirements and constraints to the team. In subsequent sessions, each specialist presents the proposed option or solutions for his/her domain, highlighting/discussing the implications for the other domains. Out of the debate a baseline is retained and the related values recorded in a shared database.

30.2 CDF Evolution and Achievements

The initial CDF experiment turned into a permanent and operational facility in 2000—see Fig. 30.1. Since then more than 200 potential future missions, mostly scientific, but also for Earth Observation and other Application programmes, have been assessed at pre-Phase A level using the facility to date—see Fig. 30.4.

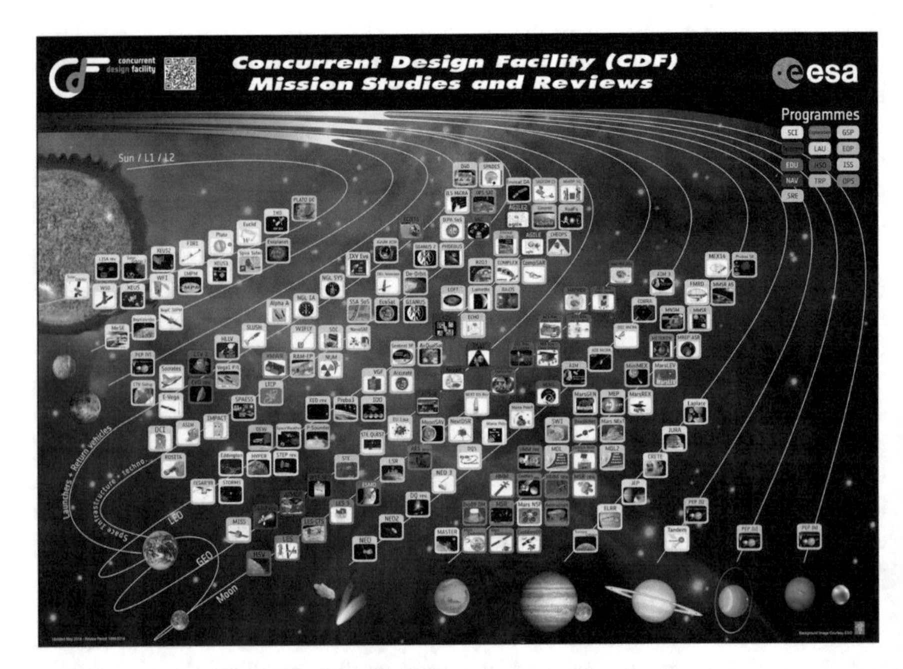

Fig. 30.4 Missions studied in CDF in the last 15 years on request of ESA programmes; mission destination mapped onto the solar system Courtesy of ESA

The ESTEC Concurrent Design Facility is today a state-of-the-art facility equipped with a network of computers, multimedia devices and analysis and design tools, which allows a team of experts from several disciplines to apply the concurrent engineering method to the design of future space missions. It facilitates a fast and effective interaction of all disciplines involved, ensuring consistent and high-quality results in a much shorter time.

The CDF is used to assess the technical, financial and programmatic feasibility of potential future space missions and related spacecraft concept (e.g. internal pre-Phase A or Level-0 assessment studies). Since the beginning, the CDF has been effectively used for:

- new mission concept assessment
- new space vehicle preliminary design
- space system options evaluation and trade-offs
- new technology validation at mission/system level
- technology infusion and derivation (e.g. as input to technology road-maps).

The concurrent engineering approach used in CDF is based on the following elements:

- Centralised Model Based architecture, including
  - central data-base (Project Data Repository and Data Exchange among disciplines),
  - domain-specific tools and data-bases,
  - Engineering models of each individual domain, with input/output interfaces towards domain tools and central data-base,
  - Web services interlink, to allow remote participation.
- Multi-disciplinary team,
- Communication and rendering systems (multi-media wall, video-conference, network, 3-D printing, stereoscopic vision, CAD, simulation).

The operational experience has shown that application of the Concurrent Engineering method, supported by appropriate informatics tools, can greatly improve the efficiency of mission design activities, both reducing the duration of a typical pre-Phase A study to a few weeks on average, and producing a higher quality design with a higher level of detail than was produced by traditional methods.

The performances achieved for a typical study design phase are:

- Reduced duration in time: 3-6 weeks to be compared to 6-9 months, hence a factor 4 reduction in time.
- Reduced Cost for the Customer: half of the cost of previous approaches, hence a factor 2 reduction in cost.

Other benefits include:

- Increased number of studies per year, hence better service provided by the Agency to the supported programmes and Space communities.

- Improvement in quality, providing quick, consistent and complete mission design, including technical feasibility, as well as programmatic, risk assessment and cost evaluation.
- Capitalisation of corporate knowledge for further reusability, thanks to the centralised approach.
- Remarkable reduction of engineering changes observed in subsequent phases of the project life-cycle.

The use of CD in the ESA CDF has demonstrated many advantages, and there are many further benefits that CE could bring when applied to later phases of the project life-cycle.

Through the use of the centralised approach of the CDF, a data base of knowledge has been accumulated within the organisation. The CDF results provide detailed information to ESA decision-making bodies to aid their assessment of which mission are selected for further industrial implementation. Furthermore this more detailed assessment of the new potential missions allow critical issues to be discovered and highlighted well in advance in the project life-cycle, and consequently this reduces the risk of engineering changes being required later or mistakes occurring.[2,3]

More recently the CDF infrastructure has been used to perform scientific requirements definition and consolidation, prepare specifications, payload instrument conceptual design, industrial work reviews, System of Systems architecting and coordinate international project work. Finally, it is also being used for training and educational purposes.

The ESA CDF is used extensively by most of the ESA programmes, as well as by the ESA Institutional partners. The results and experience are regularly exchanged and shared with the international space engineering community, for which the CDF represents a reference for the application of Concurrent Engineering methodologies.

The CDF has grown not only in number of studies, but also in facility-architecture (see Fig. 30.1), range of software tools and models, number of teams and team-members, and diversity of applications and customers. For further information about ESA CDF evolution and achievements, consult the CDF Web site.[4]

Following few years of successful application, ESA CDF started sharing its know-how about CE, as well as the software and the models that had been developed in the facility, with the rest of the Community. This included practitioners in National Agencies and institutions, European space industry and Academia.

Today, thanks to ESA's initiative, CE is quite well known and applied in the European space sector.

For ESA this has been a further achievement because it complies with one of the mandates of the Agency, which is to support the European industry and partners to become more competitive on the international market.

---

[2]Bandecchi, M. et al. (2001). The ESA concurrent design facility (CDF): Concurrent engineering applied to space mission assessments (2001).

[3]Bandecchi, M., & Biesbroek, R. (2004). The ESTEC concurrent design facility—proceedings of concurrent engineering workshop 2004 (CEW04), ESTEC, Noordwijk (NL).

[4]ESA portal. CDF Web site: www.esa.int/cdf.

ESA's promotion of CE has reached also non-space sectors, and in particular those companies whose final products are complex systems, with design challenges similar to those of a spacecraft or a space mission. For instance, yachts or racing cars (e.g. F1) or nuclear plants are all custom design complex systems, characterized by high performances, need of high reliability, and in general rather expensive. In all these cases the benefits and performances of the concurrent approach have shown to be very similar.

30.3 CDF and Education

ESA's Concurrent Design Facility (CDF) is built to facilitate collaborative working and knowledge exchange. Intended for mission engineering, it has also proved to be a valuable resource for education, also in collaboration with universities, academia and ESA's Education Office.

With the support of ESA a number of European universities have taken the initiative to organize lectures and courses in Concurrent Engineering and some of them to create facilities based on the model of ESA CDF. Figure 30.5 provides a map of the universities in Europe that have built their own educational design facility. Figure 30.6 shows two examples: the International Space University (ISU) in Strasbourg, France and the Ecole Polytechnique Federale de Lausanne (EPFL) in Switzerland.

ESA CDF and Education Office have been collaborating on a new facility called the ESA Academy Training and Learning Centre (TLC), see Fig. 30.7. The training

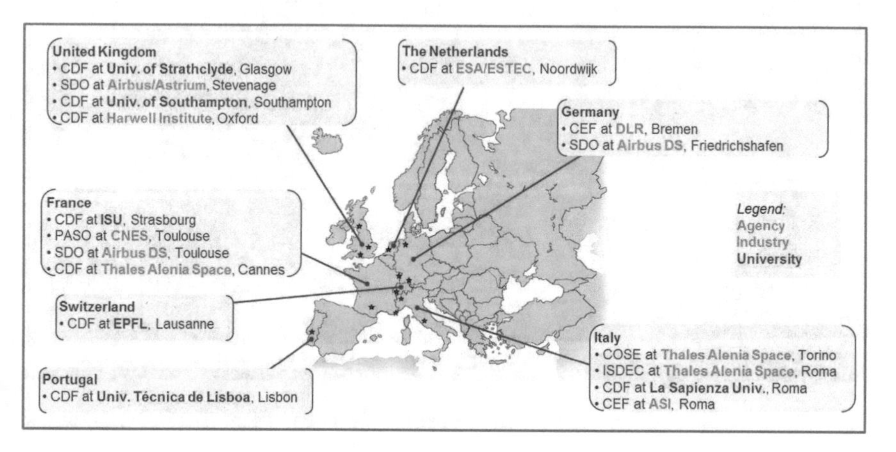

Fig. 30.5 Space design centres created in Europe, based on concurrent design and following the model of ESA CDF. Similar centres have been created in National Space Agencies, Industries, Universities, as well as in non-space organisations Courtesy of ESA

Fig. 30.6 Examples of CE/CD education centres built following the model and using the software of the ESA CDF. The pictures represent the International Space University (ISU) in Strasbourg—France and the Ecole Polytechnique Federale de Lausanne (EPFL)—Switzerland Courtesy of International Space University (F) and Ecole Polytechnique Federale de Lausanne (CH)

Fig. 30.7 ESA Academy Training and Learning centre—Redu (B)—March 2016 Courtesy of ESA

room, located on the ESA Redu Centre in Belgium, has been set up with the intention to use, for some training courses, the Concurrent Engineering techniques and the

Open Concurrent Design Tool (OCDT).[5] The CDF system engineers have been working closely with ESA Education colleagues on all aspects of the room, from layout to specifications for the hardware and software on the desks as well as the traditional horseshoe shape of desk layout and the central display screens found in (almost) all CDF type of facilities.

The ESA Academy Training and Learning Centre is a facility of the ESA Education Office dedicated to hosting training courses for university students. It aims at complementing the standard academic education in space-related disciplines offered in the ESA Member and Cooperating States' universities by transferring knowledge, know-how, and standards in all fields of ESA expertise. In this frame, ESA Education Office asked the support of ESA CDF to develop an Educational CDF, to be able to teach students how to use concurrent engineering to develop a satellite.

This collaboration with ESA Education Office has been very successful and the results and the feed-back obtained in the first 2 years of activity have demonstrated the value of the "hands on" approach to concurrent engineering in a real environment with real models, coached by ESA experts.

Thanks to these initiatives, ESA expects to see many more engineers with a background knowledge of concurrent engineering entering the space market in the near future.

The fame of CDF went beyond the borders of Europe; one example of educational institution that was built in another continent, but was inspired by the ESA CDF is the Victorian Space Science Education Centre in Melbourne, Australia, whose building and infrastructure are illustrated in Fig. 30.8.

Fig. 30.8 Victorian Space Science Education Centre (VSSEC), Melbourne—Australia; an example distant from Europe of an education design centre built following the model of the ESA CDF Courtesy of Victorian Space Science Education Centre (VSSEC)—Melbourne—Australia

[5]ESA portal. CDF Web site News: Teaching_OCDT_in_ESA_REDU_Centre http://www.esa.int/Our_Activities/Space_Engineering_Technology/CDF/A_New_Education_Facility_Teaching_OCDT_in_ESA_REDU_Centre.

The ESA CDF hosts regular workshops and educational initiatives whereby teams from all over Europe collaborate on designing small satellites. The teams can be physically relocated in the facility or collaborate remotely: each institution or team is given a specific subsystem to work on, coordinating their progress via online means. The CDF network infrastructure allows team members of various disciplines to interact together in real time, while also receiving advice from ESA experts. Various mission factors can be analysed in conjunction, from structural and mass considerations to subsystem compatibility, instrument selection, communication solutions, even project planning and cost projection.

The CDF, in collaboration with the ESA Internal University, holds two internal System Engineering courses per year, to enable domain specialists to understand the interdisciplinary relations to get an insight into the problems associated with system design and to enable them to act as systems engineers within the CDF environment.

Finally, the facility is employed to familiarize engineering teams from new ESA partners and cooperating states with mission design and Agency standards and practices.

30.4 Interoperability Based on a Concurrent/Collaborative Approach

To face an increasingly competitive environment within a globalization context, one of the strategies is to establish partnerships with other companies specialized in complementary domains. Such an approach, primarily based on optimization of the value chain, is called "Virtualization of the Enterprise".[6] Enterprises relying on virtualization, sub-contracting and outsourcing have to coordinate activities of all the partners, to integrate the results of their activities, to manage federated information coming from the different information systems. The adopted organization, which is considering as well as the internal and external resources, is called "Extended Enterprise".

In such complex emerging networked organizations, it is more and more challenging to be able to interchange, to share and to manage internal and external resources such as digital information, digital services and computer-based processes.

Following the example of the ESA CDF, in the last few years, several centres have been created in Europe and today more than 20 facilities (see Fig. 30.5) are operational and potentially interoperable with ESA CDF and with each other.

Firstly, a number of European national agencies created their facilities. Figure 30.9 illustrate (a) the Centre d'Ingénierie Concourante (CIC) of the French Space Agency (CNES), opened in 2005 in Toulouse, (b) the Concurrent Engineering Facility (CEF)

[6]Nicolas Figay, N., Ferreira da Silva, C., Ghodous, P., & Jardim-Goncalves, R. Resolving interoperability in concurrent engineering, chapter from book Concurrent Engineering in the 21st Century Foundations, Developments and Challenges pp. 133–164.

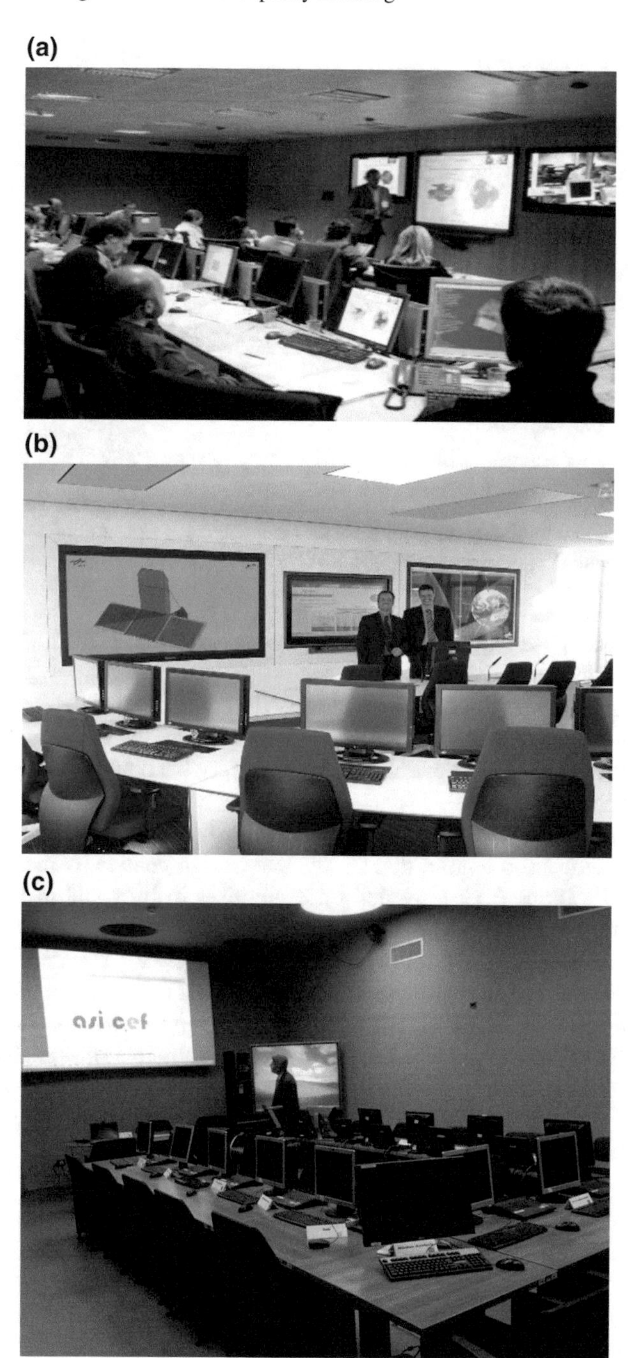

Fig. 30.9 Design centres based on concurrent engineering at European national agencies, following the example and standards of ESA CDF. **a** CNES CIC (Toulouse—F)—Nov. 2005 **b** DLR CEF (Bremen—D)—Dec. 2008 **c** ASI CEF (Rome—I)—proto: July 2008, facility: Nov. 2013 Courtesy of CNES (Toulouse—F), DLR (Bremen—D), ASI (Rome—It) respectively

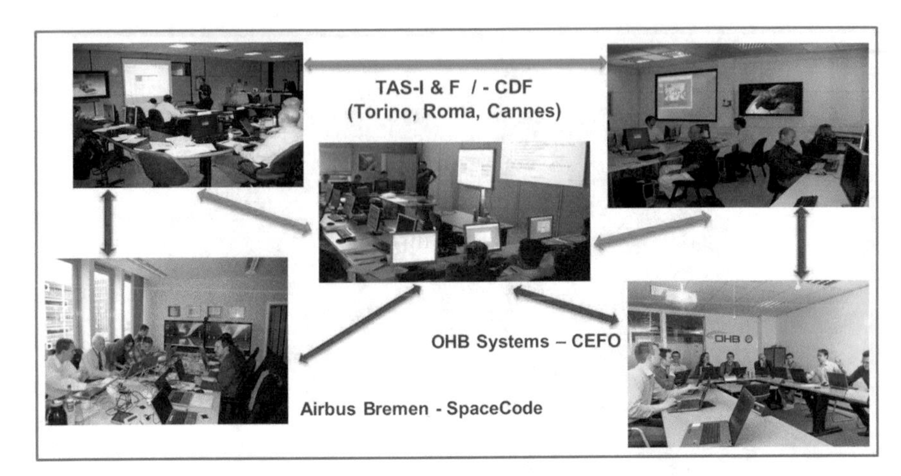

Fig. 30.10 Example of interoperation across the borders of organisations/design facilities using the CE approach and following the ESA processes and standards Courtesy of TASinI (Torino & Rome—It), TASinF (Cannes—F), OHB Systems (D), Airbus (Bremen—D)

of the German Space Centre (DLR) in Bremen, (c) the CEF of the Italian Space Agency (ASI) in Rome.

Following the successful implementation of the concurrent/ collaborative methodology within one single organisation, it was natural to try the next step towards collaboration across the borders of enterprises, towards what could be defined a "global virtual design facility".

An example of remote concurrent design is shown in Fig. 30.10 where the approach is being used among three CDF like design centres, named Concurrent Engineering Facilities (CEF), located on 3 premises in Italy and in France of the same company, Thales Alenia Space (TAS), one of the Prime contractors of ESA. The teams of specialists, virtually collocated and complementary to each other, perform design activity on-line, in real-time. Furthermore other partners can join the design activity, at different stages, from different companies and remote location/countries.

Via the CDF, ESA has proposed and is promoting initiatives that enable and facilitate remote concurrent engineering and interoperability of similar facilities sharing data models, processes and tools. The following paragraphs illustrate in turn some of these initiatives.

30.5 Standard Data Model—ECSS-E-TM-10-25

"Organizations are looking for new business relationships, and the exchange of information and documents with new partners is often incapable of being executed automatically and in electronic format. This is principally due to problems of incompatibility with the information representation adopted by the software applications they are working with".[7]

One of the first initiatives of the CDF was therefore the proposal to standardize the data model used in CDF, and adopt it in all the newly created design centres. Such an operation was conducted within the European Cooperation for Space Standardisation (ECSS), established in 1993, is an organisation which works to improve standardization within the European space sector. The ECSS frequently publishes standards, which contractors working for ESA must adhere to.

As a prelude to the true standard, a Technical Memorandum (TM) was created under the E-10 "System engineering" branch in the ECSS series of standards, handbooks and technical memoranda, namely the ECSS-E-TM-10-25 "System Engineering—Engineering Design Model Data Exchange (CDF)".

This TM facilitates and promotes common data definitions and exchange among partner Agencies, European space industry and institutes, which are interested to collaborate on concurrent design, sharing analysis and design outputs and related reviews. This comprises a system decomposition up to any level and related standard lists of parameters and disciplines. Further it provides the starting point of the space system life cycle defining the parameter sets required to cover all project phases, although the present TM only addresses Phases 0 and A. This TM is intended to evolve into an ECSS Standard in the near future. In conjunction with related development and validation activities, this Technical Memorandum should be regarded as a mechanism for reaching consensus prior to building the standard itself.

The Technical Memorandum provides the basis for creating interoperable Concurrent Design (CD) centers across the European space community:

- Allowing semantically consistent data exchange between CD centers,
- Enabling and supporting joint real-time collaborative design activities involving multiple CD centers.

The initial objective of the TM is thus to act as a reference for the creation of new CD centers or upgrade of existing ones.

[7]Morris, E. J., Levine, L., Place, P. R., Plakosh, D., & Meyers, B. C. System of system interoperability—technical report—April 2004—Software Engineering Institute—Carnegie Mellon University.

30.6 ESA Open Concurrent Design Tool (OCDT)

A second initiative has been the development, partially under ESA contract and partially as an internal development, of a software package to enable efficient multi-disciplinary concurrent engineering of space systems in the early life cycle phases, named the Open Concurrent Design Tool (OCDT, for short).[8]

The OCDT implements the standard semantic data model (i.e. the formal UML model) defined in Annex A of the ECSS-E-TM-10-25 Technical Memorandum, as well as the Web services API (application programming interface) defined in Annex C of the TM.

OCDT is a client/server software package. The client is an add-in for Microsoft Excel®, that is integrated with Excel® to perform simple analysis and simulation. Other client tools for engineering analysis and simulation can also be integrated, through the use of OCDT adapters. The OCDT server consists of a front-end web-services processor and a back-end PostgreSQL database system for the persistent storage of OCDT shareable data. The server is able to support concurrent teams of more than 30 users and synchronising their engineering models. Typically each user would represent a different domain of expertise (discipline) in a concurrent design environment.

OCDT is being used in the ESA CDF as the main modelling tool for all the internal studies.

Supported with a Community Portal for developers and end users, the OCDT package is distributed under an *ESA community open source software license* (i.e. free of charge), available for use and further development to users that qualify as a member of the OCDT Community.[9]

30.7 International Conference SECESA

The European Space Agency (Technical and Quality Management Directorate and Concurrent Design Facility), in collaboration with European universities organizes, on a biannual basis, the Conference on Systems Engineering and Concurrent Engineering for Space Applications (SECESA, for short).[10]

The conference aims to foster networking and exchange of ideas, experiences, lessons learned and future trends in the area of Systems and Concurrent Engineering

---

[8] de Koning, H. P., Ferreira, I., Gerene, S., Beyer, F., Vennekens, J., Pickering, A. (2014). Open concurrent design tool (OCDT): ESA community open source ready to go!. In *ESA Conference Paper*, SECESA'14, October 2014, Stuttgart (Germany).

[9] ESA OCDT Community Portal: https://ocdt.esa.int/login?back_url=%3A%2F%2Focdt.esa.int%2F.

[10] SAGE Journals—Concurrent engineering research and applications journal special issue SECESA First Published March 15, 2018 Guest editorial by: Massimo Bandecchi, Ilaria Roma (ESA) http://journals.sagepub.com/doi/full/10.1177/1063293X18762834.

for agencies, companies, organizations, universities and institutes involved in space and related activities.

SECESA offers the Community of practitioners the opportunity to remain informed and coordinated, on these subject.

Every edition of SECESA is hosted at a different European University premises, and co-organised by a European university chosen via a formal selection process.

This approach enables the Universities to showcase specific areas or projects where students are excelling at utilising SE and CE approaches as well as accomplishing the Agency educational objective of inspiring young generations of engineers with new methodology and approaches for Systems Engineering, by facilitating their participation to the technical sessions.

2018 was the year of Strathclyde University in Glasgow (Scotland), where the Department of Mechanical and Aerospace Engineering has a Concurrent Design Facility, named the Concurrent and Collaborative Design Studio (CCDS), located in the same building where the conference was held.[11]

30.8 System of Systems Architecting

With Ref. to,[12] "a System of Systems should be distinguished from large but monolithic systems by the independence of its elements, their evolutionary nature, emergent behaviors, and a geographic extent that limits the interaction of their elements to information exchange".

For some time the Agency has been addressing mission objectives that require the exploitation of disparate space and ground systems acting as a System-of-Systems (SoS). Such complex mission architectures would further exploit the benefits of space systems, for research, e.g. Space Exploration, or for operational use, providing new services and integrated applications, e.g. Global Monitoring for Environment and Security (GMES), Space Situation Awareness (SSA) and others.

The identification of such integrated mission architectures for crisis response was the subject of GIANUS, a user focused and service driven design study in ESA CDF.

The complex concept of Global Integrated Architecture for iNnovative Utilisation of space for Security, or GIANUS in short, is a SoS architecture combining the services of a number of Earth Observation, Telecommunication, Navigation and Launcher independent assets and capabilities, see Fig. 30.11a. The study was aimed at preparing the definition of future European infrastructure and the associated technology developments.[13]

In the original intentions, the GIANUS architecture should have provided a comprehensive set of services to the European Union and its Member States in efficiently

---

[11]ESA organized International Conference SECESA'18, web site: https://atpi.eventsair.com/QuickEventWebsitePortal/secesa-2018/secesa.

[12]Maier, M. W. (1998). Architecting principles for system of systems.

[13]ESA. (2009). GIANUS initiative, CDF Study report, CDF-95(A), December 2009.

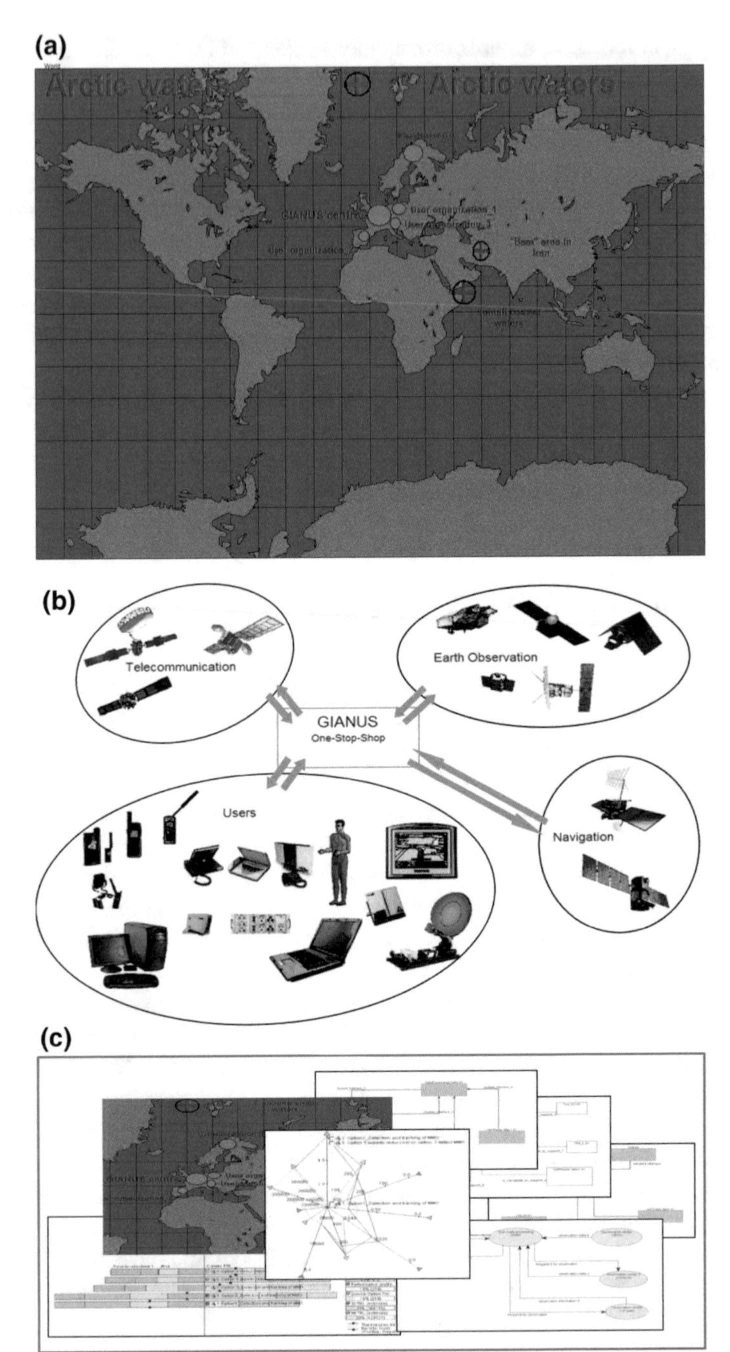
(a)
Arctic waters
Arctic waters
GIANUS centre
User organization_1
User organization_3
User organization_2
"Bam" area in Iran
(b)
Telecommunication
Earth Observation
GIANUS
One-Stop-Shop
Users
Navigation
(c)

◂**Fig. 30.11** **a** The 3 scenarios studies in GIANUS CDF study: (i) The Arctic waters are being targeted for monitoring and identification of ships and sea routes. (ii) The Somalian coastal waters are being targeted for monitoring and tracking of pirate ships. (iii) Bam area in Iran is targeted for monitoring earthquakes and to provide assistance during the consequences of these earthquakes. **b** System of Systems architecture—Example of GIANUS CDF Study. Service oriented, generating a macro service by integration of a number of heterogeneous and independent space assets/services. **c** Results, evaluation and comparison of enterprise architectures for the 3 GIANUS scenarios, using trade-off tools Courtesy of ESA

addressing their security needs. This architecture could potentially support the monitoring and management of security-related events within and outside EU borders, providing European and national security actors with guaranteed and timely access to space-based services both of a strategic and operational nature. It would support the full spectrum of internal and external civil security missions in the context of EU policies, including border surveillance, maritime surveillance, illegal activities monitoring, non-proliferation and treaties monitoring, as well as natural disaster, technological accident and man-made crises management.

The GIANUS initiative consisted in several elements, namely: a space infrastructure, a set of services based on this infrastructure, associated technology developments and a user consultation process.

GIANUS was expected to simultaneously deliver routine monitoring/surveillance services and crisis management services in the following fields:

- Border Surveillance,
- Maritime Surveillance,
- Illegal activities monitoring,
- Non-proliferation and treaties monitoring,
- Natural disasters,
- Technological accidents,
- Man-made crises (terrorism, piracy, illegal immigration, conflicts, etc.).

This architecture could potentially support the monitoring and management of security-related events within and outside EU borders, providing European and national security actors with guaranteed and timely access to space-based services both of a strategic and operational nature. It would support the full spectrum of internal and external civil security missions in the context of EU policies, including border surveillance, maritime surveillance, illegal activities monitoring, non-proliferation and treaties monitoring, as well as natural disaster, technological accident and man-made crises management.

End users for potential services would therefore include at national level, civil protections, fire brigades, police authorities, customs and coast guards, as well as interior ministries, ministries of defence, ministries of foreign affairs.

The ESTEC Concurrent Design Facility CDF was requested to perform a System-of-Systems study in order to provide a design and the assessment of architectures for crisis management in different crisis phases based on a selected set of case scenarios.

The CDF-GIANUS team reflected the integrated nature of the activity with experts based in four ESA establishments (HQ, ESOC, ESRIN and ESTEC) coming from eight different directorates. These experts applied common Concurrent Engineering (CE) principles as well as specific SoS design tools to establish mission architectures capable of efficiently delivering the user services. Such tools had been already tested in a preparatory assessment for some functions of the Space Situational Awareness (SSA) architecture.

The study focused on three reference scenarios (i) a major earthquake, like the one of 2003 in Bam, Iran, (ii) counter-piracy in the Strait of Malacca and (iii) maritime surveillance in the Arctic sea e.g. a hypothetical tanker accident), see Fig. 30.11b.

Based on these three scenarios, different assets and capabilities were combined and integrated. To gain an insight on the capabilities of different actors, four options with different portfolios of assets and their associated performances were created. To provide additionally an overview on the development and change of the options over time three different timeframes (2015, 2020 and 2025) were selected.

For the specific first GIANUS CDF assessment at horizon 2015, the service provision was tentatively schematised through a single One-Stop-Shop (OSS) approach (see Fig. 30.11a). With the development of integrated applications, GIANUS would have progressively supported the establishment of an operational OSS able to provide the full spectrum (both at strategic and operational level) of coordinated space-based services required for crisis management, based on the needs certified by the users.

These objectives requested the implementation of a whole new type of study that enables ESA and the CDF to design, analyse and improve SoS in a concurrent way. This meant the development of an Integrated Architecture Model within the CDF as well as the selection and development of supporting models, tools and methodologies to derive architectures and analyse SoS from different "view points" such as operations, system, technology as well as to identify gaps and opportunities. The Open Group Architecture Framework (TOGAF) and its Architecture Development Model (ADM) with its various phases was identified as the method providing the best flexibility to be tailored for ESA and CDF needs.

The new tools and modified CDF methods enable the team operating in the facility to provide the customer with an evaluated set of architectures, identified potential additional services and trade-offs among the options.

The outcome of the SoS studies is a complete assessment of the various technical views, as well as those associated with the operations, programmatics, cost, risk and simulation, see Fig. 30.11c.

The GIANUS exercise and subsequent SoS applications showed that the CDF is an excellent place to perform System of Systems feasibility studies and shall play a major role in SoS architecting activities, since the concurrent design approach elements interlink perfectly with the TOGAF-ADM approach.

In the frame of their cooperation ESA and the European Defense Agency (EDA) conducted in cooperation the Capabilities Package Assessment (ICPA) study using the Concurrent Design Facility and its expertise in System of Systems (SoS) architecture to implement the tool and to assess the value of Intelligence, Surveillance &

Reconnaissance (ISR) capabilities in support of Common Security and Defence Policy (CSDP) operations.[14] ESA's CDF has thereby been the focal point where space and non-space experts came together to elaborate and validate ISR architectures and assessment models.

30.9 Conclusion

The use of concurrent engineering has become embedded in the application of space system engineering, and there are now many concurrent design facilities around the world. The ESA CDF has now been in existence for 20 years and its involvement in all phases of a space programme development continues to grow. The method has enabled significant improvements in overall performance by reducing assessment and conceptual design study duration and cost, and by allowing a considerable increase in the number of studies that can be performed. At the same time, the quality has improved and the output of the CDF has become part of any follow-on industrial activity. The ESA CDF is now an essential tool for ESA decision making, and in the associated risk management processes.[15]

At the same time a number of other centres have been created in Academia, Industry and National institutions, following the example of the ESA CDF. A substantial effort has been made to supply these facilities with the means (such as standard data models, open source tools) to become interoperable; today they can be connected and support joint and collaborative activities, providing high efficiency and performance.

The education and training of Concurrent Engineering Design have progressed in parallel and today many technical universities and engineering faculties teach courses and offer lectures in CE on a regular basis. The use of educational facilities is becoming more and more popular for hands-on exercises and practice, above all for in designing small and affordable space systems (e.g. Cubesats).

The usage of CD, complemented by appropriate tools and enterprise architectures frameworks, applied to the definition and evaluation of System of Systems has shown remarkable advantages and performances.

Along with Education and Interoperability, System of Systems architecting could represent invaluable assets and synergies for capacity building in the XXI century!

---

[14]ESA portal. CDF web site: http://www.esa.int/Our_Activities/Space_Engineering_Technology/CDF/CDF_Supports_an_EDA_Contracted_Study.

[15]Spacecraft Systems Engineering. In P. Fortescue, G. Swinerd, & J. Stark (Eds.), Fourth Edition book, Chapter 20, pp. 643–666.

Chapter 31
The Open Universe Initiative

Paolo Giommi, Gabriella Arrigo, Ulisses Barres De Almeida, Massimo De Angelis, Jorge Del Rio Vera, Simonetta Di Ciaccio, Simonetta Di Pippo, Sveva Iacovoni and Andrew Pollock

Abstract The almost universal availability of electronic connectivity, web software, and portable devices is bringing about a major revolution: information of all kinds is rapidly becoming accessible to everyone, transforming social, economic and cultural life practically everywhere in the world. Internet technologies represent an unprecedented and extraordinary two-way channel of communication between producers and users of data. For this reason, the web is widely recognized as an asset capable of achieving the fundamental goal of transparency of information and of data products, in line with the growing demand for transparency of all goods that are produced with public money. This chapter describes the "Open Universe", an initiative proposed to the United Nations Committee on the Peaceful Uses of Outer Space (COPUOS) with the objective of stimulating a dramatic increase in the availability and usability of space science data, extending the potential of scientific discovery to new participants in all parts of the world.

31.1 Introduction

The almost universal availability of electronic connectivity, web software, and portable devices is bringing about a major revolution: information of all kinds is

P. Giommi (✉) · G. Arrigo · M. De Angelis · S. D. Ciaccio · S. Iacovoni
Italian Space Agency (ASI), Rome, Italy
e-mail: paolo.giommi@asi.it

G. Arrigo
e-mail: gabriella.arrigo@asi.it

U. B. De Almeida
Centro Brasileiro de Pesquisas Físicas, Rio de Janeiro, Brazil

J. Del Rio Vera · S. Di Pippo
United Nations Office for Outer Space Affairs, Vienna, Austria

A. Pollock
Sheffield University, Sheffield, UK

S. Ferretti (ed.), *Space Capacity Building in the XXI Century*, Studies in Space Policy 22,
https://doi.org/10.1007/978-3-030-21938-3_31

rapidly becoming accessible to everyone, transforming social, economic and cultural life practically everywhere in the world. Internet technologies represent an unprecedented and extraordinary two-way channel of communication between producers and users of data. For this reason the web is widely recognised as an asset capable of achieving the fundamental goal of transparency of information and of data products, in line with the growing demand for transparency of all goods that are produced with public money.

In the field of space science, almost all existing data sets have been produced through public funding, therefore they should be considered a public good and become openly available to anyone at a certain point in time. In particular, high-level calibrated data products, like images, spectra and similar products, should be available in a transparent form, that is usable by all. To ensure a fair scientific reward and to protect the intellectual property of the teams that conceive, design, build and operate the instruments that generate the data, this should happen according to clear rules.

The benefits of openness and transparency have been widely emphasised. They are so large and evident for both users and data providers that even scientific space data generated through private funds should aim at transparency.

Much has been done in recent years, especially in space astronomy, to offer open access, user-friendly platforms and services, demonstrating how natural the evolution is towards an increasingly transparent and inclusive ecosystem of tools and services. However, despite the recent progress there is still a considerable degree of unevenness in the services offered by providers of space science data.

Further efforts are necessary to consolidate, standardise and expand services, promoting a significant inspirational data-driven surge in training, education and discovery. Such a process, leading to a much larger level of availability of space science data, should be extended to non-scientific sectors of society.

To respond to these needs, at the fifty-ninth session of the United Nations Committee on the Peaceful Uses of Outer Space (COPUOS), the Government of Italy, working closely with the Italian Space Agency (ASI), proposed the "Open Universe Initiative".

In this paper we describe the main principles behind the Open Universe, and illustrate some of the features of the first version of a web portal that is under development at ASI.

We also briefly address the potentially very large socio-economical returns of the initiative, whose costs are easily sustainable and certainly marginal with respect to the total investment that is made to produce space science data.

The far reaching vision of the Open Universe Initiative and the potentially global reach, which extends the benefits of space science to large sectors of society, including emerging and developing Countries, call for a wide international cooperation under the auspices of the United Nations with activities fully integrated into the UN Space2030 agenda.

31.2 The Open Universe Initiative

"Open Universe" is an initiative under the auspices of COPUOS with the objective of stimulating a dramatic increase in the availability and usability of space science data, extending the potential of scientific discovery to new participants in all parts of the world and empowering global educational services.

Following the initial proposal at COPUOS by Italy in 2016, the initiative was included among the activities to be carried out in preparation of the fiftieth anniversary of the first United Nation Conference on the Exploration and Peaceful Uses of Outer Space (UNISPACE+50).

"Open Universe" will ensure that space science data will become gradually more openly available, easily discoverable, free of bureaucratic or administrative barriers, and usable by the widest possible community, from professional space scientists (several thousands of individuals) to citizen scientists (potentially of the order of millions) to the common citizens interested in space science (likely hundreds of millions).

The services delivered by existing space science data producers have significantly improved over time, but they are still largely heterogeneous, ranging from basic support reserved to a restricted number of scientists, to open access web sites offering "science-ready" data products, that is high-level calibrated space science data that can be published without further analysis by professionals with suitable knowledge.

"Open Universe" will implement a method to improve the transparency and usability level of the data stored in current space science data archives, and will urge the data producers to increase their present efforts so as to extend the usability of space science data to the non-professional community.

After a number of public discussions and international meetings, Open Universe is now being defined in detail under the leadership of the United Nations Office Of Outer Space Affairs (UNOOSA) in close cooperation with the Government of Italy and in collaboration with other participating countries.

31.3 The ASI Open Universe Portal

In an effort to make progress towards the implementation of the principles put forward by"Open Universe", the Italian Space Agency is developing a prototype web portal for the initiative that aims to become a multi-discipline, multi-provider open space data service.

The first version of the ASI Open Universe portal has been released on the occasion of the United Nations/Italy workshop on Open Universe held in Vienna in November 2017, and is openly available at the link: openuniverse.asi.it.

The main aims of the portal are:

1. Develop the first prototype of a multi-discipline multi-provider space science website that aims for data transparency.

2. Concentrate, in a single web page, the potential of accessing space science data and information from several data archives and related information systems (e.g. catalogues, bibliographic services, etc.)
3. Facilitate new types of scientific research based on data-intensive analysis.
4. Stimulate discussion among experts and users to collect suggestions on how to reach the goals of the “Open Universe” initiative.
5. Help define the requirements for a new generation of “user-centred” integrated space science data archives that could be used in principle by anyone having access to touch-screens or an equivalent technology of the future.
6. Explain and demonstrate the potential of “Open Universe” to the non-space science professionals (e.g. museums, education sector, common citizens).
7. Provide links to a large number of services that give access to space science data services that could be used to evaluate the level of transparency provided.

This web site is not initially meant to be a software system where all space science data can be accessed, viewed and fully analysed in a homogeneous and integrated way. It is rather a web site, where a wide range of services developed independently by different space data providers and archive sites can be found in the same place. It is a sort of space science data “shopping mall” where professional scientists and common citizens alike can go, visit, and use (shop) the many web services (each clearly identified by its “brand” or developers logos) available next to each other, with the peculiarity that all the services know what the user is looking for as soon as she/he enters the mall.

Figure 31.1 shows the front page of the Open Universe portal. It provides general

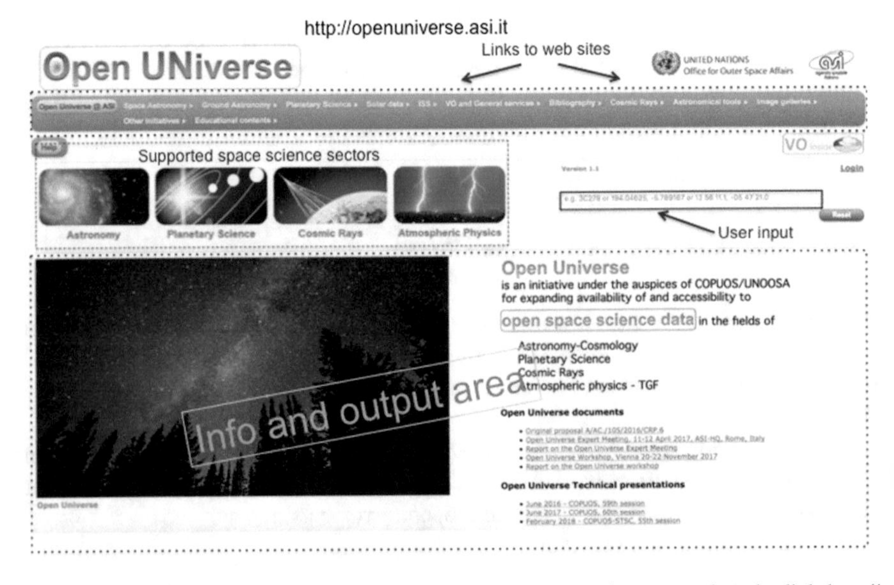

Fig. 31.1 The front page of the Open Universe portal that can be accessed at the link http://openuniverse.asi.it

information about the initiative, documentation for the user, several links to space science related sites, and an input area (marked as "User input") where the user can specify his/her requests about astronomical sources, planets, cosmic ray particles etc. The response of the portal following a request is directed to the area marked as "info and output area".

In building this portal we took into account requirements that maximise usability and accessibility, such as comprehensibility, interactivity, web readiness, and openness. We also cared about avoiding information overload, which is sometimes present in web services in the form of unnecessary or redundant information, and we added synthetic indices.

The grey area in the top part of the portal above the *user input* and highlighted by a red rectangle marked as "links to web sites" gives access to links to the major existing web pages offering data, services or information about space science. Sites are grouped in several different categories, namely: Space astronomy, Ground astronomy, Planetary science, ISS (International Space Station), Virtual Observatory (VO) and General services, Bibliography services Cosmic Rays, Astronomical web tools, Astronomical image galleries, Open Software, and Educational content.

Four space science disciplines are supported, namely **Astrophysics, Planetary Science, Cosmic Rays and Atmospheric Physics**. Each discipline corresponds to an icon, as shown in the "supported space science sectors" box of Fig. 31.1. Clicking on one of the four icons activates the corresponding space science area and the "user input" entry point provides appropriate suggestions for their use. For example, the Astronomy section is activated, then the examples in the input area are set to: 3C279 (if entering the name of an astronomical source) or "194.046, −5.7892" or "12 56 11.1, −05 47 21.0" (if using different types of sky coordinates). If the "Planetary science" icon is activated the examples in the entry area are set to "Moon or Mars", etc.

The front page of the portal is where the user has his/her first contact with the system and it is very important that the interaction is as smooth and natural as possible. Great care will be devoted to this requirement. To improve the existing version (V1.1), in the short term we intend to provide a series of video tutorials explaining how to use the system. In the mid/long term we plan to develop new interfaces, based on machine learning techniques, that will make the portal as user-centred and effective as possible, in line with our principle that all the existing barriers to data must be as low as possible, ideally removed.

31.3.1 The Open Universe Portal for Astronomy

As a practical example of the use of the Open Universe portal we show the case of a request for an astronomical source.

By entering in the user's input area the name of a known astronomical source (e.g. M101, or Crab Nebula, Andromeda, 3C273, to name just a few famous astronomical objects) the system activates the astronomy part of the "Open Universe" portal giving access to services offering astronomical data services (Fig. 31.2). A series of icons appear in the mid part of the page allowing the users to activate services of different types (each identified by a contour of a specific colour[1]), as provided by several existing major web sites. More general information is also provided by services like Wikipedia, NED, and SIMBAD located at the top right part of the portal. In the example of Fig. 31.2 after entering the name of the galaxy M101 the user clicked on the service "Aladin Lite" from the Centre de Données astronomiques de Strasbourg (CDS), causing the image of M101 to appear in the bottom part of the web page.

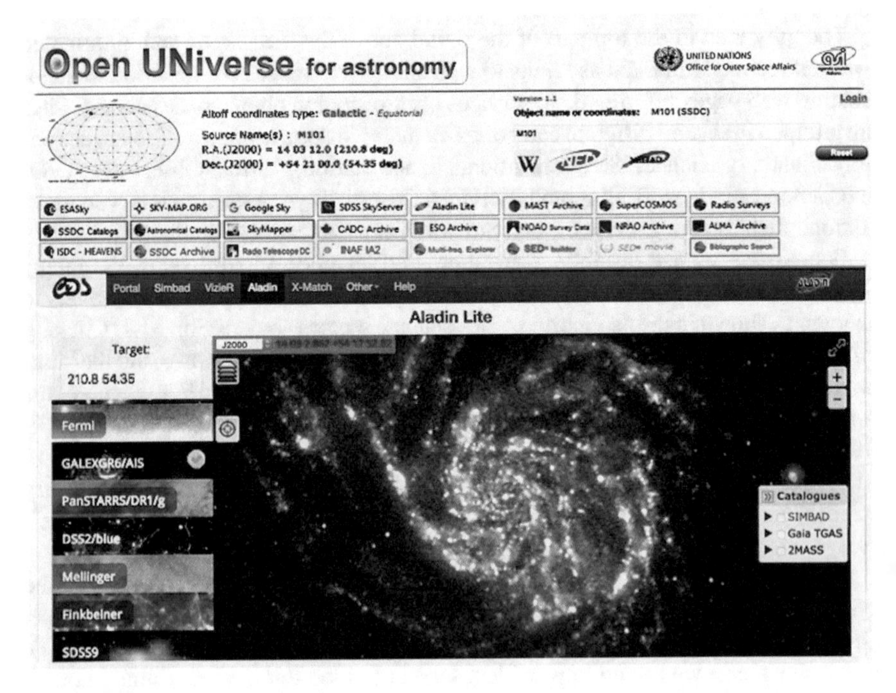

Fig. 31.2 The Open Universe portal for astronomy: the case of the galaxy M101. The picture shows the image of M101 at ultraviolet frequencies, one of the options provided by the Aladin-Lite tool, a service of CDS in Strasbourg

[1]Blue colour corresponds to services providing astronomical images (from sky surveys at different frequencies), green colour is for astronomical catalogues, red is for data archives, light blue is for sites providing bibliography services, and yellow is used for multi-frequency, data intensive services, that is web sites providing data from several satellites and telescopes.

31.4 Cost/Benefits Considerations on the Open Universe Initiative

31.4.1 *Social and Economic Benefits*

The Open Universe initiative aims at largely improving the level of transparency, and therefore accessibility and use of space science data both in the scientific community and in non-scientific sectors of society. In both cases, socio-economic benefits may be considerable, in the first case in the form of an increased number of publications in scientific journals, and in the second case in the form of, for example, better education and a larger discovery potential.

A detailed cost-benefit analysis of the Open Universe initiative is premature and certainly beyond the scope of this chapter. In the following paragraph we limit ourselves to a few considerations on the social and economic impact of the initiative.

The following is a non-exhaustive list of benefits of the Open Universe initiative:

- Improve services to professional scientists;
- Facilitate cross-discipline, multi-experiment and data-intensive research;
- Increase the efficiency in the production of knowledge from the same data;
- Enable more citizen science activities;
- Improve/promote scientific thinking in society;
- Inspire young people to pursue scientific careers;
- Improve the quality of education;
- Extend the potential of scientific discovery to non-professional scientists, and in principle to everyone interested in Space Science.

Cost-benefit analysis of research infrastructures is a difficult matter and it is rarely attempted, mostly because of the unpredictability of future economic benefits of science. Nevertheless, it has been recently conducted on some large scientific projects, the most important one being the Large Hadron Collider (LHC), the world's largest particle accelerator, showing that the benefits may be extremely large. For instance, Florio, Forte & Sirtori, (2016[2]), showed that the economic value of the two widely used open access software ROOT and GEANT4 developed and maintained within LHC activities can be estimated to be 2.8 billion Euros.

Expanding the use of space science data to large sectors of society may be achieved by lowering the barriers of usability of space science data to the level of every citizen, (that is offering web-ready or transparent data products) rather than following the more frequently used approach of educating a small fraction of citizens (usually students, researchers or citizen scientists) to be able to perform complex analyses or data processing on low or intermediate level products like raw or un-calibrated data.

The benefits are clearly potentially very large and will depend on how the initiative (and other related activities, e.g. the Virtual Observatory) will evolve.

[2]Florio, Forte & Sirtori, Social benefits and costs of large scale research infrastructures, 2016, vol. 112, issue C, 38-53. arXiv:1603.00886.

31.5 The Role of International Cooperation

International cooperation, usually implemented through inter-governmental agreements, plays a large role in the development and the operation of space hardware, especially in very large facilities such as the International Space Station (ISS). Coordination activities, like for example the International Exploration Coordination Group (ISECG) are also crucial for the implementation of highly ambitious future space projects, such as human and robotic exploration of the solar system, and the development of bases on the Moon and Mars. These initiatives clearly hold a large technical, political and economical value.

There are no similarly large and structured examples of international cooperation projects in space science data management, processing and archiving. However, especially in the astronomy sector, some important examples of cooperation have emerged in the form of spontaneous aggregation of international research institutions and individuals. Perhaps the most notable one is the cooperation that led to the development of the FITS format. This data format was originally designed in the 1970s for the exchange of radio telescope images, and over the years, and through the collaboration of experts from space and ground based telescopes operating in all energy bands, evolved into a standard capable of supporting any type of astronomical data type, and endorsed by all major organizations, for example NASA, ESA and The International Astronomical Union (IAU).

Today the FITS format is not only a consolidated asset for space science and astronomy, but also a valuable resource that can be used for other purposes. One important example is the use of FITS for the preservation of high-resolution images of ancient documents to ensure that present and future generations will have simple access to the books of the Vatican library.

Another important example of cooperation is the International Virtual Observatory Alliance (IVOA), which, since 2002, focuses on the development of standards and services to the astronomical community to support good data management and interoperability.

The IVOA currently comprises 21 programs from Argentina, Armenia, Australia, Brazil, Canada, Chile, China, Europe, France, Germany, Hungary, India, Italy, Japan, Russia, South Africa, Spain, Ukraine, the United Kingdom, and the United States and an inter-governmental organization (ESA).

Open Universe, with the coordination of UNOOSA, wishes to largely expand the current level of international cooperation in space science data services with large benefits for the scientific community, and other sectors of society.

31.6 Sustainable Costs

What are the costs of achieving the goals of the Open Universe Initiative? Often the answer to this question is "large" or "too large". This is the perception usually expressed by many decision-makers and some people operating in the data archiving sector. On close inspection, however, we must note that these costs are certainly negligible compared to the overall investment that has been made to produce space science data, and very small compared to the value that the initiative would generate for our society.

In fact, we argue that the costs associated to Open Universe can be easily sustainable if the following principles are followed.

For **future space scientific missions**, minor adjustments of agencies' cost-to-completion models are necessary to reach the goals of Open Universe: Every year about 15 Billion Euros are spent to produce space science data (not including the sector of Earth observation). By implementing policies at decision making level, that ensure that the final data products meet the Open Universe openness and transparency requirements including the costs of proper data handling, in the overall budget since the very beginning, the cost of reaching transparency would be a tiny percentage of the investment made for the overall project.

For **past, current and future data** the following general principles should be followed to ensure cost-efficiency:

- Avoid duplication of efforts;
- Foster collaboration and coordination among data centres;
- Make use of existing high quality infrastructure and data services (e.g. full use of IVOA standards when possible);
- Develop innovative tools (e.g. openuniverse.asi.it);
- Use new paradigms (e.g. distributed analysis);
- Take full advantage of new technologies.

31.7 Conclusion

Since the advent of the first web-based digital archives offering on-line open astronomical data services in the early nineties, much has been done in the direction of offering space science data to an ever increasing number of users, from the small community of scientists involved in the experiments that produced the data to several thousands of non-specialists researchers. This progress, however, has been strongly discipline dependent, with the astronomy sector leading the way, with other space science disciplines moving at a much lower speed, and in many cases still restricting the use of the expensive data they produce to the small number of scientists belonging to the projects teams.

Much work is still ahead of us to meet these goals. The ASI prototype that has been described above represents a first step along this path. Even in the astronomy sector, today's best digital archives provide in most cases calibrated data and the associated software suitable for higher level scientific analysis that must be carried out by expert users. The rest of the world is still largely excluded from the utilisation of the data. It is too early to predict the detailed work that is necessary to achieve all goals of "Open Universe". It is however easy to predict that the associated extra cost is only a tiny fraction of the amount of around 15 Billion Euros that is spent every year in the world to generate scientific space data.

The Open Universe initiative may play a strategic role in the near future by improving the use of digital technologies to maximize transparency and enhance web interfaces with applications that may go beyond the field of space science and astronomy.

From the time table point of view, 2016 has been the year of the proposal to COPUOS[3]; while 2017 was dedicated to ample discussions, with the organization of an expert meeting in Rome[4] and a workshop dedicated to Open Universe open to the international community that took place at the United Nations in Vienna.[5]

In 2018 Open Universe entered a more operational phase, within the UN Space2030 agenda, with the coordination of the United Nations through UNOOSA, in close cooperation with Italy, the scientists and the institutions that collaborated in the initial phase, together with all the UN member States that support it.

[3]http://www.unoosa.org/res/oosadoc/data/documents/2016/aac_1052016crp/aac_1052016crp_6_0_html/AC105_2016_CRP06E.pdf.

[4]http://wwwdev.openuniverse.asi.it/documents/ou_documents.php.

[5]http://www.unoosa.org/oosa/en/ourwork/psa/schedule/2017/workshop_italy_openuniverse.html.

Chapter 32
Space and SATCOM for 5G—European Transport and Connected Mobility

Stefano Ferretti, Hermann Ludwig Moeller, Jean-Jacques Tortora and Magali Vaissiere

Abstract Profound change lies ahead for the transport sector, both in Europe and in other parts of the world. A wave of technological innovation and disruptive business models has led to a growing demand for new mobility services. Transportation systems develop towards connected mobility. The choice of communication technology will depend on the location, the type of service and cost efficiency. The user should be unaware of the communication technology used, and in the 5G ecosystem, satellite communication should be integrated, as future technologies, into the hybrid communication mix. The Space Strategy for Europe underlines the role satellites can play to provide cost-effective solutions to improve connectivity for Europe's digital society and economy as part of the future 5G networks, where numerous applications and services using space data will also require uninterrupted connectivity. It encourages the uptake of space solutions by integrating space into future strategies addressing, for example, autonomous and connected cars, railways, aviation and unmanned aerial vehicles. The paper provides the major findings and recommendation stemming from the June 2017 ESPI-ESA Conference on "Space and SATCOM for 5G", identifying cooperative frameworks to encourage the interworking of satellite and terrestrial technologies in support to the transport sector. It also provides perspectives of the respective business and institutional communities and the roles and approaches of the diverse groups of stakeholders involved in the development of satellite based 5G networks and of future transportation vehicles. It then focuses on markets where satellite communications and other space applications provide strong value-add and complementarity to terrestrial 5G solutions, in automotive, aeronautics, maritime, trains and other domains. The chapter finally discusses the ESA strategy in the context of a European support to satellite communications in 5G, including possible public-private partnership models.

S. Ferretti (✉) · J.-J. Tortora
European Space Policy Institute (ESPI), Vienna, Austria
e-mail: stefano.ferretti@esa.int

H. L. Moeller · M. Vaissiere
European Space Agency, Paris, France

S. Ferretti (ed.), *Space Capacity Building in the XXI Century*, Studies in Space Policy 22,
https://doi.org/10.1007/978-3-030-21938-3_32

32.1 Introduction

Profound changes lie ahead for the transport and communications sectors, both in Europe and worldwide. While technological innovation and disruptive business models spur the demand for new mobility services, transportation systems are developing towards connected mobility. These changes and how we should respond to them are addressed by a number of recent initiatives. The European Commission, the European Space Agency, and the European Space Policy Institute are all taking an active role, in coordination with initiatives at industrial and national level, including the transport, communications and space sectors to develop joint strategies.

This is being done through a number of consultation events with decision-makers and the associated development of formal strategies and action plans. The latter include the European Strategy on Cooperative Intelligent Transport Systems[1] and the Space Strategy for Europe,[2] which underlines the role satellites can play in providing cost-effective solutions within future 5G networks, where many users will require both space data and uninterrupted connectivity.

ESA convened, in April 2017, a 5G Round Table in Paris on a possible 'Satellite for 5G' initiative to determine how the space industry could join forces to work with the terrestrial sector in areas of mutual interest, such as the definition of interoperability, network orchestration and spectrum. The outcome of this activity materialized in an ESA/Industry Joint Statement, signed at the Le Bourget Air Show in June 2017.

A technical report was issued as a first step towards an ESA initiative as a united European effort, supported by an initial set of 16 industry stakeholders. It was concluded that the choice of communication technology (e.g. cellular, fibre, satellite) should be transparent to the end user and based upon location, type of service and cost efficiency, with satellite communications integrated into the future hybrid 5G-communication mix. The transport sector has been identified as a key user market to be addressed, also referred to as 'vertical', with inherent strength of SATCOM in areas including connected cars, trucks, ships, trains, airplanes and unmanned aerial vehicles. While this activity was initiated by the space community, the next step represents an opening to other communities, in particular to the transport sector and terrestrial operators (Fig. 32.1).

To this end, ESA and the European Space Policy Institute (ESPI) co-hosted a conference in Brussels on 27–28 June 2017, where both policy and technology aspects related to 5G and future space-based solutions were discussed. This conference provided a high-level dialogue platform, to reflect upon the evolution of the space telecommunications sector, the new connectivity strategy and in particular its interaction with the transport sector. The event included the participation of the European Commission (EC), with top-level representation from DG-CONNECT, DG-GROW

---

[1] A European Strategy on Cooperative Intelligent Transport Systems, a milestone towards cooperative, connected and automated mobility. COM. (2016). 766 Final, 30 November 2016.

[2] Space Strategy for Europe. COM. (2016). 705 Final, 26 October 2016.

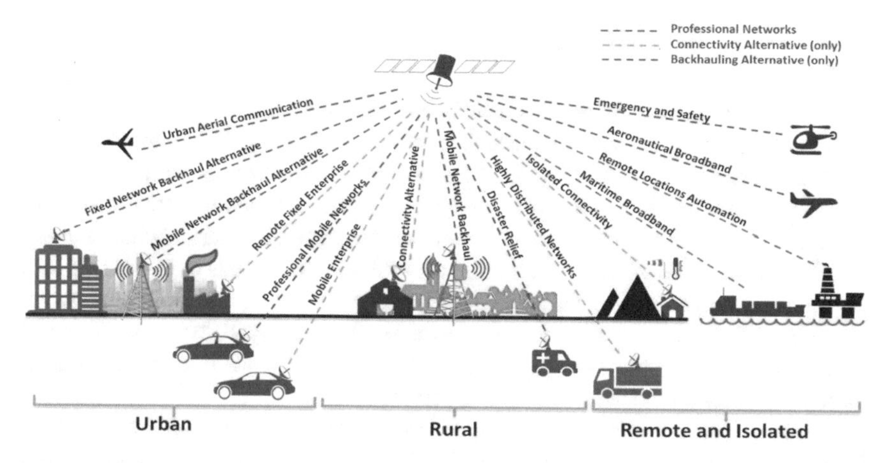

Fig. 32.1 Selected use cases for satellite communication in 5G. *Source* Booz Allen, SES, Fraunhofer FOKUS, EURESCOM

and DG-MOVE, providing a balanced discussion platform, which involved institutional players, the space sector and the transport sector, and that is key to a successful outcome.

Accordingly, the agenda focused on the markets in which satellite communications and other space applications provide strong value-add and complementarity to terrestrial 5G-solutions. Finally, the participants discussed the user and market needs on the demand-side and the possible solutions on the supply-side, with the aim to outline a way ahead including public-private partnership models (Fig. 32.2).

Fig. 32.2 ESPI-ESA conference contributors

32.2 Policy and Strategy Perspectives

Strategically two distinct domains have been identified, related on one hand to the infrastructures and on the other hand to the user segment (vertical). The infrastructural element is characterized by the contraposition of satellite and terrestrial segments. Since the digital revolution is posing new requirements on the infrastructure quality of service as well as on cost efficiency, new business models emerge, where satellite is at the convergence of fixed and mobile technologies and needs to operate on joint architectures and software. The second element is the user market segment, which will require more robust and more performing telecom capacities soon. As a first step, it is thus advisable to engage user stakeholders to work together as soon as possible towards new developments and services.

In this context, it is crucial that Space gets involved in the elaboration of the 5G architecture as well as service development from the outset, to allow satellite to become integral part of verticals. In this respect, it seems necessary to further elaborate some key policy considerations at times when the 5G-infrastructure trial roadmap is being defined. This shall ensure future interoperability of systems with appropriate technological developments and relevant standardization. Equally, the applications sector shall evolve accordingly to provide the users with efficient seamless services ready to be fully adopted, yielding proper return on investments for the 5G-stakeholders.

With the support of satellite, 5G shall also help to bridge the digital divide. However, after twenty-five years of operations of Global System for Mobile Communications (GSM), only some 60% of the population is actually covered. This means that neither market forces nor terrestrial solutions alone have been able to provide full coverage to all citizens. This coverage issue will be even more challenging with 5G, which requires higher density networks. In this respect, cellular and fibre optic networks alone cannot be expected to provide an adequate solution. Therefore the complementarity of satellite and terrestrial networks appears as a most promising option to ensure ubiquitous coverage, provided that cost-effective terminals are developed to equip vehicles, buildings and remote stations.

32.2.1 Space, 5G and Transport

Nowadays all transportation systems are developing towards connected mobility, and the challenge ahead is to meet the conditions for adequate coordination among three different sectors - space, telecommunications and transport—to enable new connectivity capacities for the future European transport systems.

At the moment, stakes are high for the space sector to secure a significant role in 5G. Inherently, SATCOM is particularly adapted to mobility applications, but a dedicated effort to engage with non-space players and vertical user segments is still required to ensure that these merits can be turned into practical, effective and affordable services as part of a seamless user experience. In order to develop and successfully deploy such services, the space sector needs to focus specific efforts in the areas of technologies, availability of spectrum, demonstration of capabilities via trials, and finally, application of the most appropriate standards to seamlessly integrate space assets into a hybrid communication mix.

Space telecommunications also need to overcome their current situation, perceived as operating in a niche market, in order to integrate into the broader picture of connected mobility, ensuring that space demonstrates its benefits and also gets the "critical mass" to gain the necessary political support from governments and acceptance by the terrestrial stakeholders. To achieve this objective, European institutions have the key role to raise awareness among the main stakeholders (e.g. space agencies, national authorities, regulatory and standardization bodies and policy makers), and to propose concrete measures to help ensure that satellite becomes a fundamental component of future 5G infrastructures.

In this respect, the European Commission underlined the key role that satellites can play to provide cost-effective solutions and uninterrupted connectivity for Europe's digital society and economy. In fact, the Space Strategy encourages the uptake of space solutions by integrating space into various future strategies addressing other sectors. Therefore, current and future strategies of the European Commission are structuring contributions to the European innovation agenda, aiming at the creation of new business opportunities and innovative services for the European citizens. There are high stakes and opportunities for the space sector in this fast-evolving framework, which need to be tackled jointly by manufacturers, space and ground operators, to gain a common understanding of the impact of the 5G-revolution on the satellites business and to identify the roadblocks potentially preventing terrestrial operators to effectively rely on a space segment.

The European Commission is currently putting considerable efforts on the digital transformation as a key priority for the evolution and growth of Europe. Creating a digital society also means inclusion of all through the network, and this will hopefully mean in practical terms the provision of a 100 Mbit connection by 2020 across Europe. According to the European Commission, this objective could be best achieved working together with all actors in the framework of Public Private Partnerships. The 5G ecosystem is characterized by transformative technologies at large, which will impact not only the mobile telecommunications, but many other sectors such as mobility, logistics and health, among others. This is a great challenge Europe must win, particularly in relation to the creation of a single market, which today is still highly fragmented (Fig. 32.3).

To fully enable this transformation, some key legislative steps still need to be completed, such as essential reforms for the utilization of the spectrum. The relationship with the United States on 5G deployment still requires further harmonization efforts, to build credible scenarios for the utilization and possible sharing of the spectrum, since two different frequencies were selected (EU 26 GHz vs US 28 GHz). Spectrum allocation continued to be a major discussion point at the World Radiocommunication

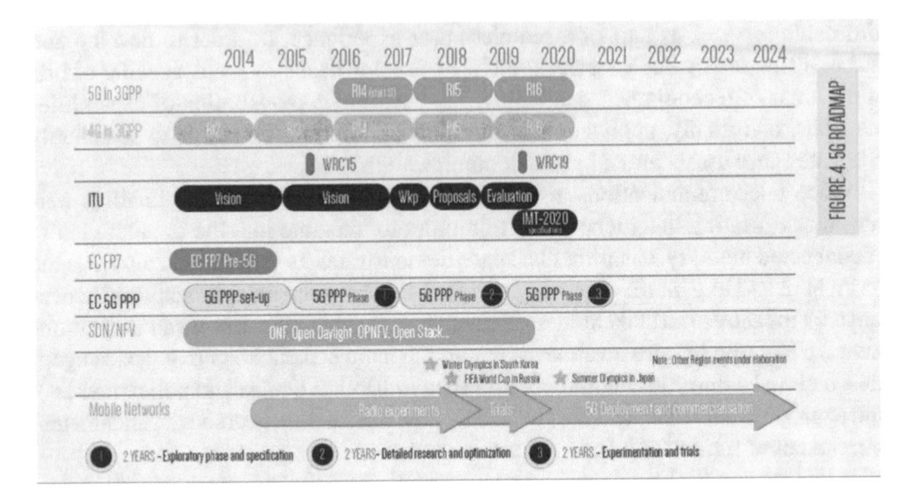

Fig. 32.3 5G terrestrial roadmap

Conference (WRC) in 2019, as maintaining access to higher frequencies is necessary to ensure the growth of the satellite sector and its services while acknowledging investments in existing and future space systems in all fields of space applications.

Some open key elements for properly engaging with the transport sector are the development of dedicated ground antennas, the forging of global alliances around 5G and the considerable work with ITU to converge on the same frequencies globally. Space has already proven to be instrumental in the aviation and maritime sectors. Now it is time to ensure that Space will contribute also in other sectors to the new 5G ecosystem. The Space community is already adopting this new digital transformation, through its new strategy (i.e. Space 4.0) (Fig. 32.4).

32.2.2 *The Transport Perspective*

The transport sector is facing a significant transformation towards a sustainable and intelligent evolution, and space can play a key role in enabling it, by fully exploiting the potential of satellites within the 5G connectivity ecosystem. Connected transport has the remarkable advantages of improving road safety, reducing its environmental impact, while ensuring a higher degree of social inclusiveness and favouring cross-border initiatives.

The European Union aims at creating the conditions for a strong industrial leadership, by supporting research and development activities towards cooperative, connected and automated mobility. Europe is already promoting this evolution of the transport sector by increasing citizens' awareness and is planning to support the

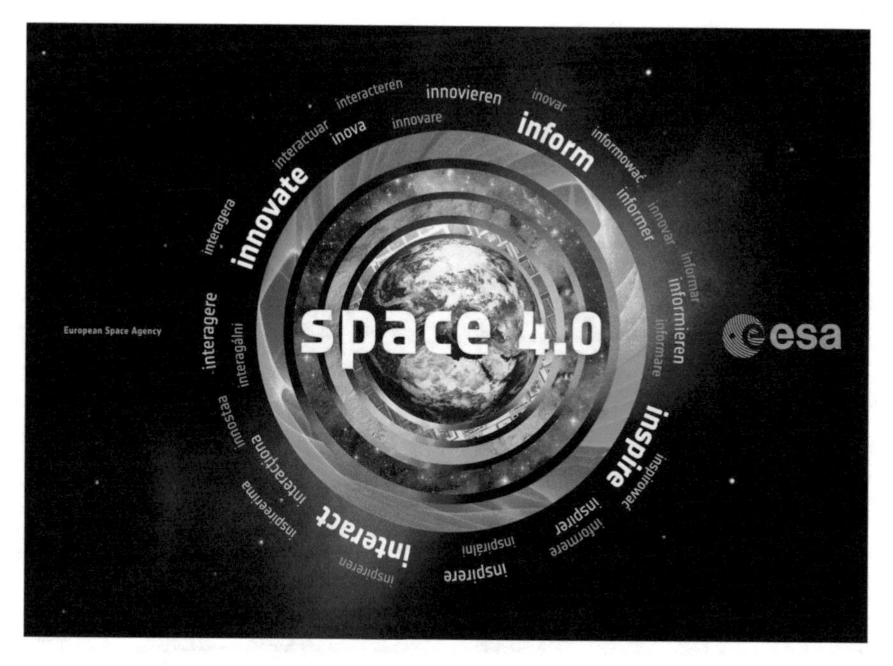

Fig. 32.4 ESA Space 4.0

roll-out of 5G based automotive connectivity applications in the 2020s. The European Commission DG-MOVE foresees to implement vehicular cooperation and automated driving on most roads in the 2030s, deploying a highly coordinated and fully automated transportation in the 2040s (Fig. 32.5).

These plans require availability of advanced and cyber-secure technologies, providing ubiquitous coverage also in remote and rural areas, to ensure a seamless automated driving experience particularly along highways. This new context will require significantly increased coverage and therefore satellite will represent an essential element in the new 5G ecosystem.

32.2.3 The ESA Perspective

ESA approaches 5G from the perspective of an R&D agency with the objective to support industrial competitiveness and the development of new space technologies and solutions. This primarily concerns satellite communications as a future component of seamless 5G communications infrastructures and services. It is also relevant for space applications of navigation and earth observation, providing space-based information in support to 5G.

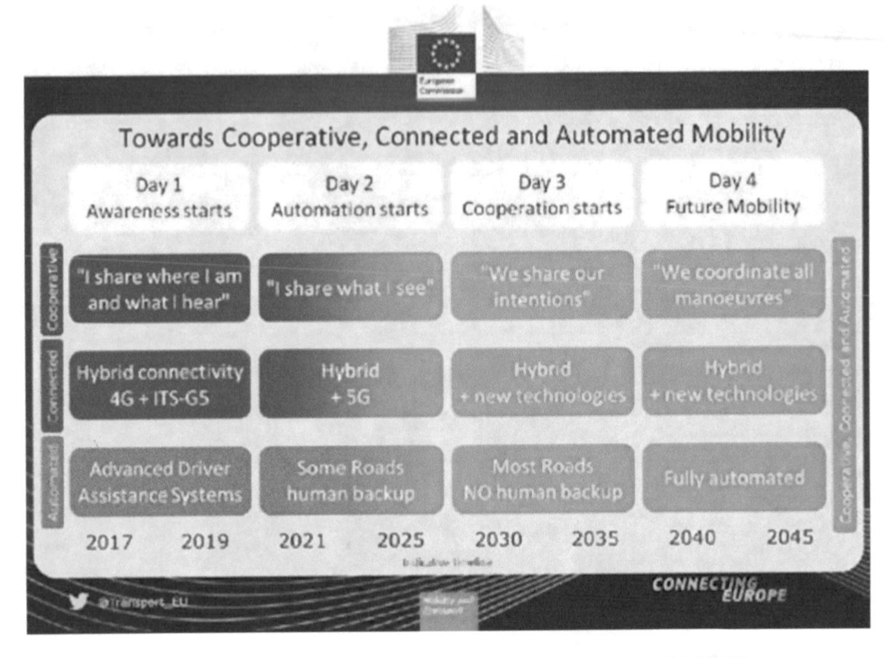

Fig. 32.5 Cooperative, connected and automated mobility. *Source* EC DG-MOVE

It has to be noted that 5G comes with new challenges for ESA, as 5G primarily originates in a non-space environment. Telecommunications today are largely dominated by cellular operators and by the deployment of fibre optic networks. This fact may also explain why space and satellite communications in other world regions for now are little or not represented in the global 5G effort, an opportunity for Europe to take a leading role. The convergence of terrestrial and satellite 5G solutions has been initiated in Europe, in particular thanks to the initiative undertaken by the European Commission as part of the H2020 and 5G PPP. However, at present only one of 40 projects include a strong satellite component.

Now ESA, together with industrial partners, wishes to build upon and reinforce that European initiative, for the benefit of the European and Canadian satellite and terrestrial industry, in close collaboration with the European Commission and other key stakeholders such as the ones represented by the 5G Infrastructure Association. The ESA approach is industry-driven, in line with ESA's strategy in satellite communications based on PPP models. This is also increasingly frequent in the New Space age. ESA therefore is consolidating a programmatic framework, in consultation and agreement with its Member States, to provide support to industry for the innovation and competitiveness challenges of 5G (Fig. 32.6).

The ESA Directorate for Telecommunications and Integrated Applications launched the 'Satellite for 5G Initiative' for the period 2018–2021 and beyond. ESA will work together with operators, service providers and manufacturers, in synergy

Fig. 32.6 ESA 'Satellite for 5G' initiative

and aligned with the EC 5G-PPP and 5GIA roadmaps as well as national initiatives, to implement a three-pronged strategy:

- SATCOM-5G Service Trials, with focus on vertical user markets where space can contribute with its inherent benefits, i.e. in transport, public safety and media and entertainment
- R&D support activities in technology and services development, e.g. user antennas/hybrid receivers, resource management, interoperability demonstrations including access to or provision of 5G test beds
- Outreach activities aimed at promoting awareness of satellite in 5G, including informing policy makers on the infrastructural key role of satellite next to fibre and cellular networks (Fig. 32.7).

As a first step, ESA and 18 signatories from the space industry issued the Joint Statement on Satellite for 5G. First trials may be implemented shortly, based on existing space and ground segment assets while a second phase may support new developments and incremental levels of SATCOM-5G integration. This shall allow to synchronise with trials within the expected 5G-PPP timeframe and the anticipated

Fig. 32.7 Signatories of the Joint Statement on Satellite for 5G

industrial deployment of 5G, and support first trials, pilot projects by 2020 and full operational deployment towards 2025. At present, 13 expressions of industrial interest for 5G developments and trials have been received by ESA, with 29 specific activities being proposed from industries from 14 different European countries and Canada. The Joint Statement and the resulting programme framework is open for further participants to join, in particular from the downstream/user sector as well as from the telecom sector beyond space, including start-ups and SMEs. The above will support the definition of a joint way ahead regarding the ESA programmatic framework including PPP models, the EC 5G Action Plan, as well as National and EC Research activities.

In conclusion, at ESA the development of 5G will be a key element to maintain and increase the competitiveness of the European SATCOM industry, which is the commercial foundation on which also other sectors of space build on. As the core SATCOM market of TV broadcasting is in decline, the success in Space 4.0 and 5G will be essential for a strong industry base to continue and provide solutions in other fields such as space imagery, positioning and Navigation. Therefore, strategically 5G is not only key to Europe's satellite communications industry but also to the space sector at large. A joint European effort is required to secure Europe's place in the global market and digital society.

32.3 Conclusion

To effectively respond to the needs of European Transport and connected mobility, three sectors have to join forces, i.e. the transport sector, the telecommunications sector and space. Space and in particular satellite communications have inherent strength and already support connected mobility in the maritime and aeronautics domains, expanding into connectivity for trains, trucks, cars and UAVs.

Equally, three groups of stakeholders need to work in synergy to secure Europe's future in the 5G era, i.e. industry in manufacturing and operations/service provision, national initiatives and institutional cooperation at European level. For the latter, the Space Strategy for Europe and the European Strategy on Cooperative Intelligent Transport Systems provide an excellent framework for European Commission and ESA initiatives (Fig. 32.8).

In this endeavour, a new challenge for space will be the establishment of new three-party partnership models, between space, non-space, and the user markets. As a first step, ESA's engagement in a Joint Statement with space industry opens the door for other stakeholders to join. The challenges will be technological as well as political. Technical solutions will require architectures and new services to appear to the user as one, seamlessly hiding the underlying infrastructures be it satellite, cellular or fibre. Politically, satellite still needs to find its recognition next to the current focus on investing in fibre, as it is the only mean to overcome the digital divide and ensure coverage to all European citizens, well beyond the 60% provided today by GSM and terrestrial networks.

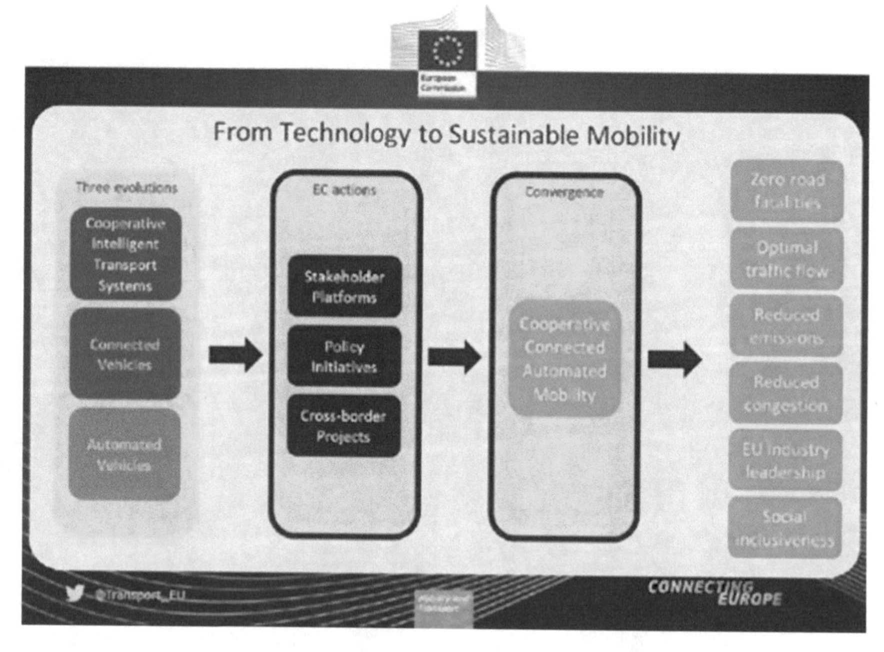

Fig. 32.8 From technology to sustainable mobility. *Source* EC DG-MOVE

5G deployment may be expected in the 2020–2025 timeframe, while automated driving and a highly coordinated and fully automated transportation may become reality in the 2030s and 2040s. Therefore, the initiatives by the European Commission and as elaborated by ESA are timely. They are open to industry-driven partnerships and focus on trials. The coming years will see the emergence of the digital society, as supported by the European Commission, and the transport sector as well as the ambitions of Space 4.0 will be a key component in this new development enabled by 5G.

This paper was presented at the 68th International Astronautical Congress, 25–29 September 2017, Adelaide, Australia. www.iafastro.org.

Part V
Concluding Remarks

Chapter 33
Building Space Capacity Through Synergies: Three Programmatic Frameworks

Stefano Ferretti

33.1 Introduction

The outcomes of the conference "Space2030 and Space 4.0: synergies for capacity building in the XXI century", co-organized by the European Space Policy Institute (ESPI) and the United Nations Office for Outer Space Affairs (UNOOSA), with the support of the European Space Agency (ESA), and the contributions in this volume at large, point to the timeliness of an in-depth reflection upon the various perspectives and strategies, that could be implemented in future space programmes, with a focus on space capacity building. They include examples of successful programs, research, entrepreneurial activities, and future developments, mapping out available options and identifying the ideal conditions for their successful implementation. Innovative frameworks, partnerships and collaborations in the space capacity building ecosystem are explored, posing special attention to improving the dialogue with civil society and other sectors to make them aware of the potential of space, and to the creation of new mechanisms to identify, collect and process user needs, in order to design, implement and fully exploit future space programmes.

Together, the different contributions point to potential synergies between the UN agenda Space2030, the ESA Space 4.0 strategy and various national space strategies, focusing on four thematic priorities of UNISPACE+50: Global partnership in space exploration and innovation; Strengthened space cooperation for global health; International cooperation towards low emission and resilient societies; Capacity building for the XXI century. Through these themes, the interplay and dependencies amongst key actors are identified, and a special emphasis is placed on future approaches of the diverse groups of stakeholders involved, leveraging the existing space infrastructure, institutions and networks while reinforcing and expanding their scope and effectiveness in ensuring that space becomes an important driver for sustainable

S. Ferretti (✉)
European Space Policy Institute (ESPI), Vienna, Austria
e-mail: stefano.ferretti@esa.int

S. Ferretti (ed.), *Space Capacity Building in the XXI Century*, Studies in Space Policy 22,
https://doi.org/10.1007/978-3-030-21938-3_33

development. For example, future integrated services will capture new citizen needs and target sustainable development goals, creating unprecedented opportunities for Europe worldwide. This will be enabled by Copernicus, Galileo and by the emergence of new satcom infrastructures for 5G, serving new markets by providing broadband connectivity to rural areas. Space is therefore increasingly becoming the link among systems of systems, and its enabling function may represent a key element actively contributing to a sustainable future on Earth.

The core innovation of this book is the wide nature of the disciplines addressed, with views and perspectives from a wide geographical distribution of actors. The user has been placed at the centre, to identify concrete needs and achieve goals in a clear and efficient manner. The importance of demonstrating best practices and showcasing available space-based services resulted in the elaboration of recommendations, that can be considered in the creation of roadmaps and in the implementation of future space programs. It is the scope of this chapter to summarise some of these recommendations for a way forward which addresses the challenges and opportunities envisaged for the future. Therefore, for each thematic priority, a relevant program is presented, stemming from a synthesis of the corresponding contributions together with critical reflections.

33.2 Innovation and Exploration

New innovation models are increasingly spread across sectors and disciplines, including space, which is becoming an integral part of many societal activities (e.g. telecoms, weather, climate change and environmental monitoring, civil protection, infrastructures, transportation and navigation, healthcare and education). In particular, Space services are nowadays becoming a substantial part of the value chain in various sectors and disciplines.

In the context of UNISPACE+50 thematic priority 1, space exploration is considered as "involving humans and robots venturing beyond Earth". Space exploration encompasses a range of activities and missions aiming at extending human presence beyond Earth (i.e. including space stations occupation and operations). It also focuses on better understanding the existence of humankind on Earth from the beginning of time, while advancing the scientific knowledge of the Universe and the search for other forms of life on other celestial bodies. This includes preparation, demonstration and operations of both human and robotic missions.

Directly or indirectly, space exploration activities contribute to new knowledge and technical innovations, stimulating the creation of tangible benefits on Earth. As a result, these activities have contributed to different aspects of everyday life, ranging, for example, from solar panels to implantable heart monitors, from cancer therapy to light-weight materials, from water purification systems to improved computing systems and a global search-and-rescue systems. These actions have a wide potential to deliver substantial benefits to all countries, both by developing new scientific knowledge and by spinning-off space technologies in the areas of health and medicine, public safety and transportation. Innovation and knowledge derived from space exploration, contribute also to foster economic growth and to deliver high returns on investment, creating new business models and legal instruments, and leveraging international cooperation and collaboration at various levels (e.g. governmental, non-governmental and private sector). Furthermore, the concept of innovation can be more widely adopted within other space systems and services, offering higher performance while lowering costs.

If the objective of space capacity building is to expand upon the number of spacefaring nations, enhancing their capacity to exploit the benefits of space assets and participate in the greatest exploratory quest of our time, the ideal starting point is the development of space technologies in these countries. Thanks to the exponential and disruptive character of recent innovations, the miniaturisation of satellite technology has become a reality. In particular, it has emerged that small satellites are very important tools for capacity building, enabling access to space and the development of downstream services across the world.

Small satellite technology, including of CubeSat size, is becoming increasingly accessible, including to emerging spacefaring nations and research entities in developing nations. These emerging developments in the field of CubeSats across the world indicate a new trend in accessing and utilizing Space. For the first time, this represents an opportunity for all nations to exploit the potential of space technology, data and information for their citizens. It also poses some challenges related to space traffic management and related governance mechanisms. In fact, Cubesats are foreseen to increase the level of debris in orbit, posing crucial questions on the future access to orbits by all. Therefore, when considering the development of a Cubesat-oriented program, any country should be well informed and prepared to tackle not only technical challenges, but also legal and regulatory ones. The United Nations Office for Outer Space Affairs plays a key role in this process and is the only international body entitled to provide support in these areas, in addition to its key role as secretariat of the United Nations COPUOS.

UNOOSA could play a key role in supporting the adoption by its Member States of innovative tools and methods which are currently available for capacity building in the XXI century. It could also facilitate the collection, processing and baselining of requirements linked to actual needs which have been identified in the field among end users. For example, methodologies such as open innovation platforms can serve the purpose (e.g. the recently released ESA Open Space Innovation Platform "OSIP"). Once these needs and requirements are clearly identified, UNOOSA can leverage Concurrent Design Facilities around the world, linking them both to its Regional Centres and to National Institutes, as very well outlined by DLR and ESA in Chaps. 6 and 30. Designing together with the experts in the Member States will allow a much greater efficacy of capacity building, versus more traditional one-off training programs. It will not only provide opportunities to learn on the job, but also to create useful relationships at working level with the space agencies around the world that are part of this potential future programmatic framework.

When approaching this, UNOOSA and interested space agencies could also create an international Young Trainee Program, where the partner country could send motivated young graduates to institutions around the world, to gain the necessary skills to design, test, integrate and operate a CubeSat. UNOOSA could both leverage the many and highly relevant partners (already signatories of MoUs) and engage with new ones. UNOOSA could then coordinate the training and following project phases via digital tools, enabling an interactive and efficient resources management, following the examples of other space organizations (e.g. Open Service Innovation, Digital Platforms and Organizations). This should be accompanied by basic project management training that could be delivered for example via MOOCs and Microsoft Project tools. The implementation and progress of such a program could then be tracked making use of digital platforms for actions tracking and videoconferences for progress meetings, providing timely support when needed via a Space Applications Help Desk (leveraging on the European Union's excellent model for Copernicus which is now starting to prove very effective). The project plan and budget should include costs for setting up the partnerships, a travel budget and funds for capacity building activities in the target countries. The communication links and local set-up could be part of ODA activities with the World Bank though Space component activities (e.g. as per the ESA-World Bank cooperation along these lines). In addition, it is proposed that innovative new space companies could contribute in kind to some of these activities both in the EU and in the USA. A realistic timeline of one year for project definition, implementation and CubeSat design could be initially foreseen. A follow up activity may include manufacturing and testing, with a more variable timeline, depending on the facilities and capabilities in the country. For the first CubeSat that each country is developing, part of these activities could take place at the premises of an institute in a country with a strongly established space program, like in the case of the Bhutan/Japan cooperation of 2017/2018. The final objective would be to have an operational system in orbit, including the downstream applications workflow, fully functional in two years since the project kick-off. This would include the CubeSat launch, which could also make use of the KIBOCUBE partnership with UNOOSA (see Chap. 8).

In this quickly evolving context, cooperation projects should be established that could encompass partnerships with national and regional research institutes, space agencies, academia, IGOs, NGOs and space industry, including new space actors. The baseline for a cooperation in this area could include specialized training and design/development facilities for small satellites (both at technical and managerial levels). A joint implementation plan with the various stakeholders and partners, could include the recruitment of engineers from emerging spacefaring nations to be seconded at specialized facilities and space agencies in order to become acquainted with the latest technologies and methodologies. In addition, the use of modern digital technologies and concurrent engineering methodologies could enhance the effectiveness of such a cooperation, allowing for remote teamwork in any phase of the design, development, manufacturing, integration and launch, operations and applications/services implementation.

33.3 Space for Global Health

The United Nations Office for Outer Space Affairs (UNOOSA) and the World Health Organisation (WHO) have conducted several workshops and conferences on space for global health, where they have examined the role of Earth Observation/Remote Sensing, Telecommunications and Positioning and space-based medical research, in improving global health efforts. One of the key outcomes was the recommendation to develop a dedicated curriculum on space and global health, that could build up on the work conducted by the COPUOS Expert Group on Global Health and on various initiatives and programs at Space Agencies' and Member States' level. UNOOSA would be in a privileged position in channelling know-how from the supply to the demand side, and user needs in the other direction, supporting capacity building in the field, while liaising with space agencies and NGOs which would then establish a permanent and efficient link within such a programmatic framework.

As very well outlined by Ramesh S. Krishnamurthy from the World Health Organization (WHO), in order to define an effective space for global health framework, it is important to ensure a sufficient level of integration of space science and technology with health systems, strengthening efforts in the context of United Nations One Health, Agenda 2030 Sustainable Development Goals (SDGs) and Universal Health Coverage. It is also crucial to set-up closer collaborations, so that at national-level, health authorities and other relevant space actors could leverage the benefits of space science, technology and data for health gains. All the 194 UN Member States could strongly benefit from such a framework and the six UN regional offices could support the implementation of activities, acting as a bridge between space actors and the various national institutions and health actors.

The United Nations concept of One Health underlines the strong interconnectedness of human health, animal health and the ecosystem. In fact, more than 60% of human infectious diseases are of zoonotic origin, as well as at least 75% of emerging ones. In the perspective of the United Nations Agenda 2030, Good Health and Well Being (SDG3) is closely linked to all the SDGs: for example prioritizing the health needs of the poor (SDG1), addressing the causes and consequences of all forms of malnutrition (SDG2), preventing diseases through safe water and sanitation for all (SDG6), ensuring equitable access to health services through UHC based on stronger primary care (SDG10), protecting health from climate risks and promoting health through low-carbon development (SDG13), supporting the restoration of fish stocks to improve safe and diversified healthy diets (SDG14), promoting health and preventing diseases through healthy natural environments (SDG15).

In order to advance national health-related SDG targets, it is important to strengthen national capacities for utilizing space science and technologies. In Fig. 33.1, from left to right, a cross sectional view of the relevance of space science to public health is depicted, as envisioned by WHO. For each United Nations member state, there is a national Sustainable Development Goals agenda, which is closely linked to Good Health and Well Being (SDG3) and to the other SDGs as

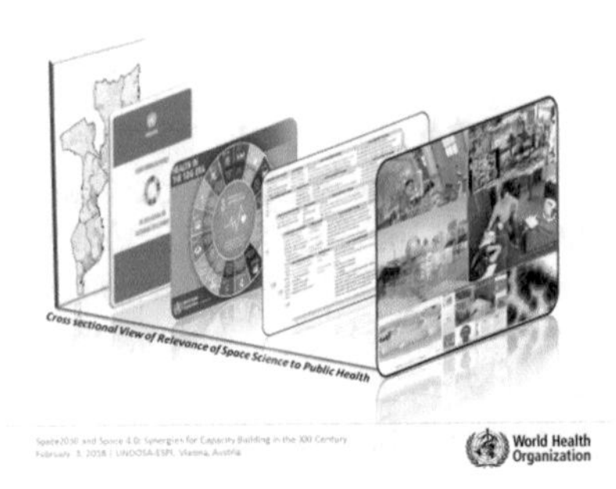

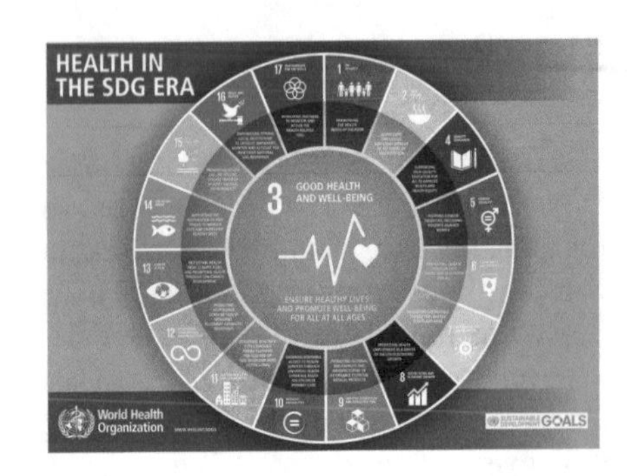

Fig. 33.1 Cross sectional view of relevance of space science to public health, as envisioned by WHO

discussed above. Each country also has a certain capacity in terms of available technologies, methodologies and approaches, which ultimately translate into actions. A key question is therefore how to best move along this path, in order to effectively apply space science and technology in the public health context. The key recommendation is to start such a program of activities focusing on three main areas: (1) Space science and technology for epidemic intelligence; (2) Space science and technology for health emergencies; (3) Shaping the research agenda on the benefits of space science and technology to public health.

The crucial point is then to effectively translate successful demonstration projects in these areas into adopted best practices, implemented with a systematic approach at national or regional level. It is therefore necessary to create a *Framework for Country Capacity Development for utilizing Space science and technology for advancing health-related SDGs*, which should include at least three components: **National readiness** for using Earth Observation (EO) data in conjunction with routine health systems data; **Multi-sectoral engagement** for establishing an Earth Observation data utilization environment in the national context; **Alignment** of stakeholders, strategies, and efforts, especially through the public-private partnerships that occur in the regions of interest (Fig. 33.2).

In terms of *national readiness*, and according to this conceptual framework, it is considered that any given country would have, as a first dimension to be explored, a certain policy environment, which could be more of less established to allow a proficient use of space science, data and technology. A second dimension to be considered is related to the capacity to utilize EO data and its associated computing environments. According to this methodological approach, any given country could be mapped in the context of space and health capacity, ranging from experimentation, early adoption, developing and building capacity, scaling up routine use until mainstreaming.

When looking at *multi-sectorial engagement* it is important that countries have appropriate coordination mechanisms among the actors involved. Coordination is essential to developing and sustaining data analytics capacities at national and sub-national levels. For example, partners within the Health Information Landscape could

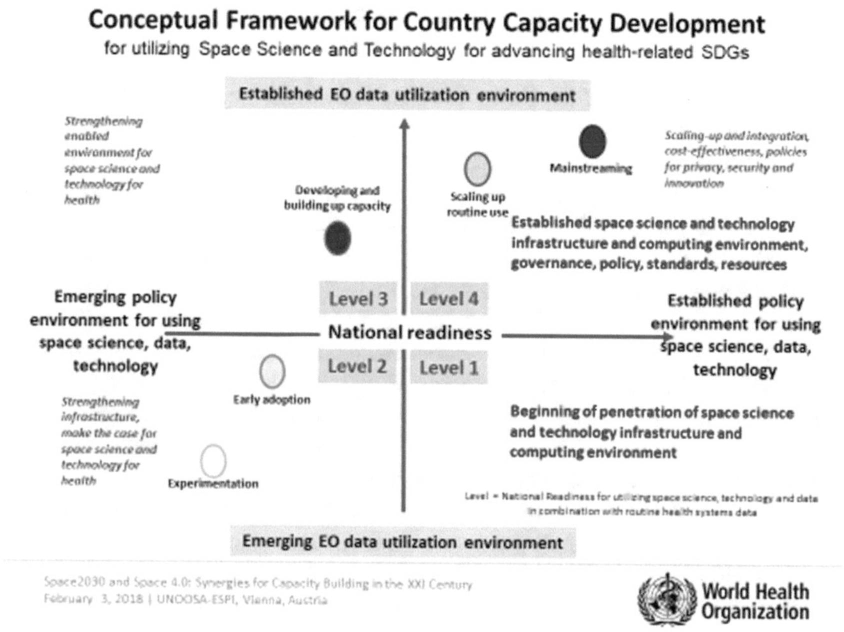

Fig. 33.2 Conceptual framework for Country Capacity Development for utilizing space science and technology for advancing health-related SDGs. *Source* WHO

be ministries (Health, Finance, Education, Labour, Telecommunications, Infrastructure, Science and Technology), academia, the private health sector, donors and implementing partners.

All these activities can take place only if there is an *alignment of stakeholders*, within the activities related to the SDG3 targets. To achieve better health outcomes, it's important that the value chain of solutions, hopefully driven by public-private partnerships, includes all the key actors, from governments to patients, hospitals and practitioners, from community to small and medium enterprises, from technical experts to entrepreneurs. But it is also important that countries align the many different strategies which are impacted by this framework (e.g. Health Information Strategy, mHealth, Health Management Information System, public health Emergency Operations Centres, Information and Communication Technologies, Human Resources capacity, etc.).

In conclusion, such a programmatic framework could be particularly effective in case integration of space science and technology to health systems is strengthened and more widely practiced in the context of One Health, UN Agenda 2030 and UHC. At national-level, closer collaboration between Health authorities and relevant space actors is essential to leverage the benefits of space science, technology and data for health gains.

33.4 Climate Change and Resilient Societies

In 2015, Member States of the United Nations endorsed 3 major global frameworks, all central to the success of the 2030 Agenda for Sustainable Development: The Sendai Framework for Disaster Risk Reduction 2015-2013; the Sustainable Development Goals; and national commitments to limit the impacts of a changing climate (COP21). At the core of these agreements is the recognition that countries needed to strengthen their capabilities in accessing and using precise and relevant data, as well as developing their capabilities to use the best technologies available for their interpretation and integration in planning and decision-making processes. The role of UNOOSA in coordinating the space contribution to the achievement of the Global Agendas (2030 Agenda, Sendai framework, COP21) has been widely recognized and praised. The need for the creation of a permanent and solid exchange between the supplier and the provider side of the sustainable development equation is becoming of primary importance, as climate change is dramatically affecting all countries of the world. Earth Observation and Integrated Applications could well serve the purpose, combined with Big Data management techniques and new technologies associated to it, such as Machine Learning, Artificial Intelligence, Cloud processing and Thematic Exploitation Platforms that will support obtaining useful and actionable information from the massive amounts of available raw data.

There are many different ways in which space technologies can support sustainable development. Space has demonstrated a wide span of societal benefits in the last 50 years, and given the recognition of its potential to allow for sustainable development on Earth it's time to fully exploit the opportunities offered by Earth Observation, Navigation, Telecommunications, Science and Exploration. For example, Earth Observation technologies enable the independent monitoring of different sustainable objectives especially in relation to climate, environment and pollution. Indeed, half of the Essential Climate Variables can only be measured from space. Moreover, telecommunication technologies can enable connectivity in remote areas, empowering populations and contributing to education and healthcare. Finally, space technologies, that are generally efficient, resilient and frugal, may have applications on Earth that foster sustainability through a more efficient use of resources. New actors (for example the World Health Organisation) and stakeholders (such as Non-Governmental Organisations) are becoming interested in innovative solutions that can create economic value while improving living conditions in a truly sustainable manner.

Therefore, the first recommendation for future space capacity building activities in this area, would be to set up an international task force in interested countries, identifying key climate-related issues and defining a strategic plan for cooperating with leading space institutions on a global scale. There are two areas that are highlighted here, where space capacity building will be increasingly important going forward, because of the potential of space assets to provide key monitoring and forecasting capabilities, and even more importantly, for their potential to create sustainable economic value through innovation and employment, in a sustainable development perspective. These areas are agriculture and ocean/coastal environments.

In the field of agriculture, much progress is being achieved with smart farming, both in forecasting and production. Droughts and flooding are nevertheless posing challenges to this progress, as climate change effects become increasingly visible. Increases in temperature and limited availability of water are the main issues affecting food security, in addition to natural disasters that are difficult to predict. The ECMWF organization is working hard to improve forecasting capabilities and is also the provider of the Climate Change Service of the European Union. In future, more countries could partner with this organization, making use of the advanced supercomputing capabilities that it is currently implementing. UNOOSA and the World Meteorological Organization (WMO) already collaborate and could enforce this approach in support of all countries. The same applies for the Charter on Space and Major Disasters that could be expanded to include aspects related to the Agenda 2030 and new climate change challenges. UNOOSA could further expand its Sustainable Development Profile activities to the impacts of climate change, providing dedicated solutions for the emerging agricultural challenges in the space solutions compendium, while supporting countries in their implementation. Moreover, whereas several organizations are working on developing applications with space data (Group on Earth Observation, space agencies, private companies including the field of IT, and new space entrants), for developing and emerging nations it is difficult to have access to or even know about these possibilities. It's even more difficult for end users, for example farmers, to understand the best use of this data. UNOOSA plays a key role in simplifying and streamlining these processes, leveraging capacities globally and linking the needs of the users to the supply side, therefore capacity building efforts should take advantage of these capabilities.

A second challenge related to climate change concerns the oceans, which are increasingly suffering due to pollution, as well as endangering coastlines due to rising sea levels. Several dedicated space missions are already collecting data on these phenomena, and more efforts are currently being put in place. A good example is the recent agreement between UNOOSA and the AIR centre (an international distributed scientific network to share science and technology responding to the Global Atlantic scale challenges, enhancing the sharing of scientific & technical resources and creating sustainable local value). For countries that might be impacted by sea level rise concerns, early warning systems should be put in place and a collaborative system could be created, including the integration of data from satellites, drones and in situ measurements. There several systems already existing, however fragmentation and lack of access to these by many countries is an issue. UNOOSA could leverage its global reach and position, in addition to the agreements and partnerships it holds with technical institutions, to ensure that this data is made available to all. This type of initiative should also focus on building and maintaining a historical dataset. Indeed, in the case of natural disasters, the International Charter "Space and major disasters" data is provided soon after the disaster for real-time response management, whereas if historical records of such events were saved in an accessible database, these could contribute to research and evaluation of trends.

Moving to the coastlines, development of Smart Cities and especially Smart Port Cities, where information technology and digital integration is used to achieve a high level of efficiency from a logistical, environmental and security point of view, is one of the priorities for Europe and therefore ESA is also investing in this topic. It would be interesting to extend the above-mentioned approach towards large cities in developing countries on the coast, since they will have the fastest growth in the coming decades. Dedicated services for sea level rise monitoring, coastal erosion and pollution will become increasingly valuable as climate change will broaden its effects on the populations living near the coasts. For these reasons, a global conference on the topic, involving different sectors and technical expertise to showcase the potential of space-based data to tackle these challenges and to anticipate the trends, is also recommended. The following steps could be the expansion of agreements and the leverage of the UNOOSA partnerships with research centres to various locations around the world, in all the seas and oceans. In addition, the EC GSA and UNOOSA partnerships could be leveraged in this context, providing key information. Space programmes like Copernicus offer solutions and best practices that could prove successful also in addressing the four pillars of UNISPACE+50 (space society, space economy, space accessibility, space diplomacy). Providing visible socio-economic benefits, through its free, full and open data policy, it also allow for strategic long-term plans at political level. It is important to further disseminate the value adding potential of space for society, and in this sense it crucial to identify innovative ways of advertising these benefits, through the involvement of a large number of partners at all levels. This approach should be applicable to the various national and international initiatives that are contributing to mitigate climate change, and cooperation between these initiatives is vital since the consequences of climate change are of global nature. UNISPACE+50 has been the occasion to move forward in this direction, through thematic priorities 6 and 7 and the four pillars of Space2030. More specifically, various initiatives were addressed, such as the Capacity Building Network, the Space Climate Observatory, encouraging women to pursue education in STEM, Space for Development Profile (SDP); Space Solutions Compendium (SSC) and Global Space Partnership for SDGs.

Now, in order to fully exploit the potential of satellite information and services, the space community needs also to ensure that potential customers and business partners are aware of the precious treasure that flies above. Several entrepreneurial endeavours have already successfully combined space technologies and sustainable development goals. The most notable example is Planet. They have revolutionised the Earth Observation market through their deployment of small satellites with frequent revisit times, offering interesting data for a range of markets such as energy, agriculture, maritime transport and insurance. Their data is also actively being used for the support of sustainable development objectives. In the words of Will Marshall, Planet's CEO, addressing the UN Sustainable Development Summit: "*According to our analysis, Planet Labs' imagery can be used to directly (or indirectly) advance 15 of 17 SDGs, and help measure more than 70 of their related targets. I'm going to mention three examples: First, monitoring deforestation, part of Goal 15: Today, we discover evidence of deforestation only after our forests are gone. Planet Labs' imagery can enable us to monitor forests every day, to see illegal logging and enable proactive intervention. Second, combating climate change, which is Goal 13: Our imagery can monitor climate change with up-to-date data on the state of the world's ice caps and carbon stocks. Third, ending hunger and establishing food security, which is Goal 2: Our imagery can measure the health of crops in every farmer's field around the world, and provide vital information to them to increase crop yield. I'm pleased to say Planet Labs is committing to make our data available and accessible to efforts aligned with meeting the SDGs*."

There are several international and private organisations actively working to support entrepreneurial efforts. The European Commission has recently allowed using satellite data to report on agriculture field usage according to the Common Agricultural Policy benefits system, ensuring less expensive and more comprehensive reporting on whether subsidised crops are being grown as expected. This has opened up new business opportunities for firms able to convert Sentinel satellite data from the Copernicus program into meaningful information towards this goal. Several accelerators and incubators are actively focusing on creating businesses from space assets.

One model is that of the Copernicus Accelerator of the EU, that looks for promising ideas that use data from the Sentinels, and offers them both funding and pairing with industry experts who will mentor the founders as they move from the first conceptualization of their idea to market entry. The accelerator doesn't offer incubation facilities, but rather relies on local incubators to include the start-ups into their programs, which is both a quality signal for the start-ups themselves, and a fast way to cover large geographic regions at minimal cost. This could therefore be an interesting model to promote entrepreneurship at a low cost across the world. Another currently existing model is applied in the European Space Agency Business Applications and Technology Transfer programs. The technology transfer program relies on a partnership with a series of pre-selected incubators, placed in ESA member states, that select and manage start-ups that are using space technology on Earth. Each selected start-up receives funding which is provided by ESA and the host country in equal measure. The ESA Business Applications program, that seeks out ideas that use integrated applications from space assets, similarly foresees a cost-sharing scheme with the country in which the company is registered, without the actual incubation.

On a global scale, the UNOOSA Regional Centres for Space Science and Technology Education are another resource that can provide support and capacity building along the model of the ARCSSTE-E centre in Nigeria (Chap. 26). The centre already set up a series of technology hubs, mentoring programs and social media platforms. They also reach out to young students in high schools and universities. These are essential first steps towards creating an entrepreneurial ecosystem, where a solid network and a steady stream of talents are paramount in creating the conditions for new ideas to flourish.

33.5 Conclusion

The capacity building programs that were outlined in this chapter point to a series of building blocks that need to be developed if countries around the world can achieve progress in becoming spacefaring nations and capitalising on the benefits that space technologies have to offer.

The first component is access to new technologies developed thanks to disruptive trends such as digitalisation, miniaturisation, the advent of new materials and concurrent design techniques. Knowledge of these technologies must be accompanied by a general knowledge of the space sector and its key technological solutions, also in relation to sustainable development, as well as the main requirements for setting up a space project and launching and managing space assets. This, in particular, requires combining theoretical teaching at all levels of education, with practical experiences that motivated individuals can develop while working with international partners, space companies, and space agencies around the world, willing to participate in capacity building efforts.

Another essential component is a development of rules and regulations as well as appropriate policies that can foster the development of new business models around space technologies. A third is a knowledge of the key market trends underpinning the space sector and the downstream applications sector, that can help steer countries in identifying a strategy going forward, to combine the concrete needs of their populations with solutions from space, in a sustainable development perspective. Finally, the appropriate funding needs to be mobilised, using innovative schemes such as cost-sharing or by pooling together unused resources, to kick-start space activities. All these activities require the joint participation of a wide range of actors, that find themselves represented in various ways in this volume. It is the hope of the editor that the concrete ideas and stories presented in the previous chapters will inspire more individuals, companies, agencies, governments and NGOs to work towards space capacity building efforts in the XXI Century, for the benefit of all humankind.

Editor and Contributors

About the Editor

Stefano Ferretti co-authored and edited this book as Resident Fellow of the European Space Policy Institute (ESPI) in Vienna, Austria, the leading European think tank for space policy, while working as an ESA Space Policy officer seconded from ESA/HQ in Paris, France, from 2015 to 2018. His main research interests are governance, innovation and future space-based services, and at ESPI he has initiated and managed the "Space for Sustainable Development" programme of activities, which included the 10th ESPI Autumn Conference in 2016 and the *Yearbook on Space Policy 2016: Space for Sustainable Development* published by ESPI at Springer; the ESPI contributions to UNISPACE+50, which included the ESPI-UNOOSA-ESA Conference *Space Capacity Building in the XXI century* in 2018; and various high-level Space policy researches, dialogue platforms, seminars and conferences in the European and global context, addressing sectors such as health, telecommunications and transport, which included the ESPI-EU-ESA Conference "Space and SATCOM for 5G: European Transport and Connected Mobility" in Bruxelles in 2017 and the participation to the UN COPUOS Expert Focus Group for Space and Global Health (EFG-SGH) based on the UNISPACE III Recommendations 6. Previously, he covered various positions at the European Space Agency, working as energy manager and infrastructure technical officer at ESA/ESRIN, and as International Space Station payloads project manager at ESA/ESTEC, coordinating the development of scientific experiments in microgravity with international space agencies (NASA, JAXA and Roscosmos), research institutions and industries. Before joining the Agency, he worked at Thales Alenia Space, where he managed development activities of the International Space Station Node 3 module, covering the various engineering and manufacturing phases of the flight hardware. For this work, he received an award from the NASA International Space Station vehicle office in 2006. Prior to that, he carried out academic and industrial research activities at NASA, and during ESA parabolic flight campaigns, for which he received the International Astronautical

S. Ferretti (ed.), *Space Capacity Building in the XXI Century*, Studies in Space Policy 22,
https://doi.org/10.1007/978-3-030-21938-3

Federation Napolitano Award in 2002. He holds a Ph.D., with a dissertation on Innovative Technologies for Space Habitats, a Master's in Mechanical Engineering from the University of Bologna and a Master's of Space Studies from the International Space University. He attended executive programmes in space policy and law, innovation and entrepreneurship and leadership, at George Washington University and MIT. He has authored several articles, reports and papers in the fields of space policy and law as well as science, engineering and technology. He is a member of the Professional Engineers Association (*Ordine degli Ingegneri*) of Italy, since 2005, of the International Astronautical Federation Committee on Space Applications, and he represented ESPI for three years at the United Nations Committee on the Peaceful Uses of Outer Space (UN COPUOS).

Contributors

Anas Abdulaziz received his Ph.D. in Environmental Biotechnology from Cochin University of Science and Technology, India. Currently, he is a Senior Scientist at the CSIR-National Institute of Oceanography in India. His major areas of research are host-microbe interactions in marine ecosystems, and the ecology of environmental Vibrios and their reservoirs. He has also worked on the applications of probiotics and immunostimulants to protect crustaceans from infections by Vibrios. He has used semiconductor Quantum dot nanoparticles for the detection of Vibrio harveyi. He maintains a Marine Microbial Reference Facility, supported by the Ministry of Earth Sciences (MoES, Govt of India), for health indicator bacteria from coastal waters of India.

Aranka Anema is an epidemiologist and public health scientist from Harvard University and holds research fellowships at Harvard Medical School and Boston Children's Hospital (USA). She has designed and led independent impact evaluations for major bilateral and multilateral agencies including the WHO, USAID, European Commission, World Bank Group, United Nations World Food Programme (WFP) and United Nations Children's Fund (UNICEF) spanning operations across 50 countries. She also advised the United Nations and XPRIZE Foundation on evidence-based policies and programs. She is the Co-Founder of GlobalHealthPx.com, a Canadian advisory that supports government, academia and corporate agencies on using evidence-based science and technology. She is member of the Expert Focus Group for Space and Global Health (EFG-SGH) based on the UNISPACE-III Recommendation 6.

Gabriella Arrigo is the Director of International Relations at the Italian Space Agency (ASI), where she started her space professional career in 1991. After a degree in Political Sciences at the University of Messina (1984), she specialized in international humanitarian law at the University of Geneva (CH) and in international relations at the Italian Society for International Organization (SIOI) in Rome. She was elected Vice President of the International Astronautical Federation (IAF) by the General Assembly in 2017 in Adelaide (Australia). Under her portfolio, which focuses on IAF science and academic relations for the years 2018–2020, she implemented the International Space Forum 2018 in Buenos Aires (Argentina) "The South America and Caribbean Countries Chapter" and the International Space Forum 2019 in Reggio Calabria (Italy) "The Mediterranean Chapter" to involve Academy and University in space activities. She is a member of the Italian delegation to the European Space Agency (ESA) in the International Relations Committee (IRC) since 1992 and at the United Nations Committee of the Use of Outer Space for Peaceful Purposes (COPUOS) since 1994. She is ASI representative, Chairperson or advisor in several intergovernmental committees, international councils, task forces and working groups, as well as in the Boards of the European Space Policy Institute (ESPI) in Vienna and of the International Space University (ISU) in Strasbourg. She also coordinates the Italian participation in the Global Space Forum of the OECD. As an expert in space policy and space affairs, she negotiated numerous international space agreements on behalf of the Italian Space Agency or in support of the Italian government. In 2009, convinced of the interdisciplinary nature of the space system and of the necessity to train officials and young professional, capable of representing space interests in international contexts, she created the first Master's course in Space Institutions and Policies at SIOI in Rome, where she is the Director and teaches space policy and international relations. In 2017 she became Scientific Advisor to the SIOI President. She also teaches international space relations and models of space governance in different master's courses in Space disciplines in Italy. She is coordinator of numerous space research projects in Italy and abroad. Since 2007 she is member of the Academic Council of the Mario Gulich Institute of High Space Studies at the University of Cordoba (Argentina). In 2018, in Bremen, the International Academy of Astronautics (IAA) awarded her Academic in the "Social Sciences" section. She is author of several articles and essays in international space policy issues, a distinguished speaker and moderator in international conferences and seminars around the world. She is a great promoter of space activities under different forms of communication.

Ajit Babu MBBS, MPH is Professor of Internal Medicine at St Louis University, a staff physician at the St. Louis VA Medical Center, USA, a research scientist with Centre for Disease Modeling, York University, Canada and an American Board-Certified practitioner in Internal Medicine and Preventive Medicine (Clinical Infomatics). He volunteered as a visiting professor in health informatics (telehealth) at the University of Addis Ababa, Ethiopia. He has over 35 international journal publications and is an associate Editor for Indian Journal of Medical Informatics. He has served as Vice-President of the Indian Association of Medical Informatics and was part of iHIND workgroup on health informatics, reporting to the past scientific advisor on innovation to the Prime Minister of India. From 2004-2008, he was founder and director of the Centre for Digital Health at the Amrita Institute of Medical Sciences in Kerala, India. He has been an invited expert on biomedical informatics for the Indian Council of Medical Research (ICMR) and was principal investigator on a dengue project funded by the Department of Science and Technology, Government of India, using geospatial tools for early warning and disease surveillance. In 2007, Dr Ajit became a member of the United Nations Action Team (AT6) which has evolved into the Expert Focus Group for Space and Global Health (EFG-SGH) based on the UNISPACE-III Recommendation 6.

Werner Balogh is Chief of the Satellite Data Utilization Division in the Space Programme Office of the World Meteorological Organization (WMO). He previously worked as Chief, ad interim, of the Space Applications Section of the United Nations Economic and Social Commission for Asia and the Pacific (ESCAP), as Programme Officer and Associate Expert on Space Applications with the United Nations Office for Outer Space Affairs (UNOOSA), as Programme Manager with the Austrian Space Agency and as International Relations Officer with the European Organisation for the Exploitation of Meteorological Satellites (EUMETSAT). He holds Dipl.-Ing. and Dr. degrees in technical physics from the Vienna University of Technology, an M.Sc. degree in space studies from the International Space University and an MA degree in international relations from the Fletcher School of Law and Diplomacy.

Massimo Bandecchi joined ESA in 1989, where he is heading the Systems and Concurrent Engineering section in the Directorate of Technology, Engineering and Quality, at ESTEC in The Netherlands. He has over 30 years of experience as software and system engineer, applied to the design, development, testing and operations of several space projects. In 1998 he was one of the co-founders and main lead of the ESA Concurrent Design Facility (CDF), used by the Agency and its partners for the assessment and preliminary design of future space missions. He authored several papers and publications and is a lecturer on Concurrent Engineering methodology and applications.

Ulrike M. Bohlmann is a German Attorney working for the European Space Agency since 2002. She joined ESA's Strategy Department in 2015, where she manages relations with the Agency's Member States. From 2002 to 2015, she held positions in ESA's Legal Department, where she advised the Agency on legal solutions concerning the carrying out of space programmes, provided legal advice on the interpretation and application of the Convention and negotiated international cooperation agreements. She represented ESA at the UNCOPUOS Legal Sub-Committee and several other international conferences. In the institutional field, she managed legal disputes and represented ESA in these proceedings. She has published a book on commercial space activities and intellectual property, a number of book chapters in the field of space law and general public international law, as well as numerous conference papers and articles in law and policy journals. Before joining ESA, she held the position of full-time Senior Research Associate at the Institute of Air and Space Law of the University of Cologne, where she also earned her doctoral degree magna cum laude. She continues to contribute to academic research projects, lectures regularly on space law and policy at University level and continues to publish regularly.

Arnaud Bossy is Vice President, Head of Consulting Europe at Frost & Sullivan. He is a strategy and transformation consultant, bringing 20 years of experience in enabling organizations to grow, in digital and global ecosystems. From developing a medium term vision to driving operational changes and delivering financial results, he has successfully orchestrated and led consulting missions across industries. Long time recognized and certified Lean Six Sigma Master Black Belt, he leverages leadership and facilitation skills as well as the complementary change management and analytical outlooks. Entrepreneur in both the start-up and corporate environments, he evolves in multicultural environments, always with enthusiasm and dynamism. Recent work and passion revolve around exploring how digital innovations impacts organizations and people, and how businesses can contribute to a sustainable environment.

Amalia Castro Gómez is a Remote Sensing Project Scientist, working at the European Space Agency (ESA/ESRIN). In her role she supports the Earth Observation educational activities within the Science, Applications and Climate Department of the ESA Earth Observation Programmes Directorate. Previously she worked as a Remote Sensing analyst at Airbus in the UK. She holds a bachelor in Environmental Sciences by the Autonomous University of Madrid, Spain and a master in Geo-Information Sciences by the Wageningen University, The Netherlands.

Rita Colwell is Distinguished University Professor at the University of Maryland at College Park and President of CosmosID, Inc. Her interests are focused on global infectious diseases, water, and health. She has authored or co-authored 20 books and more than 800 scientific publications. Dr. Colwell served as 11th Director of the National Science Foundation and Co-chair of the Committee on Science, National Science and Technology Council. Dr. Colwell is a member of the National Academy of Sciences, Royal Swedish Academy of Sciences, Royal Society of Canada, Royal Irish Academy, American Academy of Arts and Sciences, and American Philosophical Society. Dr. Colwell has been awarded 63 honorary degrees from institutions of higher education and is the recipient of the Order of the Rising Sun, Gold and Silver Star - Japan, 2006 National Medal of Science, 2010 Stockholm Water Prize, and the 2018 Lee Kuan Yew Water Prize of Singapore.

Clémentine Decoopman is currently the Executive Director of the Space Generation Advisory Council (SGAC). Responsibilities of the position include (but are not limited to) Operations Management, and Events Management. The Operations Management components of this role has an emphasis on administration of the organisation and communications management, including working closely with the Executive Office on the development and facilitation of opportunities for SGAC members in the space sector. The latter includes overseeing development of the SGAC brand, web and printed content, and supporting public relations and communication efforts. The Events Management component of this role is focused on overseeing and coordinating the SGAC event managers to ensure success of our global events, such as the Space Generation Congress, Space Generation Fusion Forum, and SGx. The

SGAC Executive Director also oversees the implementation of other SGAC activities including regional and local events. Clémentine also represents the SGAC at international conferences and at the United Nations (particularly at the UN COPUOS annual sessions) and contribute to the discussions by making regular statements at the plenary sessions.

Simonetta Di Pippo is the Director of the United Nations Office for Outer Space Affairs (UNOOSA), which is mandated to enhance international cooperation in space activities to promote their use for humanity. Prior to joining UNOOSA in 2014, she was Head of the European Space Policy Observatory at Agenzia Spaziale Italiana (ASI) in Brussels. Ms. Di Pippo also served as Director of Human Spaceflight of the European Space Agency from 2008 to 2011, and Director of the Observation of the Universe at ASI from 2002 to 2008, where she started her career in 1986. She has been an Academician of the International Academy of Astronautics (IAA) since 2013, and since 2016 a member of the IAA Board of Trustees. Ms. Di Pippo is also a member of the World Economic Forum Global Future Council on space technology since 2016. Ms. Di Pippo holds a Master's Degree in Astrophysics and Space Physics from University "La Sapienza" in Rome, an Honoris Causa Degree in Environmental Studies from St. John University, and an Honoris Causa Degree of Doctor in International Affairs from John Cabot University. Ms. Di Pippo was knighted by the President of the Italian Republic in 2006. In 2008, the International Astronomical Union named asteroid 21887 "Dipippo" in honour of her contribution to space activities.

Isabelle Duvaux-Béchon is in charge of the Relations with ESA Member States ensuring the coordination at corporate level of the information on the Agency's policies and actions towards the 22 ESA Member States, acting as their entry point to the Agency and aiming at optimisation of mutual current and future interests. She is also in charge of the coordination across ESA of transverse initiatives representing global challenges on Earth, with a thematic (Sustainable Development) or geographical approach (Oceans, Arctic, Antarctic, Alps, Africa, etc.), making the links between ESA and potential users, and coordinating partnerships with non-space entities. She is co-chairing the ESA Space & Arctic Task Force and the ESA Blue Worlds Task Force formed with interested ESA Member States. After 4 years in the space industry, she joined ESA in 1987 and worked in various areas: Microgravity, International Space Station, Budget, Education, Corporate Planning, Finance, Studies, Advanced Concepts. Isabelle Duvaux-Béchon is an engineer with a degree from Ecole Centrale Paris (specialisation in Air & Space Engineering) and an auditor of the French Institute of Advanced Studies in National Defence (IHEDN). She is a Commander in the citizenship reserve of the French Navy and a full member of the International Academy of Astronautics.

Funmilayo Erinfolami is a scientific staff of the African Regional Centre for Space Science and Technology Education in English (ARCSSTE-E), Nigeria. She obtained her Bachelors and Masters Degree in Engineering Physics (Materials Science) at the Obafemi Awolowo University, Ile-Ife, Nigeria. She has a Postgraduate Diploma (PGD) in Basic Space and Atmospheric Science; Space Studies Program (SSP14) at the International Space University, Montreal Canada, and Space Life Sciences at PUCRS. She is involved in capacity building through space education outreach to schools across different grades (elementary to university level); and as a faculty member of ARCSSTEE, she teaches a Space Weather Module and co-lectures in the Disaster Management Module of the PGD programme. Other activities include outsourcing and organizing space activities around the country. In addition to teaching and organising space programmes, she is currently serving as the secretary of the CSSTE consortium under the GMES & Africa Project for Flood Monitoring in West Africa. She is a member of the African Geophysical Society and of the Materials Society of Nigeria.

Hayley Evers-King is an Earth observation scientist who has worked across the satellite data value chain, from sensor validation, to algorithm development, and applications in both research and operational settings. She has experience in a wide range of marine and transitional water applications including water quality, harmful algal bloom detection, ocean carbon and heat content, and climate model validation. She has a keen interest in growing the use of satellite data through capacity building and science communication, and applying the data to support sustainable development.

Grazia Maria Fiore graduated in Political Sciences form LUISS University in Rome and she has over ten years of experience in research, public affairs and communications. She was previously Research Officer within the UNDP REDIVU project, aimed at fostering university social responsibility in Latin America, and Project Manager at the European Public Law Organization, managing EU-funded projects in the fields of rule of law and human rights. At Eurisy, she implemented an extensive analysis on the uses of satellite applications by European public administrations. She is currently responsible for managing the Eurisy Space for Cities initiative.

Grinson George Ph.D. in Marine Science from National Institute of Oceanography, Goa, India, started his career as a teacher of Fishery Science in the Maldives. He then moved to the Central Island Agricultural Research Institute in Port Blair, India and then to the Central Marine Fisheries Research Institute, Kochi, India. Since 2000, Dr. George has worked on the assessment of fisheries' resources using remote sensing, Geographical Information Systems, numerical modelling and oceanographic observations. He hosted the prestigious Jawaharlal Nehru Science Fellowship granted to Prof. Trevor Platt FRS by the Govt. of India, during which he collaborated with Prof. Platt and Dr. Shubha Sathyendranath of Plymouth Marine Laboratory, UK. He has 40 publications in peer-reviewed, high-impact journals, 6 books as author or co-author, 3 books as editor or co-editor, 13 book chapters and more than 100 conference abstracts to his credit. He is reviewer to some ten international journals. Dr. George has a long-standing experience with capacity building and user-engagement, and has organised many national, as well as two international, workshops. He delivers lectures in various community awareness programmes on a voluntary basis.

Paolo Giommi is a senior astrophysicist currently working at the Italian Space Agency (ASI), within the international relation unit. His scientific experience goes back to the early 1980s when he was a research fellow at the Harvard-Smithsonian Center for Astrophysics. He then worked for ESA from 1982 to 1995 and for ASI from 1995 until now. His scientific expertise is mostly in the area of extragalactic astrophysics, in particular in the field of blazars, where is one of the world's experts. He has played a significant role in multi-frequency, multi-messenger data science, as senior scientist of the European Space Information System ESIS (1991–1995), the director of the BeppoSAX scientific data center (1995–2000) and of the ASI Science Data Center (2000–2016). He is now the Italian coordinator of the United Nations Open Universe Initiative. He is author or coauthor of several hundred scientific publications with an H-Index of over 100.

Beth Healey is a British Medical Doctor (MD) who overwintered in Antarctica as a research MD for the European Space Agency. Stationed at Spaceflight Analogue Concordia, 'White Mars', she researched the effects of this extreme environment, including the isolation, inaccessibility, altitude, and low light levels on the physiology and psychology of the overwinter crew. It is believed that this research will help inform us of some of the challenges future astronauts embarking on long duration spaceflight will face. With an interest in extreme environments, she has previously worked as part of logistical and medical support teams for ski mountaineering expeditions and endurance races in a variety of polar environments. She is currently working as an emergency medical doctor in the Swiss Alps.

Barbara Imhof is an internationally active space architect, design researcher and educator. Her projects deal with spaceflight parameters such as living with limited resources, minimal and transformable spaces, resource-conserving systems; all aspects imperative to sustainability. Barbara Imhof is the co-founder and CEO of LIQUIFER Systems Group, an interdisciplinary team comprising engineers, architects, designers and scientists. As project lead, she currently works on the Gateway project, designing the habitat module for the next International Space Station in a lunar orbit. She also led projects such as SHEE, a Self-deployable Habitat for Extreme Environments, the first European simulation habitat. The SHEE habitat became part of another LIQUIFER project named MOONWALK, developed to test human-robot collaborations in two space simulation missions in Rio Tinto and subsea. Another most recent project she led for LIQUIFER was RegoLight, advancing the Additive Layer Manufacturing technology of solar sintering to build ISRU habitats on the moon. In addition, she pursues projects in the field of biomimetics and closed-loop systems such as Living Architecture and GrAB–Growing As Building—which looks at growth principles in nature and their proto-architectural translations towards self-growing buildings. LIQUIFER's partners and clients belong to internationally renowned institutions and space agencies.

Antarpreet Jutla is Associate Professor of Civil and Environmental Engineering at West Virginia University. He received his BS and MSc in Agricultural Engineering from Punjab Agricultural University, India; a M.Sc. in Civil and Geological Engineering from University of Saskatchewan, Canada and an Ph.D. in Civil and Environmental Engineering from Tufts University, Massachusetts. His research focuses on issues of predictability of water-related infections using satellite remote-sensing data. His research group – Human Health and Hydro-environmental Sustainability Simulation Laboratory – is engaged in utilizing satellite remote sensing to understand issues of water governance as well as distribution of food nutrition for human populations. He is an Editor of GeoHealth, a new flagship journal started by the American Geophysical Union to integrate health (environmental, human and ecological) with geosciences.

Gemma Kulk was trained as a marine biologist and specialises in the photophysiology and primary production of marine phytoplankton. She combines laboratory experiments, in situ measurements and ocean-colour remote sensing to study the response of marine phytoplankton to changing environmental conditions. She has enjoyed working in various oceanic regions, including the North Atlantic Ocean, Mediterranean Sea, Arctic Ocean and West Antarctic Peninsula. In addition to her scientific work, Dr. Kulk is an experienced project manager and has been involved in teaching at secondary schools and universities. At Plymouth Marine Laboratory, Dr. Kulk currently investigates the role of lake ecosystems in cholera epidemics in the South-West of India.

Toshiyuki Kurino joined the World Meteorological Organization (WMO) Space Programme Office as chief of the Space-based Observing System Division in July 2016. As a part of his duty in the WMO, Mr. Kurino is promoting WMO's initiative for improving capabilities of WMO Members to utilize meteorological satellite observations. Mr. Kurino began his career at the Meteorological Satellite Center (MSC) of the Japan Meteorological Agency (JMA) which he joined in 1981 as a scientist for meteorological satellite data processing and applications development for satellite meteorology, disaster risk reduction, and climate monitoring. For the last ten years in JMA, Mr. Kurino supervised Japan's meteorological satellite programme, including coordination to enhance satellite data utilization by domestic and international users.

Veronica La Regina is the Director for Global Engagement Europe & Asia at NanoRacks Space Outpost Europe srl (Turin, Italy). Before joining NanoRacks she was Strategy and Business Development officer at RHEA, and appointed at the European Space Agency (ESA/ESTEC) as Business Innovation Expert of the Commercial Partnerships' initiative for space exploration in the Directorate of Human Space Flight and Robotic Exploration. She previously worked for the ESA Technology Transfer Programme Office and at the Department of International Relations of the Italian Space Agency (ASI). Before that, she worked at the European Space Policy Institute in Vienna, at the International Space University in Strasbourg, in Telespazio SpA in Rome, at the Wave Energy Centre in Lisbon and as researcher at different universities in Europe and in the USA. She has been a climate-KIC coach since 2012. She holds a Master Degree in System Engineering from the Electronic Engineering Department of Tor Vergata University in Rome (Italy), a Master Degree in Space Policy, and a full Graduation in Law. She also pursued PhD Studies in Economics. In 2017 she has been awarded as a Leader by the Italian Branch of the Business Professional Women's Network.

Antonio Martelo is the head of the Concurrent Engineering Facility (CEF) and a space systems engineer in the German Aerospace Center (DLR), since 2014. His present work focuses on Concurrent Engineering-study operation as team leader, Concurrent Engineering process development, the responsibility for the CEF, and the acquisition and execution of projects and studies, as well as the dissemination of results and external communication. He holds degrees in Telecommunications Engineering from the European University of Madrid (Spain), Space Systems Engineering from ISAE-Supaero (France), and in Radio-astronomy and Space Science from Chalmers University of Technology (Sweden).

Nandini Menon holds a doctorate in Marine Science from Cochin University of Science and Technology, Kerala, has twenty years of post-doctoral research experience and eighteen years of teaching experience. After working in various academic positions at her alma mater for more than a decade, she joined the Nansen Environmental Research Centre India (NERCI), where she is the Deputy Director. A marine biologist by training, Dr. Menon has devoted her time to marine ecology, pollution, and invertebrate taxonomy, recently diversifying to aquatic bio-optics and phytoplankton dynamics. As investigator of national and international projects, she has been responsible for directing research on marine ecosystems from various perspectives. She has 42 scientific and popular publications to her credit. Dr. Menon contributes actively to outreach programmes through popular lectures and through international winter schools and workshops that she organises for young scientists. She has visiting faculty positions at a number of universities in Kerala where she contributes to post-graduate programmes in Marine Sciences.

Roberta Mugellesi is a business applications manager in the Downstream Business Applications Department of the Directorate of Telecommunications and Integrated Applications of the European Space Agency (ESA) in Harwell, UK. She has been involved as project manager in the development of more than 40 services dealing with transport, maritime, fishing and infrastructure monitoring. Since 2011 she is supporting the ESA corporate approach to knowledge and information management in the ESA Director General services. She has received 8 awards and has more than 50 publications in peer-reviewed journals and conferences. She holds a M.Sc. in Mathematics, specialization in Automatic Calculus from the Pisa University (Italy) and an MBA in International Business from the Schiller University of Heidelberg, (Germany). She is member of the International Program Committee of the International Astronautical Federation (IAF) and member of the International Academy of Astronautics (IAA).

Chiaki Mukai M.D., Ph. D, is the first Asian female astronaut, and currently JAXA Senior Advisor and Vice President of the Tokyo University of Science. In 2017 She served as the 54th Chair of the Scientific and Technical Subcommittee of the United Nations Committee on the Peaceful Uses for Outer Space, which promotes the long-term sustainability of outer space activities and the SDGs. As a medical doctor, she flew on the Space Shuttle Columbia (STS-65/IML-2) in 1994, and on the Space Shuttle Discovery (STS-95) mission in 1998, and conducted various life science and space medicine experiments.

Engelbert Niehaus is a professor at the Institute for Mathematics and Mathematics Education at the University of Koblenz-Landau (Germany) and works on mathematical modelling of spatial patterns of risk and resources. He is head of the computer science centre at the University of Koblenz-Landau, Campus Landau. Since 2007 he is member of Action Team 6 (AT6), Action Team 6 Follow-Up Initiative AT6FUI and Expert Focus Group for Space and Global Health (EFG-SGH) based on the UNISPACE-III Recommendation 6.

Trevor Platt Fellow of the Royal Society, is an oceanographer with long experience of research, field work, synthesis, and theory, as well as leadership and coordination of international marine science and science policy. He has led oceanographic expeditions from the High Arctic to the Tropics. His scientific interests include mathematical analysis of the pelagic ecosystem; phytoplankton physiology; marine optics; remote sensing; and waterborne diseases.

Melanie Platz is a mathematician specialized in education and healthcare information communication and technology for sustainable development (ICT4D). She brings 10 years of operational experience in mathematical modelling to guide healthcare investments and predict investment risks and outcomes. Melanie's passions converge around how to leverage use of GIS-tailored and GUI-design tools for problem-based learning. She specializes in creation of education tools for ICT4D for improving healthcare delivery, and has worked extensively across sub-Saharan Africa and Central America to design, implement and evaluate professional development training tools. Melanie is a Professor at the Pedagogical University of Tyrol, Austria, after she substituted professorships in Mathematics Education at the University of Siegen and the University of Education Freiburg (Germany). Since 2012 she is a member of the Expert Focus Group for Space and Global Health (EFG-SGH) based on the UNISPACE-III Recommendation 6.

Edith Rogenhofer started to work in the humanitarian field in 1991 and since 1997 has been with MSF. She has mainly worked in projects in Africa, with a few stints in Afghanistan, Pakistan, Bangladesh and Myanmar. Most of the projects have been carried out during complex emergencies, focusing on water supply and sanitation in camp settings and rural areas. She holds an MSc in water management as well as a diploma in drilling. She is part of the MSFs GIS unit and and coordinates the work related to user requirements and user validation.

Alfredo Roma is an economist now consultant for the aerospace industry. He has been member of the Advisory Council of the European Space Policy Institute in Vienna and Italian Delegate at the European Space Agency. He has been member of the RPAS Steering Committee, created by the European Commission for the integration of civil drones in to the common airspace. He has been National Coordinator for the Galileo Project at the Italian Prime Minister's Cabinet. From September 1998 to July 2003 he has been Chairman of ENAC, the Italian Civil Aviation Authority, and President of ECAC (European Civil Aviation Conference). From 1975 to 1992 he has been untenured professor of business finance, Faculty of Economics, at the University of Modena, Italy.

Luciano Saccani joined Sierra Nevada Corporation's Space Systems in 2014 and now holds the position of Senior Director for International Business Development to address business acquisition and customer relations around the world. During his career, Mr. Saccani worked in all major areas of space business including human spaceflight, space transportation, remote sensing, telecommunications, and navigation. Before joining SNC, Dr. Saccani worked in the US from 1992 to 2013 for the Space Subsidiary of Finmeccanica. In 2001 he was appointed CEO of Alenia Spazio North America. During his career he managed business development and strategies in Japan and North America and played a primary role in the development, negotiations and signature of key alliances between European and US industries. Dr. Saccani holds a Degree of Doctor in Electronic Engineering from the University of Rome La Sapienza, in Italy, a Post Graduate Diploma in Business Administration from the Edinburgh School of Business, in Scotland, UK and attended the Stanford University Advanced Management College. Dr. Saccani is citizen of both the United States and Italy.

Francesco Sarti is the Scientific Coordinator of the Education and Training Activities within the Directorate of Earth Observation Programmes at the European Space Agency. He graduated with a Master Degree in Electrical Engineering at the University of Rome La Sapienza and then conducted research at the CNR. In 1990 he joined ESA at ESOC, working on mission analysis, precise orbit determination and attitude control, then moving to ESTEC in the Netherlands. He attended the SSP1996 of the International Space University in Vienna, completed a Master in Applied Remote Sensing and Image Processing and obtained a Ph.D. on "optical-radar remote sensing for the monitoring of surface deformation" from the University Paul Sabatier in Toulouse, France. He worked at CESBIO and at CNES, as Project Manager for the International Charter on Space and Major Disasters, conducting R&D activities for remote sensing applications to disaster management and natural risk monitoring, interferometric monitoring of several seismic areas and providing training courses in Earth Observation. After a short period at the Italian Space Agency (2001), as technical interface ASI-CNES for the cooperation COSMO-SkyMed/Pléiades, he moved to ESA/ESRIN in the Earth Observation applications area, where since 2007 he coordinates the Education and Training Activities. His publications focus on the application of radar and optical remote sensing to damage mapping, tectonics and disaster management.

Shubha Sathyendranath is an oceanographer from India (Cochin University of Science and Technology), with a doctorate from France (Université Pierre et Marie Curie, Paris), who has worked in India, France and Canada before moving to the Plymouth Marine Laboratory in the UK. Her main interests are marine optics; remote sensing; marine primary production; biological-physical interactions in the ocean; climate variability; and climate change. She is the science lead of the Ocean Colour Climate Change Initiative of the European Space Agency. She worked for many years as the Executive Director of the Partnership of Observation of the Global Oceans, devoting her time there to improving international coordination and collaboration, as well as capacity building for ocean observations. Most of her research work has had an open-ocean focus. Her interest in issues of ecosystem health in coastal and inland water bodies was stimulated by learning about the degradation of the water quality of Vembanad Lake in Kerala, India, and its implications for human health.

Hilde Stenuit works in the space sector since nearly 20 years, and joined Space Applications Services soon after obtaining a Ph.D. in plasma-astrophysics from the University of Leuven (KU Leuven). Working on several operational aspects, she coordinated operations for the Automated Transfer Vehicle (ATV) and the International Space Station experiments, having substantial experience in leading console operations. From 2005 she supported the European Space Agency as Mission Scientist for several International Space Station increments, coordinating all microgravity sciences experiments. Currently she leads the SpaceApps science team, being responsible for the business development of the ICE Cubes (International Commercial Experiments) service, which provides fast and simple access for anyone or any organization to Space, in order to conduct scientific and technology research in microgravity.

Christopher Stewart is working at ESA as a research fellow in Earth Observation exponential technologies at the newly created Φ-Lab at ESA/ESRIN. Prior to this, he completed a bachelor's degree in mathematics, and a master's in remote sensing and image interpretation, before beginning his professional career at Remote Sensing Applications Consultants (RSAC) Ltd in the UK. Since 2007 he worked for the Science, Applications and Climate Department of the ESA Earth Observation Programmes Directorate, focusing on training activities, while also completing a PhD at the Tor Vergata University of Rome with distinction in 2017, with a thesis on "Archaeological Prospection using Spaceborne Synthetic Aperture Radar". His current research activities focus on the use of machine learning with EO data to explore new science (e.g. ionosphere-lithospheric coupling), and new applications (e.g. cultural heritage). In addition, he promotes the integration of potentially disruptive technologies (e.g. quantum computing, EO workflows), organising workshops, trainings and other events.

Fuki Taniguchi received her M.A. of International Public Law from Waseda University in 2009, and joined the Japan Aerospace Exploration Agency (JAXA). She started her career in the legal division, being involved in the drafting of the "Protocol to the convention on international interests in mobile equipment on matters specific to Space assets". In 2011, she was seconded to the Ministry of Education, Culture, Sports, Science and Technology for contributing to the ISS policy making. Since 2013, she had been in charge of PR activities for the ISS programme in JAXA. She has been in her current position since April 2016.

Jean-Jacques Tortora has been Director of the European Space Policy Institute (ESPI) since June 2016. Before this, he served as the Secretary General of ASD-Eurospace. From 2004 to mid-2007, he was head of the French Space Agency (CNES) office in North America and the Attaché for Space and Aeronautics at the Embassy of France in Washington, D.C. Previously he was Deputy Director for Strategy and Programs, responsible for the Industrial Strategy of CNES, the French Space Agency. Prior to that position, he was France's representative in the ESA Industrial Policy Committee and Joint Communication Board. From 1998 to 2000, he was advisor to the French Ministry of Research for Industrial Policy Funds management, aiming at industry competitiveness support and new space applications and services development and promotion.

Chandana Unnithan Ph.D., MBusComp, MBA is a global digital health expert and an applied scientist in IoT (geo-spatial technologies, health informatics, digital health, smart healthy cities, permissioned block chain, AI) and Professor of Public Health Informatics from Australia. In her applied research programs, she has pioneered IoT adoption in some of Australia's largest public and private hospitals and city councils, advised on successful EMR implementation and mentored higher degree students in health tech implementation and public health best practices. She is an invited panellist and keynote at international research conferences with over 100 scientific publications. She is an advisory to governments on national and local level on the topics of e-health implementations, health technology integration for public health, and disaster management using GIS/remote sensing techniques in Australia and Canada. She represents Australia at the United Nations Committee on Peaceful Uses of Outer Space (COPUOS), and she is member of the Expert Focus Group for Space and Global Health (EFG-SGH) based on the UNISPACE-III Recommendation 6. She is a current expert in the WHO Digital Health Roster of experts.

Servaas van Thiel is Minister Counsellor at the EU Delegation to the International Organisations in Vienna (since 2015). He teaches at the Vrije Universiteit Brussel (since 1987) and the Wirtschafts-Universität Wien (since 2016) and occasionally sits as a judge ('Raadsheer') in the Regional Court of Appeal in the Netherlands (since 2003). He published over 140 books and articles and guest-lectured world-wide for academic, professional and public sector audiences. He previously worked for the EU Delegation in Geneva, the EU Council of Ministers and the European Court of Justice. He was Director of the Brussels LL.M Program (2003–2011). Before joining the EU, Professor van Thiel, worked for the IBFD and was an international consultant (including for the Italian Economic and Labour Council, the UN Centre for Trans-National Corporations, and the UN Economic Commission for Africa). He studied law and economics at the Universities of Nijmegen ('Radboud') and Brussels ('Vrije Universiteit') and he holds a Doctoral Degree from Erasmus University Rotterdam.

Magali Vaissiere is the Director of Telecommunications and Integrated Applications (D/TIA) at the European Space Agency since 2008 and Head of ECSAT, Harwell, UK since 2013. She has an Executive MBA from the Centre de Perfectionnement aux Affaires, a Master of Science in Electrical Engineering from the University of Stanford and a Diplome d'ingénieur de l'Ecole Nationale Supérieure des Télécommunications de Paris. She had 24 years of experience in industry, dealing firstly with ground-based radars and then with satellites, prior to joining ESA in 2005, as the Head of the Telecommunications Department within the Directorate for Telecommunications and Navigation.

Andrea Vena is heading the Corporate Development Office at the European Space Agency. Graduated in Electronic Engineering, he started his career as system engineer in Alenia Spazio, the space branch of Finmeccanica (today Leonardo). In 1995, he was appointed head of the European Union Marketing Unit, coordinating the company's activities and programmes with EU institutions. In 2000, Mr Vena joined ESA and in 2007 he was appointed head of the Corporate Strategic Planning office. Since 2016, he's heading the Corporate Development Office, in charge of establishing a corporate strategy for the development of the organisation.

Cécile Vignolles is an Engineer in Agriculture and holds a doctorate degree in Remote Sensing and Agriculture. She served from 1998 to 2001 as research assistant at the Space Applications Institute of the EU Joint Research Centre (Ispra, Italy), where she contributed to the development of the agrometeorological bulletin within the "Monitoring Agriculture with Remote Sensing" project. From 2002 to 2005, she worked at SCOT, a subsidiary of CNES (The French Space Agency) as a research engineer on R&D in Agriculture and Remote Sensing projects. Research engineer at the GIP Medias-France from 2005 to 2008, she was in charge of R&D projects in tele-epidemiology. She joined the CNES in 2009, where she is affiliated to the Directorate of Innovation, Applications and Science in the 'Earth-environment-climate' team in charge of the "Earth Observation programmes". Within this team, she is responsible of the "Tele-epidemiology" and "Forests" programmes.

Roland Walter is a senior lecturer at the astronomy department of the University of Geneva fascinated by the study of high-energy phenomena occurring around compact objects or in diffuse media. He holds a physics diploma from the University of Lausanne, his Ph.D. from the University of Geneva and was a Max-Planck fellow at the Max Planck Institüt für Extraterrestrische Physik in Garching. He works on the high-energy emission of Active Galactic Nuclei, on their contribution to the hard X-ray background and on tidal disruption events. He participated to the discovery and modelling of special types of high-mass X-ray binaries. More recently, he contributed to the understanding of the gamma-ray emission of pulsar wind nebula and colliding wind binaries. He participated in almost 500 publications, supervised or contributed to 12 Ph.D. theses, observed on many different facilities and developed analysis techniques also based on machine learning. He led the development of the data center for the INTErnational Gamma-Ray Astrophysics Laboratory (INTEGRAL) of the European Space Agency and the HEAVENS project and is a member of the Cherenkov Telescope Array and of the High Energy cosmic Radiation Detection facility.

Printed in the United States
by Baker & Taylor Publisher Services